Design of Air Pollution Control Equipment

Design of Air Pollution Control Equipment

Jay Richardson
Combustion Engineer/Owner, Falcon Combustion

Louis Theodore
Consultant, Theodore Tutorials

WILEY

Published by John Wiley & Sons, Inc., Hoboken, New Jersey.

Library of Congress Cataloging-in-Publication Data has been applied for:

Print ISBN: 9781394377688
ePDF ISBN: 9781394377701
ePUB ISBN: 9781394377695
oBook ISBN: 9781394377718

Cover Design: Wiley
Cover Image: Courtesy Omar Nasr

Set in 9.5/12pt STIX Two Text by Lumina Datamatics, Inc.

To:

Joseph Vicente Richardson Cabezas

(JR)

The village of East Williston, our home for nearly a half century.

(LT)

Contents

About the Authors

Jay Richardson

Jay Richardson received his Bachelor of Science degree from the University of Connecticut in Chemical Engineering with a Minor in Chemistry. He has over 10 years' experience in the engineering, design, construction, commissioning, and testing of combustion equipment for industrial and utility boilers, process heaters, flares, thermal oxidizers, and catalyst activators. He has designed low and ultra-low NO_x burners, overfired air, flue gas recirculation, SCR/SNCR, and ignition systems. He has authored scientific journal articles on organic chemistry synthesis, magazine articles on physical combustion airflow modeling, and presented papers at international utility conferences on topics such as fuel conversions, NO_x reduction, biomass fuels, and hydrogen combustion. He has led the R&D testing of new ultralow NO_x burners and holds a patent (2022/0381433) on an AI supported ultralow NO_x burner system.

In addition to industrial combustion, Mr. Richardson took the wildlife photographs in his son Joseph's nonfiction book about peregrine falcons (*The Peregrine Falcon: A Wonderful Species*).

Louis Theodore

Raised in Hell's Kitchen by poor Greek immigrant parents, Louis Theodore received the degrees of M.Ch.E and Eng. Sc.D. from New York University and B.Ch.E. from The Cooper Union. Dr. Theodore was a successful educator for over years at Manhattan College (holding the rank of Full Professor of Chemical Engineering) as the Graduate Program Director, researcher, professional innovator, and technical communicator. During this period, he was primarily responsible for his program achieving a No. 2 ranking by the U.S. News & World Report and particularly successful in placing students in internships, jobs, and graduate schools.

Dr. Theodore is an internationally recognized lecturer who has provided more than 200 seminars/courses to industry, government, and technical associations. He also developed and served as the principal moderator/lecturer for U.S. Environmental Protection Agency (EPA) courses on air pollution control equipment, consulted for several industrial companies in the field of air pollution control equipment, pollution prevention and environmental management, and served as a consultant/expert witness for the U.S. EPA and U.S. Department of Justice. He currently serves as a part-time consultant to Theodore Tutorials and is actively involved in air pollution control equipment, potable water/desalination activities, and geothermal energy applications.

He is the author of more than 125 text/reference books ranging from air pollution control equipment to pollution prevention to hazardous waste incineration to engineering and environmental ethics, and served as section editor of Environmental Management for the last four editions of Perry's Chemical Engineer's Handbook.

Dr. Theodore is the recipient of the Air & Waste Management Association's (AWMA) prestigious Ripperton award that is "presented to an outstanding educator who through example, dedication and innovation has so inspired students to achieve excellence in their professional endeavors." He was also the recipient of the American Society for Engineering Education (ASEE) AT&T Foundation award for "excellence in the instruction of engineering students." In addition to providing invited testimony to a presidential (Ford) Crime Commission Hearing, Dr. Theodore was honored at Madison Square Garden in 2008 for his contributions to basketball and the youth of America. He is also the author of Basketball Coaching 101 (2015) and Winning at Casinos: The Definitive Guide (2021).

Dr. Theodore is a member of Phi Lambda Upsilon, Sigma Xi, Tau Beta Pi, American Chemical Society, American Society of Engineering Education, Air & Waste Management Association, and the Royal Hellenic Society.

Preface

As its title "Design of Air Pollution Control Equipment" implies, this book—more than anything else—provides design information on air pollution control equipment from an engineers' and applied scientists' perspective. The need and requirement to treat, evaluate, and design this equipment has led to the writing of this book. Concise and easy to use, it brings together a wealth of general information on control equipment in one compact book. It offers both technical information, field data, calculation details that include design and predictive algorithms, and illustrative examples.

The book has been written not only for students but also for technical individuals, e.g., engineers and applied scientists who work in the environmental area or in health-related fields. It is a tool that may be used whenever and wherever information about air pollution control equipment is sought. Thus, it is a must for those working in the environmental arena and those involved peripherally with this field. The authors feel that no one source covers all of the information on these devices in the manner presented in this book. It is hoped that this book will serve to fill the growing information required—both academically and professionally—in this field.

Controlling the emission of pollutants from industrial and domestic sources is important in protecting the quality of air. Air pollutants can often exist as combinations of particulate matter and gases. Air cleaning devices have been reducing pollutant emissions from various sources for many years. Originally, air cleaning equipment was used only if the contaminant was highly toxic or had some recovery value. Now with legislation including the Clean Air Act as well as individual state standards, control technology efficiency and effectiveness have been upgraded and more sources are regulated in order to meet the National Ambient Air Quality Standards (NAAQS). In addition, state and local air pollution agencies have adopted regulations that are in some cases more stringent than the federal emission standards.

Selecting an air pollution control device for cleaning a process gas stream can be a challenge. Some engineers, after trying to find shortcuts, employ quick estimates for both gas flow and collection efficiency that may be the entire extent of the collector specification—an inadequate approach. The end result will often be an ineffective installation that may have to be replaced or upgraded. Treating a gas stream, especially to reduce emissions, is usually not an income producer, but the costs—both capital and operating—can be minimized, not by buying the least expensive collector but by thoroughly engineering the whole system as is normally done in the process design phase of a project.

Techniques used to control gaseous emissions are absorption, adsorption, combustion—both catalytic and non-catalytic—and condensation. Equipment used to control particulate emissions include gravity settlers (often referred to as settling chambers), mechanical collectors (cyclones), electrostatic precipitators (ESPs), scrubbers (gas-liquid contactors), and fabric filters (baghouses). The applicability of a given technique depends on the physical and chemical properties of the pollutant and the exhaust stream. More than one technique may be capable of controlling emissions from a given source. For example, vapors generated from loading gasoline into tank trucks at large bulk terminals can be controlled by using several of the above gaseous control techniques. More often, one control technique is used more frequently than others for a given source-pollutant combination. For example, absorption is commonly used to remove sulfur dioxide (SO_2) from boiler flue gas.

A brief description of the seven parts of the book's content follows. Part I provides an overview of the subject title. The next two parts, Parts II and III, provide information on design and calculation details for both gaseous and particle control equipment, respectively. In addition, a number of examples are included within both of these parts. Part IV follows with similar information for other air pollution control equipment. The epilogue (Part V) addresses miscellaneous considerations that include: risk assessment, monitoring, economies, equipment selection factors, equipment comparisons, and suggestions for purchasing equipment. The book concludes with Parts VI and VII, the appendix and index, respectively.

Reasonable care has been taken to assure the accuracy of the information contained in this design book. However, the authors and the publisher cannot be responsible for erroneous omissions in the information presented or for any consequences arising from the use of the information published in the book. Accordingly, reference to original sources is encouraged. Reporting of any errors or omissions is solicited in order to assure that appropriate changes may be made in future editions.

Last but not least, the author wishes to thank the following individuals for their contributions to this work: Dr. Walter Matystik, Emma Parente, Mary K. Theodore, and Ronnie Zaglin.

Part I

Prologue

1

The Air Pollution Problem

1.1 The Role of Water and Air

1.1.1 Water

Water on the Earth is part of a closed cycle called the hydrologic cycle, that is affected by both the energy of the Sun and the gravitational pull of Earth's Moon. This results in the tides, evaporation of water, and its condensation to form precipitation in various forms. Only 1% of the water on the Earth participates in the short-term water cycle that is used for human consumption. Most water is outside the scope of human interaction and is stored in the oceans, ice caps, and deep groundwater aquifers and moves on time scales of several thousand years or more.

The gravitational pull of the Moon plays an interesting role in the movement of water on the Earth. The tides are caused primarily by the Moon's gravity. The Sun's gravity also plays a role, but the Moon is more important because it exerts more than twice as much gravitational influence as the Sun because the Moon is so much closer. The gravitational forces of the moon essentially pull the water on the Earth into an elliptical shape. The water bulges toward the moon, and equally on the opposite side, which shrinks the water on the perpendicular sides. As the Earth rotates, it passes through these high and low bulges which we experience and high and low tides.

Water on the Earth is believed to have been brought to the surface by comets that collided with the Earth early in its history. This water is a prerequisite for all life on the Earth, and it is liquid water that is the prerequisite, not steam or ice. Thus, water must exist at temperatures constrained by the boiling and freezing points of this unique material for life to flourish.

Over the course of history, human life has revolved around access and proximity to water. Those who maintained close proximity to clean water survived to learn again and reproduce, whereas those who did not disappeared. Environmental concerns took hold at about 3,000 B.C.E. Urban areas on the Indian continent developed sanitation programs such as underground drains and public baths. Aspects of health were integrated with daily activities including personal hygiene, health education, dietary practices plus food, and environmental sanitation.

Water is the original renewable resource and carries a rich history. The water molecules contained in the fruit eaten yesterday may have fallen as rain last year in a distant place or even millennia ago by one's ancestors.

The overall amount of water on the surface of the Earth has remained fairly constant over time. The hydrologic cycle describes this movement from place to place. The vast majority (96.5%) of water on the surface of the Earth is contained in the oceans. Solar energy heats the water at the ocean surface, and some of it evaporates to form water vapor. Air currents take the vapor up into the atmosphere along with water transpiring from plants and evaporating from soil. The cooler temperatures in the atmosphere cause the vapor to condense into clouds. Clouds move around the world until the moisture capacity of the cloud is exceeded, and the water falls as precipitation. Most precipitation in warm climates falls back into the oceans or onto land, where the water flows over the ground as surface runoff. Runoff can enter rivers and streams, which transport the water back to the oceans; it can accumulate and be stored as freshwater in lakes; or it can soak into the ground as infiltration. Some of this water may infiltrate deep into the ground and replenish aquifers that store huge amounts of freshwater for long periods of time.

1.1.2 Air

Earth's atmosphere is currently composed primarily of nitrogen and oxygen, but this composition has greatly changed over time. The air composition exposed to the first primitive cells was quite different to the "normal" composition of air today. Some scientists believe that Earth's earliest atmosphere probably contained almost no free oxygen. The oxygen in today's atmosphere is likely the result of several million years of photosynthesis.

Over the history of the Earth, plants and animals have had to adapt to changes in the environment. These changes occur extremely slowly through evolution. However, when environmental changes occur more rapidly than a species' ability to adapt, some species cannot adapt fast enough and oftentimes do not survive. This is the main concern with the human effect on the environment and atmosphere. The average air temperature and composition has certainly changed over the history of the Earth, but human contributions have caused significantly faster changes compared to the natural rate of change. The concern among scientists is that Earth's inhabitants' natural adaptation capabilities might not be adequate to meet this challenge. The current concentrations of air are provided in Tables 1.1 and 1.2, listing two key physical parameters of concern in the air pollution field.

Table 1.1 Concentration of current air at sea level.

Component	Concentration (vol %)
Nitrogen (N_2)	78.0
Oxygen (O_2)	20.99
Carbon dioxide (CO_2)	0.03
Argon (Ar)	0.94
Neon (Ne)	0.0015
Helium (He)	0.0005
Krypton (Kr)	0.00005
Xenon (Xe)	0.000006
Hydrogen (H_2)	0.01
Methane (CH_4)	0.0002
Nitrous oxide (N_2O)	0.0004

Note: Not including particulate, PM, and water vapor.

Table 1.2 Physical properties of selected gases of concern.

Gas	Molecular weight	Density (lb/ft^3)
Air	29	0.079
Ammonia, NH_3	17	0.048
Carbon dioxide, CO_2	44	0.12
Carbon monoxide, CO	28	0.078
Hydrogen, H_2	2	0.006
Methane, CH_4	16	0.045
Nitrogen, N_2	28	0.078
Oxygen, O_2	32	0.089

1.2 Early History of Air Pollution

Air pollution is not a new phenomenon. Natural phenomena such as volcanoes, windstorms, forest fires, and decaying organic matter contribute substantial amounts of air pollutants. Plants and trees also emit organic vapors and particles. The Earth, for the most part, is able to keep up with natural air pollution with natural cleaning systems.

Air pollution has been an issue since the first person discovered fire. This pollution was quite insignificant at first, especially when fire was kept outdoors. Humans did not significantly affect the environment until relatively recent times. This is due to two reasons: (1) the human population has been large for only a small part of recorded history, and (2) the bulk of human-made produced air pollution is intimately related to industrialization. Humans did not begin to alter the environment until they began to live in communities.

Recognition of pollutants as health hazards has rarely resulted in pollution reduction or control; traditionally, only when personal survival is at stake has effective action been taken. Unfortunately, the control of pollutants rarely takes place prior to public outcry, even though the technology for controlling pollutants may be available.

Air pollution issues date back to the reign of the English King Edward I (1271–1307), where a protest was conducted by the nobility against the use of "sea" coal. Under later kings such as Richard (1377–1399), and later under Henry V (1413–1422), England took steps to regulate and restrict the use of coal through taxation and regulation of the transportation of coal in London. Other legislations, parliamentary studies, and literary comments appeared sporadically during the next 250 years. In 1661, a pamphlet was published by the Royal Command of Charles II entitled "Fumifugium; or the Inconveniences of Air and Smoke in London Dissipated; Together with Some Remedies Humbly Proposed." The paper was written by John Evelyn, one of the founding fathers of the Royal Society. Later, in 1819, a Select Committee of the British Parliament was formed to study smoke abatement.

Air pollution was a fact of life during the first half of the 20th century. The smoke problem in London peaked around December 1952; during this "air pollution episode," approximately 4,000 people died, primarily of respiratory problems. Similar issues were recorded around the same time period in Pennsylvania, New York City, Tennessee, St. Louis, and Pittsburgh. Additionally, details of these often-referenced episodes are available in the literature [1].

Smoke (particulates) and acid gases produced by the burning of coal have been significant air pollutants for more than 400 years. The use of oil and gas has reduced this problem somewhat; however, local natural resources (and hence costs) currently dictate the choice between coal, oil, and gas for stationary energy conversion.

The internal combustion engine in the automobile is a more recent but no less significant source of air pollution. The contribution of the automobile to air pollution was recognized in 1915; however, 30 years passed before action was taken to reduce automobile pollution. Typical smog problems are related to the automobile and industry, and the pollution problems are all concentrated in urban areas.

The average person breathes 35 lb of air each day—six times as much as the food and drink normally consumed in the same period of time. However, approximately 200 million tons of waste gases are being released into the air annually. Slightly over half of the pollution comes from the internal combustion engines of cars and other motor vehicles. Roughly 25% comes from fuel burned at stationary sources such as power-generating plants, and another 15% is emitted from industrial processes.

1.3 The Need for Air Pollution Control

Profit and protection are the two fundamental reasons for the cleaning of gases in industry. One example of profit is the use of blast furnace gases from the steel industry being used to produce steam and power. However, impurities must be removed from the gases before they can be burned satisfactorily. Another example is sulfur dioxide which can be extracted from flue gases and converted into sulfur for reuse or sale. The health and welfare of the public, in general, is often a secondary concern and only successful through strict regulation. Waste gases containing toxic constituents such as arsenic or lead fumes constitute a serious danger to the health of plant operators and the surrounding populations [2]. Other pollutants can have less direct effects to human life such as killing plants, damage paintwork, or curtains, which in the past made industrial town considerably worse places in which to live [3].

The extent to which industry cleans polluted gas streams is almost exclusively dependent on government regulations. These government limits are imposed by three main considerations: concentration levels harmful to plant and animal life; protection of the public health and welfare; and the reduction of air pollution. These considerations are not independent. For example, the legal limits on emissions are also closely related to the degree of cost needed to prevent concentrations which would damage the life of animals and plants [4].

1.4 The Environmental Movement and the Environmental Protection Agency

Franklin Delano Roosevelt was the first president to make significant strides in with air pollution control. Including major conservation projects in the New Deal reforms was due to his love of nature as much as his political ideologies. The New Deal programs created The Civilian Conservation Corps, the Soil Conservation Service, the Tennessee Valley Authority, and many more programs.

At this point in time, internal combustion engines were on the rise and industries were consuming natural resources at an unprecedented rate. Soon after World War II, in the late summer of 1962, a marine biologist named Rachel Carson, author of *Silent Spring* [5], the best-selling book about pesticide impacts on the ecosystem, opened the eyes of the world to the dangers of attacking the environment. This can be considered one of the major points where America started calling for reform on the protection of nature and constraints on environmental degradation.

The year 1970 was a cornerstone for modern environmental policy. The National Environmental Policy Act (NEPA) was enacted on 1 January 1970. The NEPA was very generic at first and not based on specific legislation, but rather on quality of life. The NEPA established the Council for Environmental Quality (CEQ), which was mandated to implement legislation on environmental protection. Earth Day was also established in 1970, and while it did not have any legislative impact, it is a reminder to this day that we need to protect our planet.

The U.S. Environmental Protection Agency (EPA) was officially approved by Congress on 9 July 1970 as part of a reorganization plan to establish a single, independent agency, separate from the CEQ. The EPA officially began operation on 2 December 1970.

For additional literature regarding the early history and the environmental movement, the interested reader is referred to the book by Philip Shabecoff, *A Fierce Green Fire* [6]. This outstanding book and Ponting's *A Green History of the World* [7] are "musts" for anyone who works in or has interests in the environmental arena.

References

1 M. K. Theodore and L. Theodore, *Introduction to Environmental Management*, 3rd edition, CRC Press/Taylor & Francis Group, Boca Raton, FL, 2021.
2 L. Theodore and R. R. Dupont, *Water Resources Management*, CRC Press/Taylor & Francis Group, Boca Raton, FL, 2020.
3 L. Theodore and R. R. Dupont, *Environmental Health Risk and Hazard Risk Assessment: Principles and Calculations*, CRC Press/Taylor & Francis Group, Boca Raton, FL, 2013.
4 L. Theodore, *Air Pollution Control Equipment Calculations*, John Wiley & Sons, Hoboken, NJ, 2008.
5 R. Carson, *Silent Spring*, Houghton – Mifflin, Boston, MA, 1962.
6 P. Shabecoff, *A Fierce Green Fire*, Island Press, New York City, NY, 2012.
7 C. Ponting, *A Green History of the World*, St. Martin's Press, New York City, NY, 1991.

2

Classification, Sources, and Effects of Air Pollution

2.1 Introduction

Air pollutants may be divided into two broad categories, *natural* and *man-made*. Natural sources of air pollutants include the following:

- Wind-blown dust
- Smoke, gases, and fly ash from forest fires
- Volcanic ash and gases
- Ozone from lightning
- Natural decompositions resulting in gases and odors
- Naturally occurring radioactivity
- Esters, terpenes, and pollens from vegetation

These sources of pollution have little, if any, effect on life. The air pollution that concerns us the most is from man-made sources and covers a wide spectrum of chemical and physical activities. In the United States alone, air pollution is generated from over 200 million vehicles, the refuse of over 300 million people, generating billions of kilowatts of electricity, and innumerable products demanded by the everyday life of society.

Air pollutants by origin and state of matter are classified below.

- Origin
  - Primary—emitted to the atmosphere from a process
  - Secondary—formed in the atmosphere as a result of a chemical reaction
- State of matter
  - Gaseous
  - Particulate (solid or liquid)

The gaseous state of mater is further broken down to true gases and vapors. Examples of true gases are SO_2, NO_x, O_3, and CO. Some examples of vapors are paint, gasoline, and dry-cleaning agents.

Particulates are defined here as solid or liquid matter whose effective diameter is larger than a molecule but smaller than approximately 1,000 μm (micron). Particulates dispersed in a gaseous medium are defined as an aerosol. Dispersed particles such as smoke, fog, haze, and dust are particular types of aerosols. Aerosols are often difficult to classify from a scientific basis. Aerosols can be classified by surface-to-volume ratio, settling rate, particle size distribution, chemical activity, optical activity, ability to absorb electrical charge, etc. Air pollution control is most concerned with the particle size and settling rate. Some important limits are that particles over 100 μm settle too rapidly to be considered dispersants, and conversely, particles 1 μm or less are considered permanent suspensions because of the slow settling rate. In an effort to avoid pedantic scientific classifications, colloquial terms such as smoke, dust, and mist are considered to be a satisfactory and sufficient method of nomenclature in this text.

Dust is typically formed by mechanical processing of solid matter into particles of smaller size of about 1 μm up to about 100 or 200 μm but can be larger. Grinding, crushing, and drilling are typical processes that create dust. Dust particles are typically irregular in shape. The particle size refers to an average particle size for a given distribution. Common industrial examples include fly ash and rock dusts, while a typical household example would be ordinary flour.

Smoke is defined by optical density and how opaque a vapor is. Smoke is most often formed from the burning of organic materials such as wood, coal, and tobacco. Smoke consists of extremely fine particles, ranging in size from less than 0.01 μm up to 1 μm. It can contain both liquid and solid constituents. The liquid constituents are tarry and spherical, while the solid constituents are irregularly shaped.

Fumes are similar to smoke but are typically formed by processes such as sublimation or condensation as well as combustion. Fumes range in particle size from less than 0.1 to 1 μm and settle very slowly, exhibiting strong Brownian motion.

Mists or fogs are typically formed either by the condensation of water or some other vapor, giving a suspension of natural fogs and mists which usually range in size between 2 and 200 μm. Droplets larger than 200 μm are classified as drizzle or rain.

Pollutants, either liquid or solid, tend to physically interact and/or chemically react with other components in the atmosphere, and its properties and effects may be changed. A substance is exposed to air as it is broken up into smaller particles, and the resulting interactions are often unpredictable. Small aerosol particles (from 0.001 to 0.1 μm) can create fog by acting as condensation nuclei. Particles less than 2 or 3 μm in size—about half (by weight) of the particles suspended in urban air—can penetrate into mucous membranes and attract and convey harmful chemicals such as sulfur dioxide. Aerosols tend to exhibit high surface activity due to their high surface area to volume ratio. Aerosols are able to facilitate chemical reactions because of their readiness to adsorption of gas molecules.

Particle size has a significant impact on reactivity. Materials that burn slowly in a pile can oxidize extremely quickly or possibly even explode when dispersed as fine particles in air. Dust explosions are often caused by rapid dispersion of combustible particles in an oxidative environment. Adsorption and catalytic phenomena can also be extremely important in analyzing and understanding particulate pollution problems. The conversion of sulfur dioxide to corrosive sulfuric acid assisted by the catalytic action of iron oxide particles, for example, demonstrates the catalytic nature of certain types of particles in the atmosphere.

As noted in the introduction, gaseous air pollutants can originate from both natural and man-made sources. Natural sources of gaseous air are described above. This text is focused on man-made pollutants. Man-made sources (such as those originating from fuel combustion and industrial processes) cover a wide range of chemical and physical activities and are the major contributors to urban air pollution.

These gaseous pollutants can be divided into separate, but not unique, groups—*inorganic* and *organic*:

Inorganic Pollutants

- Sulfur compounds such as sulfur dioxide, sulfur trioxide, and hydrogen sulfide
- Carbon monoxide and carbon dioxide

- Halogens and halides such as hydrogen fluoride, chlorine, fluorine, hydrogen chloride, and silicon tetrafluoride
- Nitrogen oxides such as nitrous oxide, nitric oxide, nitrogen dioxide, and other nitrous oxides
- Photochemical products: ozone, oxidants
- Cyanides: hydrogen cyanide
- Ammonia

Organic Pollutants

- Hydrocarbons: paraffins, acetylene, olefins, aromatics
- Aliphatic oxygenated compounds: Aldehydes, ketones, organic acids, alcohols, organic sulfides, organic halides
- Organic hydroperoxides: peroxyacyl nitrate or nitrite (PAN)

Finally, the greatest long-term need for deeper understanding of the atmospheric environment lies in a crucial area: the modes of action and effects of pollutants on humans, animals, plants, and inanimate objects. Existing knowledge, buttressed by the clear need for haste, provides the current basis for the establishment of air quality criteria. However, the ability to refine and augment such criteria and standards, to predict the effects of air pollutants, and to detect such effects at an early stage will require much more penetrating knowledge than now exists of the effects themselves and of the mechanisms of contaminant action.

2.2 Human Responses to Air Pollution

The effects of air pollution on humans are very difficult to establish; this is primarily due to the problems in simulating in the laboratory the entire lifetime of a person in a significantly shorter time period. The effects on humans generally involve the eyes and the respiratory tract.

The eye is sensitive to certain gases and to particulate matter; tearing normally flushes these foreign materials from the eyes. No lasting damage has been shown to result from air pollution.

The respiratory system removes impurities from the inhaled air, warms and humidifies the air, and then supplies oxygen to the circulatory system; the circulatory system then releases carbon dioxide and is exhaled. The three major portions of the respiratory system are the nasopharyngeal, which includes the nasal cavity and the pharynx, the respiratory ducts, and the lungs.

Particles larger than 10 μm are removed in the nasopharyngeal section. The nasal hairs block large particles. Some particles impinge upon the surface of the pharynx, which is covered with mucus, and are collected there. The mucus is then swallowed, which may lead to other problems. Some gases, adsorbed by the mucus, may cause irritation of the pharynx. The respiratory ducts are also covered with mucus; the mucus moves in the direction of the pharynx. If a large amount of foreign material becomes imbedded in the mucus, the mucus stops moving and the air can no longer be "cleaned" effectively.

The oxygen and carbon dioxide exchange between the blood and the air takes place within the lungs. Particles less than 10 μm in diameter may enter the lungs; particles smaller than 1 μm may remain in the tiny sacs within the lung, thereby decreasing its efficiency. Smaller particles may pass through the lung membrane and enter the blood stream. Phagocyte cells in the lungs engulf particles trapped in the sacs and carry them to the mucus or to the blood stream for removal from the body.

Table 2.1 Maximum breathing capacity (Liters/Minute).

	Low pollution		High pollution	
Age	**Male**	**Female**	**Male**	**Female**
All ages	125.9	85.1	119.3	83.1
30–44	151.4	97.2	144.9	93.8
45–64	112.9	84.9	110.5	78.9
65+	87.5	63.7	80.3	55.6

Gases in the air may irritate the respiratory tract, irritate the tissues in the lung, or enter into the blood stream. Once in the blood stream, the dissolved gases may be deposited in various organs of the body, where they may have an effect. Unusually severe pollution levels may impair the heart.

The effects of air pollution on man are difficult to assess. Some problems become apparent only after a lifetime, but answers are needed now. The interaction between man and pollution may be described using several classifications. Epidemiological, statistical, pathophysiological, and biological classifications have been used; in addition, a number of medical indices are convenient and meaningful pollution-problem criteria. Both mortality rates and morbidity can be related to air pollution levels. Morbidity studies include the incidence and prevalence of disease, especially respiratory symptoms.

Asthma is an example of a medical effect of air pollution. Asthma is a disease, caused primarily by pollen, which is characterized by an increased responsiveness of the trachea and bronchi to stimuli and results in a narrowing of the airway. Other medical indices include chronic bronchitis, emphysema, chronic cough, and dyspnea (shortness of breath, phlegm production). Hospital admissions have also been correlated with air pollutant levels.

Breathing capacity is affected by air pollution. This is readily seen in Table 2.1. Note the ambiguity in typical results from a medical study. Additional details on respiratory rates, etc. are given in the literature.

2.3 Human Health Effects: Oxides of Sulfur, Carbon, and Nitrogen

2.3.1 Sulfur Oxides

The sulfur oxides include sulfur dioxide (SO_2), sulfur trioxide (SO_3), the acids, and the salts of their acids. The health effects of SO_2 and SO_3 are typically treated together. SO_3 is of more concern to industrial equipment as it readily reacts with water to form sulfuric acid and can attack equipment. Sulfur dioxide is the principal sulfur oxide of interest in health concerns, occurring as a component of sulfuric acid or its salts (sulfates), and has also been extensively studied.

Indoor concentrations of SO_2 have been measured as high as an average of 182 $\mu g/m^3$ and over 1,300 $\mu g/m^3$ for shorter periods. Evidence has been shown to indicate that SO_2 can cause significant indoor pollution problems.

Considerable research has been conducted for setting a National Ambient Air Quality Standard for SO_2. SO_2 has been linked with human effects ranging from increased mortality to reduced respiratory functions. These studies reveal issues with both short- and long-term exposure but were conducted when SO_2 is accompanied by elevated concentrations of particulate matter. It should be noted that the role of SO_2 alone in causing adverse effects at the NAAQS standard levels was not established, but the evidence shows it needs to be controlled.

Sulfuric acid was found to have significantly been more potent than SO_2. The thresholds for irritant effects of SO_2 are over twice as high as sulfuric acid (1,600–2,600 μg/m^3 for SO_2 vs. 600–850 μg/m^3 for sulfuric acid) [1]. Combination of the two compounds produced irritation at a substantially lower threshold, and the effects of the combination were considered to be additive. NAPCA [1] indicated measurable respiratory function changes with exposure at levels over 3,000 μg/m^3 (1.2 ppm) for SO_2 for 30 minutes and over 350 μg/m^3 for sulfuric acid mist for 15 minutes.

The NAPCA considered that several important toxicological findings were derived from their review:

- Sulfuric acid aerosol is a considerably more potent irritant than SO_2.
- Effects of SO_2 are more pronounced at particle sizes of 3 μm or less, and particle sizes less than 1 μm were more effective at lower doses.
- Sulfuric acid mist is considerably more irritative than dry sulfate particles.
  - This indicates a more pronounced effect under conditions of high humidity.
- Eighty percent or more (by weight) of ambient suspended sulfates may be less than 2 μm in diameter and penetrate deeply into the respiratory system.
  - Significant protection is provided by nose-breathing.
- At very high concentrations, SO_2 inhibits respiratory clearance through reduced ciliary action and mucus removal.
- For the most part, the effects of sulfur oxides are related to irritation of the respiratory system.

The Threshold Limit Value (TLV) established for SO_2 by the American Conference of Governmental Industrial Hygienists [2] (ACGIH) is at the high level of 5 ppm, approximately 13,000 μg/m^3. Studies with human subjects cited in this report indicate the following effects of SO_2 with increasing concentration:

- 3–5 ppm noticeable odor.
- 8–12 ppm irritation of the mucous membranes, i.e., nose, eyes, etc.
- 13 ppm increased airway resistance.
- 150 ppm increased incidence of respiratory disease, hospitalization, deaths from all causes, etc.

Note that these levels are considerably higher thresholds than those reported earlier. TLVs are established for use in industrial contexts where people are exposed for more limited amounts of time. The NAPCA limits are applicable to the nonworkplace environment in which susceptible people such as children and elderly will be exposed for a much greater duration.

The TLV of sulfuric acid is 1 mg/m^3, since above this level it causes respiratory irritation, and over prolonged periods, it causes dental injury [3]. In a study by Amdur et al. [4] in which normal human subjects were exposed to the inhalation of sulfuric acid mist for 5–15 minutes, concentrations below 1 mg/m^3 could not be detected by odor, taste, or irritation. A concentration of 3 mg/m^3 was noticed by most subjects, and 5 mg/m^3 resulted in irritation of the respiratory tract. Workers chronically exposed to H_2SO_4 may show lesions of the skin, tracheobronchitis, stomatitis, conjunctivitis, and gastritis. There is also evidence that chronic exposure causes corrosion of dental enamel.

In summary, the combination of atmospheric sulfur oxide, its conversion products, and suspended particulates appears to present a clear health hazard at commonly realized ambient levels, although the role of sulfur dioxide alone has not been clearly established.

2.3.2 Carbon Monoxide

Carbon monoxide (CO) has been extensively studied for its toxicological effects, primarily on the cardiovascular system. This colorless and odorless gas is lethal to humans for a few minutes at concentrations exceeding 5,000 ppm. As the health effects of CO are so generally known, only a brief summary is included here, and more information can be found in the literature.

Indoor concentrations of CO are similar to those of SO_2, but there are important differences. CO is relatively unreactive and decays slowly overtime, whereas as SO_2 is more reactive and will not readily decay over time. CO can be generated by a number of indoor sources such as gas-fired appliances, boilers and furnaces, motor vehicles in attached garages, and leaky chimneys. These common sources make CO an extremely important indoor pollutant for study and control.

The TLV for carbon monoxide is 50 ppm (~55 mg/m^3). The inhalation of carbon monoxide causes asphyxiation (anoxia, hypoxia) by forming metastable chemical compounds, primarily with hemoglobin and secondarily with other biochemical constituents, which in a complex manner reduces the availability of oxygen for other cellular systems of the body. This is concerning as CO is odorless, colorless, and of similar density to air, making it extremely difficult for the human body to detect. In some respects, the presence of CO is a more serious concern to the human body than a lack of oxygen.

The effect of carbon monoxide exposure on man is enhanced by factors such as heavy labor, high environmental temperature, and increasing altitude (above 2,000 feet). Susceptibility is greatest in the aged, the very young, those with cardiac or chronic respiratory disease, and in pregnancy [5]. However, some individuals, particularly cigarette smokers, have an increased tolerance to CO.

2.3.3 Carbon Dioxide

Carbon dioxide is present in the air at ambient levels above 325 ppm [6] and is typically not considered a hazardous constituent of the air. However, CO_2 is indeed an asphyxiant at high concentrations. There have been a number of instances reported, usually occupational in nature, where CO_2 has caused fatalities. The normal symptoms of excessive CO_2 exposure as reported by Hamilton and Hardy [7] are attributable to oxygen deprivation. These authors also state that "unconsciousness and death do not occur unless oxygen is as low as 5 percent—unless the victim makes strenuous exertion, in which case death may come while there is still 8 percent oxygen" [7]. Carbon dioxide has been reported to be a "weak narcotic at 30,000 ppm; at 50,000 ppm, a 30-minute exposure produces signs of intoxication, and, at 70,000 to 100,000 ppm, unconsciousness in a few minutes" [7]. The TLV has been set at 5,000 ppm for an 8-hour exposure. The personnel of a submarine experienced chronic fatigue when CO_2 levels reached 3–15% (30,000–150,000 ppm) for several weeks [8]. In the same report, LaVerne et al. [8] stated that CO_2 is easily "eliminated from the body and brain by reflex hyperventilation within seconds after treatment, thereby preventing cumulative toxic side effects."

Carbon dioxide has been used in varying concentrations for many different therapeutic reasons for almost 50 years. CO_2 is a cerebrovascular dilator and respiratory stimulant and has been used in the fields of pediatrics and geriatrics as well as for the treatment of narcotic addiction and alcoholism [8].

The normal ambient concentration of CO_2 is approximately 325 ppm. Indoor concentrations of CO_2 have been found to be 1 to 10 times higher (300–3,200 ppm) than outdoor concentrations in industrial environments. These concentrations are less than the TLV, but recall that the TLV is based on an 8-hour period, and exposure for longer periods can be harmful. LaVerne et al. have stated, "it appears that CO_2 inhalation, whether therapeutic or nontherapeutic, can be as safe or as dangerous as the conditions created in each specific situation" [8].

2.3.4 Nitrogen Oxides

The primary concerns among the oxides of nitrogen are nitrogen dioxide (NO_2), nitric oxide (NO), nitric acid, and nitrates. Nitric oxide peak ambient levels are generally not present at high enough levels to pose a direct health hazard [9]. The evidence for respiratory disease associated with NO_2 is cited below. Nitrogen oxide reaction products have been associated with equipment corrosion and smog, but biological effects are generally not well-researched. Nitrates appear to have a more uniform particle size distribution than sulfates, but the ratio of respirable aerosols (<3.5 μm) may be smaller [10].

Data from earlier studies [11] on NO_2 concluded that the findings suggest an excessive risk of acute respiratory disease from exposure to prolonged levels of nitrogen dioxide, ranging from 113 to 395 $\mu g/m^3$, in combination with short-term exposures of 301–1,203 $\mu g/m^3$. Lingering effects from exposure of these levels may last for as long as 4 years. Short-term exposure can have significant effects even in the absence of long-term exposure. A mix of short-term and long-term exposure may also trigger chronic respiratory symptoms.

The TLV for nitric acid is 2 ppm (~5 mg/m^3) [3]. In practice, HNO_3 is usually found in conjunction with nitrogen dioxide. According to Fairhill, continued exposure to the vapor and mist of nitric acid may result in chronic bronchitis and more severe exposure to chemical pneumonitis [12]. It may also erode the dental enamel [13].

2.4 Human Health Effects: Photochemical Oxidants and Organic Pollutants

2.4.1 Photochemical Oxidants

This section covers oxidants (except for the previously discussed NO_2) that result from photochemical atmospheric reactions. The primary compounds of interest that present a potential hazard to human health are ozone (O_3), peroxyacetyl nitrates [peroxyacetyl nitrate (PAN) and peroxybenzoly nitrate (PBZN)], and aldehydes [formaldehyde (HCHO) and acrolein (CH_2CHCHO)].

Ozone and PAN are secondary pollutants in ambient air. They typically come from interactions between NO_x and the exhaust of internal combustion engines (automobiles) in the presence of sunlight. Ozone is typically not present in significant quantities in indoor environments. Ozone and PAN are highly reactive and, like SO_2, do not persist in high concentrations indoors even when outdoor concentrations are high because of rapid decay through adsorption on indoor surfaces.

There are hypothesized to be other organic compounds such as peroxides and ketones in air pollution, but the concentrations appear biologically inactive or too low to have any affect. The precursor ambient gaseous hydrocarbons are considered irritant only as an index of potential photochemical conversion [14]. However, low-concentration pollutants should not be ignored as their toxicity can have effects. As of writing, their cumulative effects of exposure have not been extensively studied. It has been demonstrated that repeated sublethal doses (one-fifth LD_{50}) of hydrogen peroxide can result in a cumulative effect and eventually in death of experimental animals [15].

Studies have shown that ozone is the major component of photochemical oxidants in ambient air and is thus used as the standard for total oxidants. The terms oxidant and ozone are often used interchangeably in this context, but this doesn't mean that other compounds don't exist. The national standard for ambient ozone and oxidants is the same. The effects of ozone and other oxidants continue to be studied, especially the long-term effects.

2.4.2 Organic Pollutants

The term "organic pollutants" used here refers to a group of manufactured organic chemicals that may enter the air including, but not limited to, aerosol sprays, vaporized liquid solvents, household cleaners, fuels, complex aldehydes, and pesticides. The term "organic pollutants" in this context does not include the photochemical oxidants which have been discussed earlier, nor does it include the biological and allergenic materials which have not been examined earlier.

This group of organic air pollutants that are nonphotochemical does not fit into neat categories of gases and particulates. They often occur in multiphase systems, and aerosols can exist simultaneously as both gases and particulates.

Many aerosol propellants contain chlorine, bromine, or fluorine, forming a general class known as halogenated hydrocarbons. These substances have been shown to cause respiratory and liver disease, as well as being a neurotoxin. Several (carbon tetrachloride, trichlorethylene, chloroform, vinyl chloride) of these chemicals have produced cancer in animals, but there is a wide range in the effects of the toxicity.

The major categories of these chemicals include:

- Solvents and fuels
- Saturated aliphatic hydrocarbons
- Aromatic hydrocarbons
- Chlorinated hydrocarbons
- Aerosol propellants
- Other solvent compounds (other chemicals of interest are the alcohols, ketones and ethers, esters, acids, and caustics)

Details on these classes of chemicals are available in the literature [10].

2.5 Human Health Effects: Particulates

Particulates are a subset of the larger class of aerosols, an aerosol being defined as a colloid system in which liquid or solid particles are dispersed in a continuous gaseous medium. Aerosols include dust, pollen, fibers, and soot emitted from natural sources and suspended in ambient air, but they also include nitrates and sulfates that are formed in the atmosphere by gaseous reactions involving SO_2, NO_2, and oxidants. Aerosols also include materials such as paints, deodorants, and hair spray that are dispersed in the atmosphere by a gaseous propellant. The particle sizes distribution of aerosols range has an extremely wide range.

Aerosols include a mix of substances that are inert and active (both biochemically and biologically). Aerosols can be completely harmless to man in appropriate quantities and can also be extremely hazardous in even small concentrations. These chemical substances may not be detectable by any of the senses, may not produce significant early physiological effects (to warn of their presence before serious injury), and may not produce clinically recognizable pathology until many years after exposure. These facts, and the continuing introduction of new materials in our environment from changing technology, make the evaluation of hazards extremely difficult and complex.

The deposition of particles involves multiple physical mechanisms including sedimentation, inertial impaction, and Brownian motivation [16]. These mechanisms are effected by various parameters such as the particle size distribution of the pollutant (size, shape, and density), patterns of breathing (exertion, stimuli, and pulmonary disease), and the anatomy of the respiratory system itself (changes with age and disease).

There are multiple ways for the human body to rid itself of inspired pollutants. These can be as simple as direct expiration but can involve more complex processes such as transportation by mucus and cilia to the pharynx where the gastrointestinal and urinary tracts will take over. The inspired pollutants can negatively affect these collection mechanisms. They may also induce pathology in the metabolism process before clearance or be deposited in body tissues where the potential for long-term action is realized. Results from the ICRP demonstrated that variations in effective particle size and solubility can greatly alter retention and exposure in some respiratory compartments [17].

Another recent study deserving citation here is an analysis of trace element retention in the human respiratory system using proton-induced X-ray emission techniques. Desaedeleer and Winchester [17] measured the respiratory retention of lead halide and chalk dust sized by cascade impactor with stages ranging from 0.25 to 4 μm. For both substances, minimum deposition was well under 50% with particles in the 0.5 μm-diameter region, and deposition was well over 50% for both smaller and larger particles. Retention results were consistent with those found using the ICRP model and with those reported by other investigators: minimal retention of particles in a size range of about 0.3–1.0 μm, with larger particles captured in the nasopharynx compartment and smaller particles lodging primarily in the pulmonary compartment.

The health hazards for several classes of particulates and other aerosols are discussed next, grouped under the headings of total suspended particles, tobacco smoke, and lead and asbestos.

2.5.1 Total Suspended Particles

The particulate matter of outdoor air is typically measured by using a high-volume pump to force air through a filter paper for a 24-hour period. The total weight is collected and measured against the volume of air passed through the filter. This is referred to as the total suspended particles (TSP) and is only characterized by its weight per volume of air. The wide availability of these measurements of TSP, usually expressed in terms of 24-hour average ambient concentrations in micrograms per cubic meter, has led to the use of this measure of particulate concentration in epidemiological studies.

It should be noted that large particles are not inhaled deeply into the lungs. This is why there was a switch from an air quality standard based on total suspended matter (TSP) to one based on particles with an aerodynamic diameter less than 10 μm (PM_{10}). Studies in the United States and Germany have related higher levels of particulates to increased risk of respiratory, cardiovascular, and cancer-related deaths, as well as pneumonia, lung function loss, hospital emissions, and asthma. Particles 2.5 μm in aerodynamic diameter have been identified as a major contributor in elevated death rates in polluted cities. One hypothesized biological mechanism is pollution-induced lung damage, resulting in declines in lung function, respiratory, and cardiovascular disease potentially related to hypoxemia.

In general, the data on suspended particulates suggest that the quantity, composition, size, and interaction of the particulates with pollutant gases must be taken into account when evaluating their health hazards.

2.5.2 Tobacco Smoke

Tobacco smoke is a well-researched topic. It is a known carcinogen and can increase the risk of other cardiovascular diseases. Tobacco smoke can also act in a synergistic manner with other compounds. These hazards have been extensively studied from the point of view of both the smoker and the bystander. Nicotine is the most prominent compound found to have adverse effects to exposure at chronic, low levels. Nicotine has been found in indoor air in concentrations as high as 10.3 μg/m^3. Others compounds of interest are NO_x compounds because of their possible conversion in the air or body.

2.5.3 Lead

The respiratory intake of lead has been of concern because of widespread generation from automotive sources. Research relative to the problem of atmospheric lead was reviewed earlier and concluded that, at present ambient levels, lead inspiration did not represent a clear-cut significant problem, but that further research was required on acceptable body burden and the related potential long-term effects. The TLV established by the America Conference of Governmental Hygienists for lead is 200 μg/m^3 [3].

Volcanic activity and airborne soil are the primary natural sources of atmospheric lead. Smelters and refining processes, as well as incineration of lead-containing wastes, are major point sources of lead. Approximately 70–80% of the lead that used to be added to gasoline was discharged to the atmosphere. Submicron-sized lead particles, which are formed by volatilization and subsequent condensation, attach to larger particles, or they form nuclei before they are removed from the atmosphere. Once they have attained a size of several microns, they either settle out or are washed out by rain.

Atmospheric lead occurs as a particulate. The particle size range is between 0.16 and 0.43 μm. In contrast to the other major air pollutants, lead is a cumulative poison. A further difference is that it is ingested in food and water, as well as being inhaled. Of that portion taken, ion, approximately 5–10%, is absorbed in the body. Between 20 and 50% of the inspired portion is absorbed. Those portions that are not absorbed are excreted in the feces and urine. Lead is measured in the urine and blood for diagnostic evidence of lead poisoning.

An early manifestation of acute lead poisoning is a mild anemia (deficiency of red blood cells). Fatigue, irritability, mild headache, and pallor indistinguishable from other causes of anemia occur when the blood level of lead increases to 60–120 μg/100 g of whole blood. Blood levels in excess of 80 μg/100 g^{-1} result in constipation and abdominal cramps. When an acute exposure results in blood levels of lead greater than 120 μg/100 g, acute brain damage (encephalopathy) may result. Such acute exposure results in convulsions, coma, cardiorespiratory arrest, and death. Acute exposures may occur over a period of one to three weeks.

2.5.4 Asbestos

Asbestos and talc have been associated with neoplasms in persons exposed to occupational concentrations, and asbestos with nonoccupational exposures [18].

Asbestos is another well-studied compound and "asbestos" is often used in a broad manner, consisting of various fibrous mineral silicates that differ in chemical composition. The hazardous effects of these compounds are similar due to their physical similarities. The three compounds of most significance are chrysotile (hydrated magnesium silicate), amosite (iron magnesium silicate), and crocidolite (sodium iron silicate). Chrysotile is the most widely used [18]. The TLV of asbestos is 5 fibers longer than 5 μm per milliliter [3].

Asbestos was heavily used since the 1920s. It is no longer used for large-scale insulation in walls and pipes, but many older sites still have significant amounts of asbestos insulation in use. The material is not hazardous while undisturbed but is extremely easily disbursed into the air when cut or torn. The clinical consequences are generally considered to be of three types:

- A potentially disabling pneumoconiosis characterized by a "restrictive" pulmonary functional pattern with reduced total lung capacity, lowered vital capacity without evidence of airway obstruction, reduced compliance, and impaired transfer factor for carbon monoxide.
- Bronchogenic carcinoma. Many believe that there clearly is a greater risk of cancer of the upper respiratory tract in asbestos workers who also smoke.
- Mesothelioma, a diffuse carcinoma which invades the pleura and sometimes the peritoneum.

2.6 Effect on Animals, Vegetation, Materials, and Visibility

2.6.1 Animals

The effects of pollutants on animals continue to be studied. Gas-aerosol systems, in particular, are an area of interest. The results of compounded pollutants are widely varying from the simply additive (synergistic), to reduced effectiveness (antagonistic) or somewhere in between [19].

These studies on animals are made difficult because the mechanism by which an animal can become poisoned can be completely different from mechanisms by which humans are affected. In humans, direct poisoning form mechanisms such as inhalation are an important route of entry. However, probably the most common exposure to herbivorous animals grazing within a zone of pollution will be the ingestion of feed contaminated by air pollutants. In this case, inhalation is of secondary importance. Furthermore, animals higher on the food chain can be affected by eating prey that has ingested pollutants that may not be harmful to the prey animals (see the effects of DDT on peregrine falcons) [19].

Air pollutants such as arsenic, lead, fluorides, and molybdenum have been shown to be taken up by vegetation and present a hazard to livestock, through ingestion.

2.6.2 Vegetation

Many air contaminants have more effect on vegetation than they do to animals. Pollution can affect vegetation through color degradation, reduced crop yield, stunted growth, or even death. Although the aesthetics of crops can be affected with no danger to consumers, consumers are typically weary of vegetables that are strange or discolored in appearance.

Among the pollutants that can harm plants are sulfur dioxide, hydrogen fluoride, and ethylene. Plant damage caused by constituents of photochemical smog has been studied extensively. Damage has been attributed to ozone and peroxyacyl nitrates, to higher aldehydes, and to products of the reaction of ozone with olefins. However, none of these cases precisely duplicates all features of the damage observed in the field, and the question remains open for debate and further study.

2.6.3 Materials

The damage that air pollutants can do to some materials is well known: ozone in photochemical smog cracks rubber, weakens fabrics, and fades dyes; hydrogen sulfide tarnishes silver; smoke dirties laundry, acid aerosols ruin nylon hose. The most important effects include discoloration, corrosion, and soiling of goods.

Because the leaf is the primary indicator of the effects of air pollution on plants, we shall define some terms and explain how the leaf functions. A typical plant cell has three main components: the cell wall, the protoplast, and the inclusions. Much like human skin, the cell wall is thin in young plants and gradually thickens with age. Protoplast is the term used to describe the protoplasm of one cell.

It consists primarily of water, but it also includes protein, fat, and carbohydrates. The nucleus contains the hereditary material (DNA), which controls the operation of the cell. The protoplasm located outside the nucleus is called cytoplasm. The cytoplasms are tiny bodies or plastids. Examples include chloroplasts, leucoplasts, chromoplasts, and mitochondria. Chloroplasts contain the chlorophyll that manufactures the plant's food through photosynthesis. Leucoplasts convert starch into starch grains. Chromoplasts are responsible for the red, yellow, and orange colors of the fruit and flowers.

Many air pollutants accumulate on and discolor buildings. Sooty material not only blackens buildings, but it can accumulate and become encrusted. This can hide lines and decorations and thereby disfigure structures and reduce their aesthetic appeal. Another common effect is the discoloration of paint by certain acid gases. A good example is the blackening of white paint with a lead base by hydrogen sulfide.

A more serious effect and one of great economic importance is the corrosive action of acid gases on building materials. Such gases can cause stone surfaces to blister and peel. Mortar can be reduced to powder. Metals are also damaged by the corrosive action of some pollutants.

Another common effect is the deterioration of tires and other rubber goods. Cracking and apparent "drying" occur when these goods are exposed to ozone and other oxidants.

Wearing apparel and household goods can easily be soiled by air contaminants. This can become expensive due to the necessity for more frequent cleaning. Also, more frequent cleaning often leads to a shorter life span for the materials involved, and new goods will have to be purchased more often.

2.6.4 Visibility

Polluted atmosphere has been shown to impair visibility by the scattering of sunlight by particles suspended in the air, rather than sunlight being obscured by materials in the air. Since light scattering and not obscured, reduced visibility due to the presence of air pollutants occurs primarily during bright sunny days. On cloudy days or at night, there may be no noticeable effect, although the same particulate concentration may exist during these times as during the sunny days.

Reduction in visibility creates several problems. The most significant are the adverse effects on aircraft, highway, and harbor operations. Reduced visibility can also cause adverse aesthetic impressions, which can seriously restrict the growth and development of any area. Extreme conditions such as dust storms or sandstorms can actually cause physical damage in and of themselves.

References

1 U.S. Department of Health, Education and Welfare, National Air Pollution Control Administration, *Air Quality Criteria for Sulfur Oxides*, NAPCA Publication No. Air AP-50, U.S. Govt. Printing Office, Washington, D.C., 1969.

2 American Conference of Government al Industrial Hygienists, *Documentation of the Threshold Limit Values for Substances in Workroom Air*, 3rd edition, 3rd printing, American Conference of Governmental Industrial Hygienists, Cincinnati, 1976.

3 American Conference of Governmental Industrial Hygienists, *TLV's for Chemical Substances and Physical Agents in the Workroom Environment with Intended Changes for 1975*, American Conference of Governmental Industrial Hygienists, Cincinnati, 1975.

4 M. O. Amdur, L. Silverman, and P. Drinker, Inhalation of sulfuric acid mist by human subjects, *Arch. Ind. Hyg. Occup. Med.*, 6, 305, 1952.

5 C. V. Mosby, *Handbook of Medical Physiology*, 11th edition, St. Louis, 1961.

6 J. Seinfeld, *Air Pollution Physical and Chemical Fundamental*, McGraw-Hill, New York City, NY, 1975.

7 A. Hamilton and H. L. Hardy, *Industrial Toxicology*, Acton Publishing Science Group, 1974.

8 A. A. LaVerne, D. J. DiMaio, and A. J. Fernandez, Occupational, accidental, explorational carbon dioxide inhalation poisonings and prevention, *PDM*, 4–5(1–12), 83–89, 1973.

9 U. S. Department of Health, Education and Welfare National Air Pollution Control Administration, *Air Quality Criteria for Nitrogen Oxides*, NAPCA Publication No. AP-84, U.S. Govt. Printing Office, Washington, D.C., 1971.

10 A C. Stern, *Air Pollution*, 2nd edition, 3 vols, Academic Press, New York City, NY, 1968.

11 I. G. French, Recent Epidemiologic Studies on Health Effects Related to Exposure to NO_x, Presented at Scientific Seminar on Automotive Pollutants. 11 Feb., 1975, at Thomas Jefferson Memorial Auditorium, Washington, D.C. Environmental Protection Agency Publication No. 600/9-75-003. U.S. Govt. Printing Office, Washington, D.C., 1975.

12 G. G. Esposito, Quantitative measure of photochemically reactive aromatic hydrocarbons in enamels and thinners, *J. Paint Technol.*, 40(520), 214–221, 1968.

13 J. B. Lynch and J. Bell, Dental erosion in workers exposed to inorganic acid fumes, *Br. J. Ind. Med.*, 4, 84, 1947.

14 U.S. Department of Health, Education and Welfare, National Air Pollution Control Administration, *Air Quality Criteria for Photochemical Oxidants*, NAPCA Publication No. AP-63, U.S. Govt. Printing Office, Washington, D.C., 1970.

15 H. E. Stokinger and D. L. Coffin, Biological effects of air pollutants, In *Air Pollution*, Vol. 1 Academic Press, New York, 1968.

16 L. Theodore, *Air Pollution Control Equipment Calculations*, John Wiley & Sons, Hoboken, NJ, 2008.

17 J. D. Brain and P. A. Valberg, Models of lung retention based on ICRP Task Group Report, *Arch. Environ. Health*, 28, 1–11, 1974.

18 I. J. Selikoff, R. A. Bader, M. E. Bader, and J. Churg, Asbestosis and neoplasia, *Am. J. Med.*, 42, 487, 1967.

19 J. Richardson, *The Peregrine Falcon: A Wonderful Species*, Snowfinch Publishing, 2023.

3

Air Pollution Regulatory Framework

3.1 Introduction

1970 was a cornerstone year for modern environmental policy. The National Environmental Policy Act (NEPA), enacted on 1 January 1970, was considered a "political anomaly" by some. The NEPA was not based on specific legislation; instead, it referred in a general manner to environmental and quality of life concerns. The Council for Environmental Quality (CEQ), created by the NEPA, was one of the councils mandated to implement legislation. The date 22 April 1970 brought Earth Day, where thousands of demonstrators gathered all around the nation. The NEPA and Earth Day were the beginning of a long, seemingly never-ending debate over environmental issues.

The Nixon Administration at that time became preoccupied with not only trying to pass more extensive environmental legislation but also implementing the laws. Nixon's White House Commission on Executive Reorganization proposed in the Reorganizational Plan 3 of 1970 that a single, independent agency be established, separate from the CEQ. The plan was sent to Congress by President Nixon on 9 July 1970, and this new U.S. Environmental Protection Agency (EPA) began operation on 2 December 1970. The EPA was officially born.

In many ways, the EPA is the most far-reaching regulatory agency in the federal government because its authority is so broad. The EPA is charged by the Congress of the United States of America to protect the nation's land, air, and water systems. Under a mandate of national environmental laws, the EPA strives to formulate and implement actions that lead to a compatible balance between human activities and the ability of natural systems to support and nurture life.

The EPA works with the states and local governments to develop and implement comprehensive environmental programs. Amendments to federal legislations such as the Clean Air Act, the Safe Drinking Water Act, the Resource Conservation and Recovery Act, and the Comprehensive Environmental Response, Compensation and Liability Act, all mandate more involvement by state and local governments in the details of implementation.

This chapter presents the regulatory framework governing air management. It provides an overview of environmental laws and regulations used to protect human health and the environment from the potential hazards of air pollutants.

3.2 The Regulatory System

Since the early 1970s, environmental regulations have become a system in which laws, regulations, and guidelines have become interrelated. Requirements and procedures developed under previously existing laws may be referenced to in more recent laws and regulations. The history and development of this regulatory system has led to laws that focus principally on only one environmental medium, i.e., air, water, or land. Some environmental managers feel that more needs to be done to manage all of the media simultaneously since they are interrelated. Hopefully, the environmental regulatory system will evolve into a truly integrated, multimedia management framework in the future.

Federal laws are the product of Congress. Regulations written to implement the law are promulgated by the Executive Branch of government, but until judicial decisions are made regarding the interpretations of the regulations, there may be uncertainty about what regulations mean in real situations. Until recently, environmental protection groups were more frequently the plaintiffs in cases brought to court seeking interpretation of the law. Today, industry has become more active in this role. Forum shopping, the process of finding a court that is more likely to be sympathetic to the plaintiffs' point of view, continues to be an important tool in this area of environmental regulation. Many environmental cases have been heard by the Circuit Court of the District of Columbia.

Enforcement approaches for environmental regulations are environmental management—oriented in that they seek to remedy environmental harm, not simply a specific infraction of a given regulation. All laws in a legal system may be used in enforcement to prevent damage or threats of damage to the environment or human health and safety. Tax laws (e.g., tax incentives) and business regulatory laws (e.g., product claims, liability disclosures) are examples of laws not directly focused on environmental protection, but that may also be used to encourage compliance and discourage noncompliance with environmental regulations.

Common law also plays an important role in environmental management. Common law is the set of rules and principles relating to the government and security of persons and property. Common law authority is derived from the usages and customs that are recognized and enforced by the courts. In general, no infraction of the law is necessary when establishing a common law court action. A common law "civil wrong" (e.g., environmental pollution) that is brought to court is called a tort. Environmental torts may arise because of nuisance, trespass, or negligence.

Laws tend to be general and contain uncertainties relative to the implementation of principles and concepts they contain. Regulations derived from laws may be more specific but are also frequently too broad to allow clear translation into environmental technology practice. Permits may be used in the environmental regulation industry to bridge this gap and provide specific, technical requirements imposed on a facility by the regulatory agencies for the discharge of pollutants or on other activities carried out by the facility that may impact the environment.

Most major federal environmental laws (perhaps unfortunately) provide for citizen lawsuits. This empowers individuals to seek compliance or monetary penalties when these laws are violated and regulatory agencies do not take enforcement action against the violator.

3.3 Laws and Regulations: The Differences

The following are some major differences between a federal law and a federal regulation, as briefly discussed in the previous section.

1. A law (or Act) is passed by both houses of Congress and signed by the President. A regulation is issued by a government agency such as the EPA or the Occupational Safety and Health Administration (OSHA).
2. Congress can pass a law on any subject it chooses. It is limited only by the restrictions in the Constitution. A law can be challenged in court if it is unwise, unreasonable, or even silly. If, for example, a law was passed that placed a tax on burping (belching), it could not be challenged in court just because it was unenforceable. A regulation can be issued by an agency only if the agency is authorized to do so by the law passed by Congress. When Congress passes a law, it usually assigns an administrative agency to implement that law. A law regarding radio stations, for example, may be assigned to the Federal Communications Commission (FCC). Sometimes, a new agency is created to implement a law. This was the case with the Consumer Product Safety Commission (CPSC). OSHA is authorized by the Occupational Safety and Health Act to issue regulations that protect workers from exposure to the hazardous chemicals they use in manufacturing processes. If those hazardous chemicals are emitted by the plant and affect the surrounding community but do not expose the workers in the plant, OSHA is not authorized to issue an order to step the practice. (Note: The EPA is authorized to regulate such practices.)
3. Laws can include a Congressional mandate directing EPA to develop a comprehensive set of regulations. Regulations, or rulemakings, are issued by an agency, such as EPA, that translates the general mandate of a statute into a set of requirements for the Agency and the regulated community.
4. Regulations are developed by EPA in an open and public manner according to an established process. When a regulation is formally proposed, it is published in an official government document called the Federal Register to notify the public of EPA's intent to create new regulations or modify existing ones. EPA provides the public, which includes the potentially regulated community, with an opportunity to submit comments. Following an established comment period, EPA may revise the proposed rule on the basis of both an internal review process and public comments.
5. The final regulation is published, once promulgated, in the Federal Register. Included with the regulation is a discussion of the Agency's rationale for the regulatory approach, known as preamble language. Final regulations are compiled annually and incorporated in the Code of Federal Regulations (CFR) according to a highly structured format based on the topic(s) of the regulation. This latter process is called codification, and each CFR title corresponds to a different regulatory authority. For example, EPA's regulations are in Title 40 of the CFR. The codified RCRA regulations can be found in Title 40 of the CFR, Parts 240–282. These regulations are often cited as 40 CFR, with the part listed afterward (e.g., 40 CFR Part 264), or the part and section (e.g., 40 CFR §264.10).
6. A regulation may be challenged in court because the issuing agency exceeded the mandate given it by Congress. If the law requires the agency to consider costs versus benefits of the regulation, the regulation could be challenged in court if the cost/benefit analysis were not correctly or adequately done. If OSHA issues a regulation limiting a worker's exposure to a hazardous chemical to I part per million (ppm), OSHA could be called on to prove in court that such a low limit was needed to prevent a worker from being harmed. Failure to prove this would mean that OSHA exceeded its mandate under the law, as OSHA is charged to develop standards only as stringent as those required to protect worker health and provide worker safety.

7. Laws are usually brief and general. Regulations are usually lengthy and detailed. The Hazardous Materials Transportation Act, for example, is only approximately 20 pages long. It speaks in general terms about the need to protect the public from the dangers associated with transporting hazardous chemicals and identifies the Department of Transportation (DOT) as the agency responsible for issuing regulations implementing the law. The regulations issued by the DOT are several thousand pages long and are very detailed down to the exact size, shape, design, and color of the warning placards that must be used on trucks carrying any of the thousands of regulated chemicals.
8. Generally, laws are passed infrequently. Often years pass between amendments to an existing law. A completely new law on a given subject already addressed by an existing law is unusual. Laws are published as a "Public Law— - —" and are eventually codified into the United States Code.
9. Regulations are issued and amended frequently. Proposed and final new regulations and amendments to existing regulations are published daily in the Federal Register. Final regulations have the force of law when published.[1]

3.4 Provisions of the 1970 Clean Air Act

Some of the major provisions of the 1970 Clean Air Act (as they appeared at that time) included the following:

1. EPA, assisted by the states, is to designate air quality control regions. These are basic geographic units in which air quality is evaluated and the control process takes place. Regional boundaries are based on considerations of climate, meteorology, topography, urbanization, and other factors affecting conditions in each area. A region can cover only part of one state, or it can include portions of several states sharing a common air pollution problem. The country has been divided into about 250 regions. As pollution patterns change, or as more information about problems is gathered, the boundaries of some of the regions may change.
2. EPA is required by law to develop air quality criteria for the major pollutants: particulate matter, sulfur oxides, hydrocarbons, nitrogen oxides, carbon monoxide, and so on. These criteria, which are issued in "criteria documents," give levels for which these pollutants—by themselves in and combination with other pollutants—are known to have adverse effects on public health or welfare. At the same time, EPA is to provide information on control technologies for each technology, the costs, and the economic feasibility of alternative control methods. EPA is required to review both criteria documents and control technology documents from time to time to revise them as new information become available.
3. A National Ambient Air Quality Standard (NAAQS) is the maximum level that will be permitted for a given pollutant. There are two kinds of such standards: primary and secondary. Primary standards are to be sufficiently stringent to protect the public health; secondary standards must protect the public welfare. EPA sets the standards after it issues the criteria document and a controlled technology document on the pollution in question. Both primary and secondary standards apply to all control regions. Each control region must satisfy the standards set forth. It is bound by federal enforcement. Radioactive effluents, however, are still regulated by the Nuclear Regulatory Commission. Ambient air quality standards are presented in Table 3.1.

1 Matystik, W., Private Communication to L. Theodore, Naples, FL, 1995.

Table 3.1 National primary and secondary ambient air quality standards in effect in 1991.

Pollutant	Average time	Primary standard	Secondary standard
	Annual mean	80 μg/m^3	—
SO_2	24 hours	365 μg/m^3	—
	3 hours	—	1,300 μg/m^3
PM_{10}	Annual mean	50 μg/m^3	Same as primary
	—	150 μg/m^3	Same as primary
CO	8 hours	9 ppm	No secondary
	1 hour	35 ppm	No secondary
O_3	1 hour	0.12 ppm	Same as primary
Nitrogen oxides	Annual mean	0.053 ppm	Same as primary
Lead	3 months	1.5 ppm	Same as primary

4. Within nine months after EPA issues the primary and secondary standards for a pollutant, each state must formulate an implementation plan to meet, maintain, and enforce those standards in each air quality control region within its jurisdiction. The states must hold public hearings on the plans, adopt them, and submit them for EPA approval. Each state plan must provide for the attainment of primary standards within three years of EPA's approval; secondary standards must be obtained within a reasonable time. If a state fails to submit a satisfactory plan, the EPA has the authority to write its own plan for that state which that state must then carry out.
5. The act requires EPA to set standards of performance for new and modified stationary sources of pollution. The standards are distinct from the ambient air quality standards described above. They constitute direct emissions limitations for all major pollutants from specified types of sources, such as cement plants, municipal incinerators, etc. All performance standards are applicable nationally to all pollution sources. However, they also apply to existing sources whenever modification, physical change or change in method of operation results in increased emissions of old pollutants, or in new emissions of new pollutants. For all existing or modified sources in the specified categories, the states are required to set state performance standards under procedures to be established by the EPA. EPA will also prescribe procedures under which the states may choose to enforce the federal standards for new modified sources. These regulations have been presented in numerous issues of the Federal Register and can be found in the Code of Federal Regulations under Title 40, Chapter 1, Subchapter C, Part 60. When the standards are established and the source of pollution identified, a method can be used to determine how much reduction is required to reach the national standard. A rather simple mathematical equation (see Equation 3.1) was devised as a general estimate of the percentage reduction necessary throughout the control region. These facts must be known first about the gaseous pollutant: (A) the existing concentration of the pollutant, (B) the air quality standard for the pollutant, and (C) the background level of that pollutant in the region (sometimes quite difficult to determine).

$$\text{Percentage of emission reduction} = ((A - C) - (B - C))/(A - C) \tag{3.1}$$

6. The EPA Administrator was charged with promulgation of standards for hazardous air pollutants regardless of whether they emanated from new or existing sources. Standards have since been promulgated for a number of hazardous materials including asbestos, beryllium, mercury, vinyl chloride, arsenic, and benzene. Enforcement of the standards is the responsibility of the EPA; however, at the states' option, implementation plans submitted to EPA may include regulatory control procedures for these materials. After EPA approval of the implementation plan, states would then be authorized to enforce the hazardous pollution standards within their jurisdiction. Source test methods were identified with the hazardous polluting standards. These test methods were the only approved means by which compliance could be determined. The regulations also specify reporting requirements for the source. Not only must operational data be maintained, but an application must be made to the EPA prior to any modification of existing sources. No new or modified source of a hazard pollutant may start operation without proper notification to the EPA.
7. EPA may require states and individual sources to monitor polluted emissions, to keep records, and to submit public reports. All such records and reports are considered to be public information, with one exception: EPA may keep confidential any trade secrets or any other information whose public availability the manufacturer has shown to be a potential harm to his or her business. Emission data, however, are specifically exempt from such protection.
8. Once standards and implementation plans are in effect, the EPA is required to oversee state enforcement. Where widespread violations indicate that the state is failing to enforce the plan, EPA may step in and enforce it. Or EPA may enforce portions of the plan by issuing orders of compliance or by bringing civil actions in federal court for violations. EPA is also empowered to sue for immediate restraint of any pollution source that is endangering the health of persons if state or local authorities have failed to abate such pollution under their own regulations. Each major source of pollution within a control region must prepare to follow a detailed step-by-step schedule of measures it will bring into accord with the implementation plan. Such individual timetables are referred to as compliance schedules. EPA requires the state to negotiate compliance scheduled with all major pollutant sources. Once negotiated, such schedules become legally enforceable and part of the state's implementation plan.
9. Any citizen may bring suit against any person or corporation alleged violating an emission standard or other limitation applicable under the Act. Citizens may also sue the administrator of the EPA for failure to perform an action required by the Act. In cases brought by citizen plaintiffs, the courts are empowered to award the cost of the litigation to such plaintiffs whenever the court determines such an award as appropriate.

3.5 Current United States Air Quality Legislation and Regulations

The Clean Air Act, enacted in 1970 and extensively amended in 1977 and 1990, currently establishes a comprehensive program for controlling and improving the air quality in the United States through both state and federal regulation. Generally, the EPA is charged with identifying air pollutants that endanger the public health and welfare and with formulating the National Ambient Air Quality Standards (NAAQS). The states are given primary responsibility for assuring air quality within their geographic areas by adopting state implementation plans (SIPs) that specify how the NAAQS will be achieved and maintained. For each NAAQS, a state must adopt an SIP that specifies emission limitations applicable to sources of air pollution and other measures designed to achieve and maintain that NAAQS. These emission limits are the basis for the air quality regulations each state enforces. Among the measures included in the SIP is a preconstruction permit program to ensure that emissions due to new or modified sources of air pollution do not cause or contribute to a violation of a NAAQS.[2]

The CAA defines the national policy for air pollution abatement and control in the United States. It establishes goals for protecting health and natural resources and delineates what is expected of federal, state, and local governments to achieve those goals. The CAA, which was initially enacted as the Air Pollution Control Act of 1955, has undergone several revisions over the years to meet the ever-changing needs and conditions of the nation's air quality. On 15 November 1990, the president signed the most recent amendments to the CAA, referred to as the 1990 CAA Amendments. Embodied in these amendments were several progressive and creative new themes deemed appropriate for effectively achieving the air quality goals and for reforming the air quality control regulatory process. Specifically, the amendments:

1. Encouraged the use of market-based principles and other innovative approaches similar to performance-based standards and emission banking and trading.
2. Promoted the use of clean low-sulfur coal and natural gas, as well as innovative technologies to clean high-sulfur coal through the acid rain program.
3. Reduced energy waste and create enough of a market for clean fuels derived from grain and natural gas to cut dependency on oil imports by one million barrels per day.
4. Promoted energy conservation through an acid rain program that gave utilities flexibility to obtain needed emission reductions through programs that encouraged customers to conserve energy.

Several of the key provisions of the act are listed below. Details on each of these provisions are available in the literature.[2]

1. Provisions for Attainment and Maintenance of National Ambient Air Quality Standards [1]
2. Provisions Relating to Mobile Sources [1]
3. Air Toxics [2]
4. Acid Deposition Control
5. Operating Permits
6. Stratospheric Ozone Protection
7. Occupational Safety and Health Act

3.6 Major Project Permitting

The Clean Air Act and associated federal and state regulations require preconstruction permits authorizing new plant construction or significant modification of existing operations [3].

3.6.1 Prevention of Significant Deterioration (PSD)

Prevention of Significant Deterioration (PSD) of all the federal laws placing environmental controls on industry (and, in particular, on new plants). perhaps the most confusing and restrictive are the limits imposed for the prevention of significant deterioration (PSD) of air quality. These limits apply to areas of the country that are already cleaner than required by ambient air-quality

---

2 Weiss, K., Personal Communication to L. Theodore, East Willison, NY, 2021.

standards. This regulatory framework evolved from judicial and administrative action under the 1970 Clean Air Act and subsequently was given full statutory foundation by the 1977 Clean Air Act Amendments.

EPA established an area classification scheme to be applied in all such regions. The basic idea was to allow a moderate amount of industrial development but not enough to degrade air quality to a point at which it barely complied with standards. In addition, states were to designate certain areas where pristine air quality was especially desirable. All air-quality areas were categorized as Class I, Class II, or Class III. Class I areas were pristine areas subject to the tightest control. Permanently designated Class I areas included international parks, national wilderness areas, memorial parks exceeding 5,000 acres, and national parks exceeding 6,000 acres. Although the nature of these areas is such that industrial projects would not be located within them, their Class I status could affect projects in neighboring areas where meteorological conditions might result in the transport of emissions into them. Class II areas were areas of moderate industrial growth. Class III areas were areas of major industrialization. Under EPA regulations promulgated in December 1974, all areas were initially categorized as Class II. States were authorized to reclassify specified areas as Class I or Class III. No area has been recategorized to Class III; so generally, industrial plants are built in Class II areas and must have no impact on Class 1 areas.

The EPA regulations also established another critical concept known as the increment. This was the numerical definition of the amount of additional pollution that may be allowed through the combined effects of all new growth in a locality (see Table 3.2). To ensure that the increments would not be used up hastily, EPA specified that each major new plant must install best available control technology (BACT) to limit emissions. BACT is determined based on a case-by-case engineering analysis and is more stringent titan on any applicable new source performance standards (NSPS).

To implement these controls, EPA requires that every new source undergoes preconstruction review. The regulations prohibited a company from starting construction on a new source until the review had been completed and provided that, as part of the review procedure, public notice should be given and an opportunity provided for a public hearing on any disputed questions.

Table 3.2 PSD air quality increments ($\mu g/m^3$).

	Class I area	Class II area	Class III area
Sulfur dioxide			
Annual	2	20	40
24 hours	5	91	182
3 hours	25	512	700
PM_{10}			
Annual	4	17	34
24 hours	8	30	60
$PM_{2.5}$			
Annual	1	4	8
24 hours	2	9	18
Nitrogen dioxide			
Annual	2.5	25	50

3.6.2 Sources Subject to Prevention of Significant Deterioration

Sources subject to PSD regulations (40 CFR. Sec. 52.21) are major stationary sources and major modifications located in attainment areas and unclassified areas. A major stationary source was defined as any source listed in Table 3.2 with the potential to emit 100 tons per year or more of any pollutant regulated under the Clean Air Act (CAA) or any other source with the potential to emit 250 tons per year or more of any CAA pollutant. The potential to emit is defined as the maximum capacity to emit the pollutant under applicable emission standards and permit conditions (after application of any air pollution control equipment), excluding secondary emissions. A major modification is defined as any physical or operational change of a major stationary source producing a significant net emissions increase or any CAA pollutant for which there are NAAQS. If existing data are not available to characterize the site, it may be necessary to collect data for up to one year before completing required environmental permit applications and assessments. The EPA or the state may accept representative existing monitoring data collected within three years of the permit application to satisfy monitoring requirements.

EPA regulations provide exemptions from BACT and ambient air impact analyses if the emissions increase that results from the modification is accommodated by historical emissions decreases at the plant site occurring within the five years prior to the project. This exemption is referred to as the "no significant net increase in emissions" provision of the applicability analysis. A complete PSD review would include a case-by-case determination of the controls required by BACT, an ambient air-impact analysis to determine whether the source might violate applicable increments or air-quality standards: an assessment of the effect on visibility, soils, and vegetation, submission of monitoring data, and full public review. The time required to complete such a review and obtain the permit or license required to complete such a review and obtain the permit or license required to commence construction can be anywhere from six months to one year, depending on project complexity, after the application is submitted to the government authorities, and the time required to prepare such an application can be as much as an additional year.

3.6.3 Nonattainment

Those areas of the United States failing to attain compliance with ambient air-quality standards are considered nonattainment areas (NA). New plants may be built in nonattainment areas only if stringent conditions are met. Emissions have to be controlled to the greatest possible, and more than equivalent offsetting emission reductions have to be obtained from other sources to assure progress toward achievement of the ambient air-quality standards. Specifically,

1. The new source must be equipped with pollution controls to assure lowest achievable emission rate (LAER), which in no case can be less stringent than any applicable NSPS.
2. All existing sources owned by an applicant in the same state must be in compliance with applicable state implementation plan requirements or be under an approved schedule or an enforcement order to achieve such compliance.
3. The applicant must obtain sufficient emissions offsets to more than make up for the emissions to be generated by the new source (after application of LAER).
4. The emissions offsets must provide "a positive net air quality benefit in the affected area."

LAER is deliberately a technology-forcing standard of control. The enabling statute stated that LAER must reflect (1) the most stringent emission limitation contained in the implementation plan of any state for such category of sources, unless the applicant can demonstrate that such a limitation is not achievable, or (2) the most stringent limitation achievable in practice within the industrial category, whichever is more stringent. In no event could LAER be less stringent than any applicable NSPS. While the statutory language defining BACT directed that "energy, environmental, and economic impacts and other costs" be taken into account, the comparable provision on LAER does not provide for consideration of economic or related impacts.

For existing pollutant sources in nonattainment areas, reasonable available technology (RACT) would be required. EPA defines RACT by industrial category.

3.6.4 Minor Source Preconstruction Permits

Commonly under the application of air pollution controls, the emissions associated with new facilities or modifications of existing facilities will be less than the emissions thresholds needed to trigger PSD or nonattainment new source review. In such rises, a minor source preconstruction permit is required from the state permitting authority, and new source performance standards (NSPS) apply.

3.6.5 Controlled-trading Program

The legislation enacted under the Clean Air Act Amendments of 1977 provided the foundation for EPA's controlled-trading program. The essential elements of which include:

- Netting provisions of PSD
- Offsets policy under nonattainment
- Banking and brokerage under nonattainment

While these different policies vary broadly in form, their objective is essentially the same: to substitute flexible economic-incentive systems for the current rigid, technology-based regulations that specify exactly how companies must comply. These market mechanisms have made regulating easier for EPA and less burdensome and witty for industry.

3.6.6 PSD No-net-increase Provisions

Under the PSD program, the affected source is the entire plant. This source definition allows a company to determine the most cost-effective way to control pollution any time a plant is modified. Emission increases associated with a new process or production line may be compensated for by historical emission decreases at other parts of the plant. As long as the net emissions increase for the entire site is below the levels identified, a PSD review is not required.

3.6.7 Offsets Policy

Emissions offsets were EPA's first application of the concept that one source could meet its environmental protection obligations by gelling another source to assume additional control actions. In nonattainment areas, pollution from a proposed new source, even one that controls its emissions to the lowest possible level, would aggravate existing violations of ambient air-quality standards and trigger the statutory prohibition. The offsets policy provided these new sources with an alternative. The source could proceed with construction plans, provided that:

1. The source would control emissions to the lowest achievable level.
2. Other sources owned by the applicant were in compliance or on an approved compliance schedule.
3. Existing sources were persuaded to reduce emissions by an amount at least equal to the pollution that the new source would add.

3.6.8 Banking and Brokerage Policy

EPA's banking policy is aimed at providing companies with incentives to find more offsets. Under the original offset policy, a firm shutting down or modifying a facility could apply the reduction in emissions to new construction elsewhere in the region only if the changes were made simultaneously. However, with banking, a company can deposit the reduction for later use or sale. Such a policy clearly establishes that clean air (or the right to use it) has direct economic value.

3.6.9 Cross-state Air Pollution Rule (CSAPR) and Good Neighbor

On 6 July 2011, the EPA finalized the CSAPR to address air pollution from upwind states that crosses state lines and affects air quality in downwind states, specifically targeting: SO_2 and NO_x. These emissions and the soot and smog they form can affect air quality and public health locally, regionally, and even in states hundreds of miles downwind. The transport of these pollutants across state borders, referred to as interstate air pollution transport, makes it difficult for downwind states to meet health-based air quality standards for $PM_{2.5}$ and ozone.

The Clean Air Act's "good neighbor" provision requires EPA and states to address interstate transport of air pollution that affects downwind states' ability to attain and maintain NAAQS. Specifically, Clean Air Act section 110(a)(2)(D)(i)(I) requires each state in its SIP to prohibit emissions that will significantly contribute to nonattainment of an NAAQS, or interfere with maintenance of an NAAQS, in a downwind state. The Act requires EPA to backstop state actions by promulgating Federal Implementation Plans (FIPs) in the event that a state fails to submit or EPA disapproves good neighbor SIPs. The Act is targeted toward power plants but affects a large number of large industrial facilities as well. EPA sets a pollution limit (emission budget) for each of the states covered by CSAPR. Authorizations to emit pollution, known as allowances, are allocated to affected sources based on these state emissions budgets. The rule provides flexibility to affected sources, allowing sources in each state to determine their own compliance path. This includes adding or operating control technologies, upgrading or improving controls, switching fuels, and using allowances. Sources can buy and sell allowances and bank (save) allowances for future use as long as each source holds enough allowances to account for its emissions by the end of the compliance period.

3.7 International Environmental Regulations

Along with an increasing awareness of the importance of the environment has come a deeper understanding that the environment is a global issue. For example, the inhabitants of the Earth share the same air (although there is minimal exchange between the Northern and Southern hemispheres), and some of the air over the United States might be over India in a few months, and vice versa. The harmful effects of air pollutants—agricultural damage, global climate change, and acid deposition—know no political boundaries.

The industrialized nations have led the way in regulating pollutant sources, mostly on national levels. However, progress is also being made in regional alliances. The cooperation between the United States and Canada to tackle the mutual problem of acid rain is just one example of the type of effort needed in this area. Numerous worldwide conferences have helped different countries establish priorities and set agendas for their pollution control programs. Another critical and farsighted step has been to include developing countries in the discussions of pollution problems and their possible solutions. Long-term solutions are possible only if pollution-reducing technology is freely shared and only if industrializing nations are provided with economic incentives to be nonpolluters.

As noted above, environmental pollution has transcended national boundaries and is threatening the global ecosystem. International environmental concerns such as stratospheric ozone depletion, the greenhouse effect, global warming, deforestation, acid rain, and mega-disasters such as the earlier devastating nuclear accident at Chernobyl and the toxic methyl isocyanate gas accidentally released by a subsidiary of Union Carbide in Bhopal, India, have set the stage to address global pollution problems. The potential effects of global environmental pollution necessitate global cooperation in order to secure and maintain a livable global environment.

Pollution that crosses political boundaries, such as acid rain, has caused friction between countries for at least a decade. Now, however, people are beginning to recognize a class of pollution problems that can affect not just one region but the entire planet [4]. Interestingly, nearly 30 years have passed since 113 governments agreed in Stockholm, Sweden, to cooperate in attacking a new threat to human welfare: the degradation of the global ecosystem from environmental pollution, overpopulation, and mismanagement of the natural resource base. Since then the awareness of how human and biological systems interlock has increased significantly, resulting in a far more sophisticated grasp of what must be done.

Is the goal of an environmentally healthy world realistic? To answer that question, one has only to examine the extent and significance of what has been accomplished to date. Governments have responded to the environmental challenge at national, regional, and global levels with a broad spectrum of institutional and programmatic initiatives. Indeed, despite many false starts, setbacks, continuing constraints, and the emergence of new hazards, some have aligned that the spirit of international cooperation has steadily grown stronger.

3.8 Future Trends

It is very difficult to predict future regulations. In the past, regulations have been both a moving target and confusing. What can be said (for certain?) is that there will be new regulations, and the probability is high that they will be contradictory and confusing. Past and current regulations provide a measure of what can be expected.

References

1 The Clean Air Act Amendments of 1990. Materials, November 15, 1990. https://www.epa.gov/laws-regulations/summary-clean-air-act
2 L. Stander and L. Theodore, *Environmental Regulatory Calculations Handbook*, John Wiley & Sons, Hoboken, NJ, 2008.
3 Stander, L., Personal Communication to L. Theodore, Raleigh, NC, 2021.
4 G. Burke, B. Singh and L. Theodore, *Handbook of Environmental Management and Technology*, 2nd edition, John Wiley & Sons, Hoboken, NJ, 2000.

4

Introduction to Air Pollution Control Equipment Technology

4.1 Introduction

Gaseous and particulate pollutants discharged into the atmosphere can be controlled. This chapter is devoted to air pollution control equipment. Following an introductory section on control technology, Section 4.4 presents a description of the five generic devices available for particulate control. These include gravity settlers, cyclones (centrifugal separators), electrostatic precipitators, wet scrubbers, and baghouses (fabric filtration). Section 4.5 reviews the various gaseous control equipment. Included in this presentation are absorption, adsorption, combustion, and condensation.

4.2 Control Technology Options

Emissions control is a constantly changing science of evaluating the known hazards of pollutants, ability of current technology to control them, the cost/benefit analysis of that technology, planning, and standard setting. This can be forward thinking as well by evaluating application of specific hardware, fuels, and materials with low emission potential. The goal of emissions control is to achieve an acceptable atmosphere in the local community and that requires a combination of measures aimed at all of the major area or point sources. Regulations often require the use of the Best Available Control Technology (BACT). Four major courses of action are available to eliminate or reduce emissions from a polluting operation:

1. Regulate the location of the operation
2. Eliminate the operation
3. Modify the operation
4. Reduce or eliminate discharges from the operation by applying control devices and systems.

Control technology is self-defeating if it creates undesirable side effects in meeting limited air pollution control objectives. Air pollution control should be considered in terms of both ecological consequences and technological systems. Ecological side effects align with the original goal of air pollution control: taking care of the environment. Issues with disposal must take into account unmanageable accumulations of contaminants. The effects of the control technology itself must also be considered. One example is injecting ammonia to reduce NO_x emissions. If the ammonia is over-sprayed, raw ammonia is injected into the air instead of NO_x. One must also consider the amount of NO_x generated in manufacturing the ammonia and transporting it to site. These may be concentrated in the collection process, such as ground water pollution resulting from landfill practices or pollution of streams from the discharges of air pollution control systems.

4.2.1 Elimination of Air Pollution Operations

Elimination is an ideal, though rarely potential, solution to air pollution control. An operation or activity can be eliminated only if it is unnecessary, or if a satisfactory alternate exists. Certain operations can be prohibited when they can be replaced by improved methods. For example, hand-firing of coal-burning installations have long been prohibited in favor of automatic fuel-feeding systems. Air pollution from motor vehicles could be drastically reduced by limiting their use if alternate, low-emission means of transportation, such as bicycles or mass transportation, were available. Especially in the United States, many communities are too large for bicycles to be feasible, and mass transportation has been largely underdeveloped and/or poorly maintained. Thus, all control strategies which involve elimination of activities of a certain type must also provide for the institution of feasible alternate means of accomplishing the ends presently served by those activities.

Sources of air pollution can be eliminated by the legislation passing of rules or regulations prohibiting a specific operation or type or design of equipment. These regulations are comparatively easy to enforce, but difficult to pass through the government.

4.2.2 Regulation of Location of Operations

Alternative approaches to air pollution control include

1. Applying zoning ordinances
   - Cause the sources of air pollution to be located or distributed to minimize the effects of air pollution on receptors downwind.
2. Imposing areal limits on emission rates
   - Derived from air quality standard requirements.

Regulatory standards may use both approaches through land-use planning, zoning controls, and the special handling of individual zone exception and land-use permit cases. The cooperation of urban planning, transportation, and zoning agencies is required in this case. They can be useful in preventing potential public nuisance problems by locating pollution sources in areas with advantages prevailing airflow or reducing emission source densities.

Enforcement personnel generally have extensive experience of the nuisance potentials of specific types of industries, the community problems that are created, and the economic and ecologic relationships of an industry to its community and environment. They can be utilized in the phases of planning, particularly in the handling of zone exception cases. In some agencies, enforcement officers are part of an interagency team that investigates zone exception cases.

4.2.3 Modifications of Operations

Improvements to control technology are often implemented through hardware, fuels, and materials. These are applied at various points in the operational cycles of processes. Some examples include a better actuator on damper and temperature-compensated flow meters, balancing fuel in a multiburner system. Continuous Emissions Monitoring Systems (CEMS) can be installed to monitor the flue gas of equipment and ensure it is operating per required regulations. Details are available in the literature [1, 2].

4.2.4 Reduce and/or Eliminate Discharges

Much of the material in this book receives treatment in the rest of this book.

4.3 Introduction to Air Pollution Control Equipment

Controlling the emission of pollutants from industrial and domestic sources is important in protecting the quality of air. As discussed in previous chapters, air pollutants can exist in the form of particulate matter or as gases. Particulate matter, or aerosols, can form fine suspensions of particles in a gaseous medium.

Air cleaning devices have been reducing pollutant emissions from various installations for many years. Originally, air cleaning equipment was used only if the effluent was highly toxic or had some recovery value. Now with more stringent legislation, control technologies have been upgraded, and more sources are controlled in order to meet the National Ambient Air Quality Standards (NAAQS). State and local air pollution agencies have adopted regulations that are in some cases more stringent than federal emission standards. These are often implemented to maintain safe living conditions in local communities when the individual administrations have relaxed regulations.

Profit and protection are the two fundamental reasons for the cleaning of gases in industry. One example of profit is the use of blast furnace gases from the steel industry being used to produce steam and power. However, impurities must be removed from the gases before they can be burned satisfactorily. Another example is sulfur dioxide which can be extracted from flue gases and converted into sulfur for reuse or sale. The health and welfare of the public, in general, is often a secondary concern and only successful through strict regulation. Waste gases containing toxic constituents such as arsenic or lead fumes constitute a serious danger to the health of plant operators and the surrounding populations. Other pollutants can have less direct effects to human life such as killing plants and damaging paintwork or curtains, which in the past made industrial town considerably worse places to live.

The extent to which industry cleans polluted gas streams is almost exclusively dependent on government regulations. These government limits are imposed by three main considerations: concentration levels harmful to plant and animal life; protection of the public health and welfare; and the reduction of air pollution. These considerations are not independent. For example, the legal limits on emissions are also closely related to the degree of cost needed to prevent concentrations which would damage the life of animals and plants.

Particulate and gaseous pollutants discharged into the atmosphere can be controlled. The five generic devices available for particulate control include gravity settlers, cyclones (centrifugal separators), electrostatic precipitators, wet scrubbers, and baghouses (fabric filtration). Gaseous control equipment includes combustion, absorption, and condensation. These control devices are introduced and discussed in the next sections in this chapter.

4.4 Air Pollution Control Equipment for Particulates

As described above, the five major types of particulate air pollution control equipment are as follows:

1. Gravity settlers
2. Cyclones
3. Electrostatic precipitators
4. Scrubbers
5. Baghouses

Each of these devices is briefly described below. More details are provided in Chapters 6–10.

4.4.1 Gravity Settlers

Gravity settlers, or gravity settling chambers, have long been utilized industrially for the removal of solid and liquid waste materials from gaseous streams. These devices are simple to construct and have low maintenance costs, low capital cost, and simple disposal of waste materials. Gravity settlers are typically constructed as long, horizontal parallelepipeds with inlet and outlet ports. The settler is a sudden expansion in the duct carrying the particle-laden gases: the contaminated gas stream enters at one end, the gas velocity suddenly decreases and larger particles settle out, and the cleaned gas exits from the other end. The particles settle toward the collection surface at the bottom of the unit with a velocity at or near their settling velocity. The main advantage of this device is that the separation force is provided by gravity (free by nature). Its use in industry is generally limited to the removal of large particles (>40 microns) and is useful with gas streams containing very high concentrations of particles. The gravity settler is typically the first of multiple collection devices in series to reduce the loading on more fine control devices.

4.4.2 Cyclones

Centrifugal separators are commonly referred to as cyclones. Cyclones are widely used in industry for the removal of solid and liquid particles (or particulates) from gas streams. Typical applications for cyclones include the following:

- Mining and metallurgical operations
- Cement and plastics industries
- Pulp and paper mills
- Chemical and pharmaceutical processes
- Petroleum production (cat-cracking cyclones)
- Combustion operations (fly ash collection)

Gravity is again used as the primary collection force, but cyclones require a significant pressure drop (compared to gravity settlers). The particulate-laden gas stream enters the unit tangentially and is forced into a constrained vortex in the cylindrical portion of the cyclone. The particles develop an angular velocity. Because of its greater inertia, it tends to move across the gas streamlines in a tangential rather than rotary direction; thus, it attains a net outward radial velocity. Due to the relatively high density of the particles, they are forced toward the wall of the unit as it rotates around the axis of the cyclone. As the particles impact the wall, they are carried by gravity toward the dust collector at the bottom the unit. The flow vortex is reversed in the lower (conical) portion of the unit, leaving most of the entrained particles behind. The cleaned gas then passes up through the center of the unit (inner vortex) and out of the collector.

4.4.3 Electrostatic Precipitators

Electrostatic precipitators (ESPs) are exceptional devices for removing particles from gas streams at high collection efficiencies. Majority of power plants with particle-laden flue gas use this technology to remove fly ash prior to discharge.

Two types of high-voltage ESPs currently being used are tubular and plate. Tubular precipitators consist of cylindrical collection tubes with discharge electrodes located along the axis of the cylinder.

ESPs that are in use at the time of writing are mostly the plate type. Plate type ESPs use flat parallel surfaces to collect particles. The plates are typically 8–18 inches apart. Discharge electrodes are located on the centerline of the adjacent plates. These electrodes create an electric field between them which imparts a static charge to the particles passing between them in the gas stream. The grounded collection electrodes attract the charged particles and separate them out from the gas stream. The particulate-laden flue gas passes horizontally between the plates. EPSs can be designed for both horizontal and vertical flows depending on site layouts. The tubes and plates are periodically rapped to remove the collected particles.

4.4.4 Scrubbers

Wet scrubbers can effectively remove both particulate and gaseous pollutants and, as a result, have widespread usage. The wet scrubbing process involves passing a contaminated gas stream through a liquid. Wet scrubbers include all the various types of gas absorption equipment (to be discussed later). The term "scrubber" will be restricted to those systems that utilize a liquid, usually water, to achieve or assist in the removal of particulate matter from a gas stream.

Wet scrubbers have the unfortunate, widely visible, side effect of producing a steam plume. Scrubber systems utilizing water in high-temperature processes (i.e., combustion) can generate a visible white plume as it leaves the stack. This plume is made of supersaturated water vapor, and while not strictly an air pollution problem, such plumes may be objectionable for aesthetic reasons. There are several ways to avoid or eliminate the steam plume such as using ESPs or cyclones that avoid adding water to the system [1–4].

4.4.5 Baghouses

Baghouses are another extremely commonplace control device in the utility and industrial sectors. The different applications use many different types of fabric filters and physical arrangement, but the method of removing collected material from the media is similar. The essential differences may be related, in general, to mode of operation, type of fabric, cleaning mechanism, and equipment used.

Baghouses can be designed as either positive or negative pressure systems. In positive pressurized systems, the flue gas typically enters through the top of the bags or through the hopper at the bottom. In negative pressure systems, the particulate-laden flue gas is forced through the inside of the bag. Negative pressure systems are generally preferred and are more common as any equipment/ductwork leaks will pull in clean tramp air from the outside instead of leaking dust-laden flue gases into the atmosphere with positive pressure systems. Furthermore, with negative pressure systems, the induced draft fan will be moving particulate-free gas and eliminates the erosion issues that are prevalent in positive pressure systems where the fan is exposed to particulates.

Baghouses can be designed for both intermittent and continuous operations. Intermittent operation is uncommon, but it is possible if operation is not continuous and can be paused to clean the bags and remove particulates. Intermittent collectors are primarily utilized in small-volume operations such as grinding, cutting, and polishing. Continuously operated baghouses are much more common in industry. Continuous operation is accomplished by arranging several separate filter compartments in a parallel flow system. One compartment at a time is isolated for cleaning, while the rest maintain operation.

Baghouses are often identified and characterized according to the method used to remove the particulate matter from the bags. Particles can be removed in various ways. These include:

- Shaking the bags
  - Fully mechanical operation
- Pulsing a short burst of compressed air
  - A bubble of air causes the bag to expand and moves down the bag, collapsing behind it. This sudden expansion and contraction loosens the particulates on the outside of the bag.
- Blowing a continuous jet of compressed air on the bags
  - Similar methodology to the air pulse, this method uses a continuous jet of air to dislodge the particulates from the bag.

A variety of materials are used for the fabrics of the bag filters. Felted type fabrics are typically more efficient dust collectors than are woven fabrics, but this is only in the short term. The filter efficiency of the bags will be essentially equal for the two materials after a layer of particulate builds up on the surface of the bags. A drawback of woven fabric filters is that when they are new, particulates can pass through them until a layer of particulates are built up on the outside of the bags. This is alleviated by prepping the bags with coatings with powders (such as limestone) prior to operation to eliminate this warm-up period.

4.5 Air Pollution Control Equipment for Gaseous Pollutants

As described earlier, the four generic types of gaseous control equipment include the following:

1. Absorbers
2. Adsorbers
3. Combustion units
4. Condensers

Each of these devices is briefly described with more extensive details in Chapters 11–14.

4.5.1 Absorbers

Absorption is a process by which a gas is dissolved in a liquid. A contaminant in a pollutant exhaust stream contacts a liquid and diffuses from the gas phase into the liquid phase. The absorption rate is enhanced by:

- high diffusion rates
- high solubility of the contaminant

- large liquid–gas contact area
- turbulent mixing between liquid and gas phases.

The liquid most often used for absorption is water because it is readily available, inexpensive, and an acceptable solvent for a number of contaminants. Removal efficiency can be increased by adding reagents to the water. Certain reagents merely increase the solubility of the contaminant in the water. Other reagents chemically react with the contaminant after it is absorbed. The absorption rate is typically much higher in reactive scrubbing, which sometimes allows for smaller systems to be used. However, these designs need to take into account precipitates that can be formed as a byproduct of the reactions.

Most wet scrubber designs will sufficiently remove contaminants that have high solubility. However, packed or plate towers are more effective if the contaminant is of low solubility due to increased mass transfer and relatively low pressure drop [5, 6]. Packed towers are one of the more common absorption devices. The packed tower simply consists of an empty shell filled with packing. Gas flowing up the tower is exposed to liquid flowing down over the packing. Plate towers are similar where the gas flows from the bottom and liquid from the top, but with the plate towers, there are horizontal plates spaced throughout the tower with small orifices from the gas to flow through [1–4].

4.5.2 Adsorbers

Adsorption is a process by which a gaseous contaminant adheres to the surface of a solid. Adsorption can take place as either a chemical or physical process. Chemisorption invoices molecules adhering to each other through chemical reactions. Physical adsorption often occurs through imbalance of electrical distribution but can also be simple impaction. These will both be described in more detail later. The major distinction is that physical adsorption is readily reversible whereas chemisorption is not.

All solids physically adsorb gases to some extent. Adsorbents have a high capacity for gas capture due to a high attraction for specific gases, and a large surface area that allows for a high level of contact between adsorbent and pollutant. Activated carbon is by far the most important adsorbent for air pollution control. Activated carbon is so desirable due to its unique surface properties. Activated carbon preferentially targets hydrocarbons and other organic compounds from the airstream, whereas other typical adsorbents preferentially adsorb water vapor. This often renders them useless and prevents them from removing pollution contaminants. The effectiveness of an adsorbent depends on the physical and chemical characteristics of the vapors such as:

- concentration in the gas stream
- system temperature
- system pressure
- humidity of the gas stream
- molecular weight of the vapor.

Activated carbon is also used to remove elemental and ionic mercury from flue gas streams, which is prevalent at coal fired at power plants.

Physical adsorption is a reversible process; the adsorbed vapors can be released (desorbed) by increasing the temperature, decreasing the pressure, or using a combination of both. Vapors are normally desorbed by heating the adsorbent with steam.

Adsorption is a useful removal technique, since it can collect very small concentrations (a few parts per million) of vapor from an airstream. The vapors are stored on the surface of the adsorbent until they can be removed. Adsorption collects the pollutants in a highly concentrated product that can be condensed, recycled, or incinerated as an ultimate disposal technique.

The most common adsorption system is the fixed bed adsorber. These systems consist of two or more adsorber beds operating on a timed adsorbing/desorbing cycle. One or more beds are adsorbing vapors, while the other bed(s) is being regenerated.

4.5.3 Combustion Units

Combustion is defined as a rapid, highly exothermic, gas-phase oxidation. The organic pollutant is burned with air and converted to carbon dioxide and water vapor. The three Ts of combustion, temperature, turbulence, and time, are the fundamentals to a complete chemical reaction. For complete combustion to occur:

- Turbulence: Each contaminant molecule must encounter an oxygen molecule. This requires turbulence and mixing.
- Temperature: The temperature must be maintained above the required activation energy of the combustion reaction.
- Time: There must be sufficient time for each molecule to reach.

These three variables are highly dependent on each other. For example, if the reaction chamber is maintained at a higher temperature, less mixing of the contaminant and combustion air or shorter residence time may be required. If there is poor mixing, a higher temperature or longer residence time must be required for complete combustion.

Combustion devices are classified here in three categories such as flares, thermal incinerators, or catalytic incinerators. Flares are direct combustion devices used to either continually dispose of small quantities of gases or rapid, emergency releases of large quantities. Flares are typically elevated (from 100 to 400 ft) to protect the surroundings from the heat and flames; however, enclosed ground flares can also be utilized. To prevent plumes of black smoke and incomplete combustion, flares can be designed for steam injection at the flare tip. The steam provides sufficient turbulence to ensure complete combustion. Flares are often found at petroleum refineries.

Thermal incinerators are also referred to as thermal oxidizers (TOs), direct flame incinerators, or afterburners. Incinerators typically contain a refractory-line residence chamber where combustion occurs. Thermal incinerators are typically designed to operate at a temperature of 700–800 °C (1,300–1,500 °F) and a residence time of 0.3–0.5 seconds to ensure complete combustion. Although combustion of the contaminant provides heat, supplementary heat from a natural gas burner (such as the FVF Burner from Falcon Combustion) is typically required to maintain the desired temperature, and that supplementary fuel is minimized to reduce costs.

Catalytic incineration is very similar to thermal incineration, but a catalyst bed is utilized to reduce the required oxidation temperature. Catalytic incinerators typically operate at 370–480 °C (700–900 °F) and are less common.

4.5.4 Condensers

Condensers utilize the natural phase change of volatile gases by cooling them back to liquids. The temperature of a vapor mixture is reduced until the partial pressure of the condensable component equals its vapor pressure. Thus, condensation typically requires low temperatures to liquefy most contaminant vapors. Additional gases with varying vaporization points hinder condensation.

Condensers are normally used upstream of other primary control devices such as incinerators, adsorbers, or absorbers. By reducing the volume of vapors that the primary control equipment must handle, the size and the cost of the primary control device can be reduced. Two examples of condensers are refrigeration units that control gasoline vapors at large gasoline dispensing terminal and in rendering plants to remove moisture from the cooker exhaust gas.

Condensers can be designed as either contact condensers or surface condensers. Contact condensers are similar to spray scrubbers and cool the vapor by spraying liquid directly on the vapor stream. Surface condensers are simple noncontact heat exchanges, normally shell and design. Coolant flows through the tubes, while vapor is passed over and condenses on the outside of the tubes. In general, surface condensers are more common because they have little to no wastewater to treat as a result. Contact condensers are more flexible, simpler, and less expensive than surface condensers, but the required wastewater disposal makes them undesirable.

4.6 Factors in Control Equipment Selection

There are a number of factors to be considered prior to selecting a particular piece of air pollution control hardware. In general, they can be grouped in three categories: environmental, engineering, and economic.

Environmental

1. Equipment location.
2. Available space.
3. Ambient conditions.
4. Availability of adequate utilities (i.e., power, water, etc.) and ancillary systems facilities (i.e., waste treatment and disposal, etc.).
5. Maximum allowable emission (air pollution codes).
6. Aesthetic considerations (i.e., visible steam or water vapor plume, etc.).
7. Contribution of air pollution control system to wastewater and land pollution.
8. Contribution of air pollution control system to plant noise level.

Engineering

1. Contaminant characteristics (i.e., physical and chemical properties, concentration, particulate shape, and size distribution—in the case of particulates, chemical reactivity, corrosivity, abrasiveness, toxicity, etc.).
2. Gas stream characteristics (i.e., volumetric flow rate, temperature, pressure, humidity, composition, viscosity, density, reactivity, combustibility, corrosivity, toxicity, etc.).
3. Design and performance characteristics of the particular control system (i.e., size and weight, fractional efficiency curves—in the case of particulates, mass transfer, and/or contaminant destruction capability—in the case of gases or vapors, pressure drop, reliability and dependability, turndown capability, power requirements, utility requirements, temperature limitations, maintenance requirements, flexibility toward complying with more stringent air pollution codes, etc.).

Economic

1. Capital cost (equipment, installation, engineering, etc.).
2. Operating cost (utilities, maintenance, etc.).
3. Expected equipment lifetime and salvage value.

The usual design/procurement/construction/start-up problems can be further compounded by any one or a combination of the following:

1. Unfamiliarity of process engineers with air pollution engineering.
2. New and changing air pollution codes.
3. New suppliers, frequently with unproven equipment.
4. Lack of industry standards in some key areas.
5. Interpretations of control agency field personnel.
6. Compliance schedules that are too tight.
7. Vague specifications.
8. Weak guarantees for the new control equipment.
9. Unreliable delivery schedules.
10. Process unreliability problems.

Proper selection of a particular system for a specific application can be extremely difficult and complicated. In view of the multitude of complex and often ambiguous pollution control regulations, it is in the best interest of the prospective user to work closely with regulatory officials as early in the process as possible. Finally, previous experience on a similar application cannot be overemphasized.

Equipment selection is also strongly dictated by the ability to comply with regulatory codes at the lowest uniform annual cost (amortized capital investment plus operation and maintenance costs). Attempts have been made to include employee concerns, neighbor/consumer goodwill, liability problems, etc. in the economic analysis, but these effects are difficult to quantify. Knowledge of particular site layouts and the application itself is essential to compare control equipment alternatives. A preliminary screening may be performed by reviewing the general advantages, disadvantages, and costs, but the final decision should be made by a qualified person.

4.7 Factors in Purchasing Control Equipment

Prior to the purchase of control equipment, experience has shown that the following points should be emphasized:

1. **Test Data**. Request the manufacturer to provide design specifications, previous performance information, and references of similar successful applications.
2. **Pilot Test**. If sufficient performance data are unavailable, review the data from small pilot model or an experience list of existing installations and field performance data in different applications (it is rare that two applications will be completely identical).
3. **Specifications**. A good set of specifications are essential to obtaining equipment that is appropriate for the intended application. Include strong performance guarantees that will meet all applicable local, state, and federal codes at specific process conditions.
4. **Emissions Bubble**. Assess the potential for emission trade-offs (offsets) and/or applying the "bubble concept." The bubble concept permits a plant to find the most efficient way to control

its emissions as a whole, rather than regulating each individual source separately. This way, emissions can be more economically reduced on certain equipment and be less stringent on aged equipment that may be more costly to reduce, while maintain the same minimum effect on the environment.

5. **Material Balance**. Carefully study a material balance of the equipment to completely understand the control equipment required.
6. **Installation Costs**. Escalating installation costs are the rule rather than the exception. Make every effort to understand any installation challenges ahead of time and use trusted and reliable contractors. Oftentimes, the lowest cost contractor will become more expensive with adders or mistakes.
7. **O&M Costs**. Give operation and maintenance costs high priority on the list of equipment selection factors. While capital costs can be large, they are only subjected once. O&M costs will affect the plant on an annual basis.
8. **Utilities.** Obtain a solid commitment from the utility supplier(s) prior to purchase.
9. **Common Components**. Make every effort to ensure that the new system will utilize common devices throughout the plant. Fuel, controllers, filters, motors, and so on can often be interchangeable on various processes. Maintaining similar brands will also assist maintenance and operation staff with cross-training and troubleshooting plant issues.
10. **Operating Manuals**. The specification should include multiple copies of a complete operating manual (with parts lists and full schematics) to have on hand. These will be invaluable down the road once the people intimately associated with the project have moved on.
11. **Delivery Schedules**. Equipment installations are often only conducted during total or partial plant shutdowns. These outages require an extreme amount of coordination and planning and typically fixed time periods. In such cases, delivery guarantees should be obtained from the manufacturers and penalties identified.
12. **Instrumentation**. Instrumentation should be included to continually monitor the performance of the air pollution equipment to alert operations when performance is deteriorating.
13. **Payment Terms**. Withhold 10–15% of the purchase price until compliance is clearly demonstrated.

References

1 L. Theodore, *Air Pollution Control Equipment Calculations*, John Wiley & Sons, Hoboken, NJ, 2008.

2 L. Theodore and A. J. Buonicore, *Air Pollution Control Equipment*, CRC Press, Boca Raton, FL, 1976.

3 L. Theodore and A. J. Buonicore, *Air Pollution Control Equipment: Selection, Design, Operation, and Maintenance*, Theodore Tutorials, East Williston, NY, 1986.

4 J. Richardson and L. Theodore, *Optimizing Air Pollution Control Equipment Performance: Operation & Maintenance*, John Wiley & Sons, Hoboken, NJ, 2025.

5 L. Theodore, *Chemical Engineering: The Essential Reference*, McGraw-Hill, New York City, NY, 2014.

6 L. Theodore and F. Ricci, *Mass Transfer Operations for the Practicing Engineer*, McGraw-Hall, New York City, NY, 2010.

Part II

Design of Air Pollution Control Equipment: Gases

Part II is concerned with gaseous air pollution control equipment. Following a chapter concerned with basic principles and fundamentals, air pollution control equipment receive treatment. The devices include:

- Chapter 5: Basic Principles: Gases
- Chapter 6: Absorbers
- Chapter 7: Adsorbers
- Chapter 8: Incinerators/Combustion Equipment
- Chapter 9: Condensers

Each of the chapters contains a description of the device, process design considerations, equipment design considerations, and three illustrative examples that are concerned with design aspects of the control equations. The chapters concludes with a “Suggested Reading” section.

5

Basic Principles: Gases

Contributing Author: Ernesto Daniel Cabezas Herrera

5.1 Introduction

This chapter provides a review of some basic concepts from physics, chemistry, and engineering in preparation for material that is covered. These basic concepts include units and dimensions, some physical and chemical properties of substances, the ideal gas law, the Reynolds number, and phase equilibria. Because many of these topics are unrelated to each other, this chapter admittedly lacks the cohesiveness that chapters covering a single topic might have. This is usually the case when basic material from such widely differing areas of knowledge such as physics, chemistry, and engineering is surveyed. Although these topics are widely divergent and covered with varying degrees of thoroughness, all of them will find later use in this book.

(Note that the bulk of the material in this chapter has been drawn from the work of Reynolds [1] and Theodore [2].)

5.2 Measurement Fundamentals

5.2.1 Units and Dimension Absolutes

The units used in this text are consistent with those adopted by the engineering profession in the United States. For engineering work, SI (Système International) and English units are most often employed; in the United States, the *English engineering* units are generally used, although efforts are still in place or underway to obtain universal adoption of SI units for all engineering and science applications. The SI units have the advantage of being based on the decimal system, which allows for more convenient conversion of units within the system. However, the English engineering units will primarily be used here. Conversion factors between SI and English units and additional details on the SI system are provided in the Appendix.

5.2.2 Conversion of Units

Converting a measurement from one unit to another can conveniently be accomplished by using *unit conversion factors*; these factors are obtained from a simple equation that relates the two units numerically. For example,

$$12 \text{ inches (in)} = 1 \text{ foot (ft)} \tag{5.1}$$

The following conversion factor can be obtained:

$$\frac{12 \text{ in}}{1 \text{ ft}} = 1 \tag{5.2}$$

Since this factor is equal to unity, multiplying some quantity (e.g., 18 ft) by this factor cannot alter its value. Hence,

$$18 \text{ ft} \left(\frac{12 \text{ in}}{1 \text{ ft}}\right) = 216 \text{ in} \tag{5.3}$$

Note that in Equation (5.3), the original units of feet on the left-hand side cancel out leaving only the desired units of inches.

Physical equations must be dimensionally consistent. For the equality to hold, each term in the equations must have the same dimensions. This condition can be and should be checked when solving engineering problems. Throughout this book, great care is exercised in maintaining the dimensional formulas of all terms and the dimensional homogeneity of each equation. Equations will generally be developed in terms of specific units rather than general dimensions (e.g., feet, rather than length). This approach should help the reader to more easily attach physical significance to the equations presented.

5.2.3 Significant Figures and Scientific Notation

Significant figures provide an indication of the precision with which a quantity is measured or known. The last digit represents, in a qualitative sense, some degree of doubt. For example, a measurement of 8.32 inches implies that the actual quantity is somewhere between 8.315 and 8.325 inches. This applies to calculated and measured quantities; quantities that are known exactly (e.g., pure integers) have an infinite number of significant figures.

The significant digits of a number are all the digits starting from the first nonzero digit on the left. Zeros at the beginning of a number are not counted as significant digits, but zeros at the end are counted if they come after a nonzero digit. If there is a decimal point, all digits to the right of the first nonzero digit, including zeros, are considered significant. In summary, do not count zeros on the left, but always count zeros on the right if they follow a nonzero digit.

370	Has 2 significant figures
370.	Has 3 significant figures
370.0	Has 4 significant figures
28,070	Has 4 significant figures
0.037	Has 2 significant figures
0.0370	Has 3 significant figures
0.02807	Has 4 significant figures

In general, whenever quantities are combined by multiplication and/or division, the number of significant figures in the result should equal the lowest number of significant figures of any of the quantities. In long calculations, the final result should be rounded off to the correct number of significant figures. When quantities are combined by addition and/or subtraction, the final result cannot be more precise than any of the quantities added or subtracted. Therefore, position (relative to the decimal point) of the last significant digit in the number that has the lowest degree of precision is the position of the last permissible significant digit in the result. For example, the sum of 3702, 370, 0.037, 4, and 37 should be reported as 4110 (without a decimal). The least precise of the five numbers is 370, which has its last significant digit in the *tens* position. The answer should also have its last significant digit in the *tens* position. The authors have *attempted* to abide by these rules.

In the process of performing engineering calculations, very large and very small numbers are often encountered. A convenient way to represent these numbers is to use *scientific notation*. Generally, a number represented in scientific notation is the product of a number (<10 but > or =1) and (multiplied by) 10 raised to an integer power. For example:

$$28{,}070{,}000{,}000 = 2.807 \times 10^{10}$$
$$0.000\,002\,807 = 2.807 \times 10^{-6}$$

A nice feature of using scientific notation is that only the significant figures need appear in the number.

One of the authors has a strong affinity for significant figures. Significant figures are frequently ignored in industry, and it can have substantial ramifications. For example, NO_x emissions (discussed in detail in Chapter 19) in certain regions are limited to a concentration of 9.0 ppm in flue gas at the time of writing. Another common unit of measure for NO_x is lb/MMBtu. When converting 9.0 ppm to lb/MMBtu for natural gas, one obtains 0.011 lb/MMBtu. The significant figures here are indeed very significant. Often times, environmental scientists writing emissions permits for plants either round or truncate the 0.011 lb/MMBtu value to 0.01 lb/MMBtu. When converted back to ppm, one obtains 8.2 ppm. This may not seem significant at first glance, but in this author's extensive experience, most industrial ultra-low NO_x burners are able to achieve 9 ppm NO_x, but 8.2 ppm remains just out of reach in a dynamic environment. So, due to an error with significant digits, this author has seen plants rendered nearly inoperable, derated, or largely frustrated. Significant digits may not seem important, but they do have real life ramifications.

5.3 Chemical and Physical Properties

5.3.1 Temperature

Whether in the gaseous, liquid, or solid state, all molecules possess some degree of kinetic energy, i.e., they are in constant motion—vibrating, rotating, or translating.

The kinetic energies of individual molecules cannot be measured, but the combined effect of these energies in a very large number of molecules can. This measurable quantity is known as *temperature*; it is a macroscopic concept only and as such does not exist on the molecular level.

Temperature can be measured in many ways; the most common method makes use of the expansion of mercury (usually encased inside a glass capillary tube) with increasing temperature. However, thermocouples or thermistors can also be employed.

The two most commonly used temperature scales are the Celsius (or Centigrade) and Fahrenheit scales. The Celsius is based on the boiling and freezing points of water at 1 atmosphere pressure; to the former, a value of 100 °C is assigned, and to the latter, a value of 0 °C. On the older Fahrenheit scale, these temperatures correspond to 212 and 32 °F, respectively. Equations (5.4) and (5.5) provide the conversion from one scale to the other.

$$°F = 1.8(°C) + 32 \quad (5.4)$$

$$°C = \frac{(°F - 32)}{1.8} \quad (5.5)$$

where °F = a temperature on the Fahrenheit scale
°C = a temperature on the Celsius scale

Experiments with gases at low to moderate pressures (up to a few atmospheres) have shown that, if the pressure is kept constant, the volume of a gas and its temperature are linearly related (Charles' law) and that a decrease of 0.3663% or $\left(\frac{1}{273}\right)$ of the initial volume is experienced for every temperature drop of 1 °C. These experiments were not extended to very low temperatures, but if the linear relationship were extrapolated, the volume of the gas would *theoretically* be zero at a temperature of approximately −273 °C or −460 °F. This temperature has become known as *absolute zero* and is the basis for the definition of two *absolute* temperature scales. (An *absolute* scale is one which does not allow negative quantities.) These absolute temperature scales are the Kelvin (K) and Rankine (°R) scales; the former is defined by shifting the Celsius scale by 273 °C so that 0 K is equal to −273 °C; Equation (5.6) shows this relationship:

$$K = °C + 273 \quad (5.6)$$

The Rankine scale is defined by shifting the Fahrenheit scale 460°, so that

$$°R = °F + 460 \quad (5.7)$$

5.3.2 Pressure

In the gaseous state, the molecules possess a high degree of translational kinetic energy, which means that they are able to move quite freely throughout the body of the gas. If the gas is in a container of some type, the molecules are constantly bombarding the walls of the container. The macroscopic effect of this bombardment by a tremendous number of molecules enough to make the effect measurable is called pressure. The natural units of pressure are force per unit area. In the example of the gas in a container, the unit area is a portion of the inside solid surface of the container wall while the force, measured perpendicularly to the unit area, is the result of the molecules hitting the unit area and swing up momentum during the sudden change of direction.

There are a number of different methods used to express a pressure measurement. Some of them are natural units, i.e., based on a force per unit area, e.g., pound (force) per square inch (abbreviated lbf/in^2 or psi) or dyne per square centimeter (dyn/cm^2). Others are based on a fluid height, such as inches of water (in H_2O) or millimeters of mercury (mm Hg); units such as these are convenient when the pressure is indicated by a difference between two levels of a liquid as in a manometer or barometer. Barometric pressure and atmospheric pressure are synonymous and measure the ambient air pressure. Standard barometric pressure is the average atmospheric pressure at sea level, 45 ° north latitude at 32 °F. It is used to define another unit of pressure called the atmosphere (atm). Standard atmospheric pressure at sea level is defined as 1 atmosphere (atm), which is equal to 14.696 pounds per square inch (psi) or 29.921 inches of mercury (in Hg). It is important to note that barometric pressure is not constant; it changes depending on weather conditions and elevation above sea level. At higher altitudes, the atmospheric pressure decreases, while changes in weather can cause fluctuations in pressure at any location.

Measurements of pressure by most gauges indicate the difference in pressure either above or below that of the atmosphere surrounding the gauge. *Gauge pressure* is the pressure indicated by such a device. If the pressure in the system measured by the gauge is greater than the pressure prevailing in the atmosphere the gauge pressure is expressed positively; if lower than atmospheric pressure the gauge pressure is a negative quantity; the term vacuum designates a negative gauge pressure. Gauge pressures are often identified by the letter g after the pressure unit; for example, psig (pounds per square inch gauge) is a gauge pressure in psi units.

Since gauge pressure is the pressure relative to the prevailing atmospheric pressure, the sum of the two gives the absolute pressure, indicated by the letter a after the unit (e.g., psia [pounds per square inch absolute]):

$$P = P_a + P_g \tag{5.8}$$

where P = absolute pressure (psia)

P_a = atmospheric pressure (psia)

P_g = gauge pressure (psig)

The absolute pressure scale is absolute in the same sense that the absolute temperature scale is absolute, i.e., a pressure of zero psia represents the complete absence of pressure, or a perfect vacuum, which is the lowest pressure that can theoretically exist.

5.3.3 Moles and Molecular Weights

An atom consists of protons and neutrons in a nucleus surrounded by electrons. An electron has such a small mass relative to that of the proton and neutron that the weight of the atom (called the atomic weight) is approximately equal to the sum of the weights of the particles in its nucleus. Atomic weight may be expressed in atomic mass units (amu) per atom or in grams per gram-atom. One gram-atom contains 6.02×10^{23} atoms (Avagadro's number). The atomic weights of all the elements are available in the literature.

The molecular weight (MW) of a compound is the sum of the atomic weights of the atoms that make up the molecule. Units of atomic mass units per molecule (amu/molecule) or grams per gram-mole (g/gmol) are used for molecular weight. One gram-mole (gmol) contains an Avogadro number of molecules. For the English system, a pound-mole (lbmol) contains $454 \times 6.023 \times 10^{23}$ molecules.

Molal units are used extensively in air pollution control calculations as they greatly simplify material balances where chemical (including combustion) reactions are occurring. For mixtures of substances (gases, liquids, or solids), it is also convenient to express compositions in mole fractions or mole percentages instead of mass fractions. The mole fraction is the ratio of the number of moles of one component to the total number of moles in the mixture.

5.3.4 Mass and Volume

The density (ρ) of a substance is the ratio of its mass to its volume and may be expressed in units of pounds per cubic foot (lb/ft^3), kilograms per cubic meter (kg/m^3), and so on. For solids, density can be easily determined by placing a known mass of the substance in a liquid and determining the displaced volume. The density of a liquid can be measured by weighing a known volume of the liquid in a volumetric flask. For gases, the ideal gas law, discussed later in Section 5.4, can be used to calculate the density from the pressure, temperature, and molecular weight of the gas.

Densities of pure solids and liquids are relatively independent of temperature and pressure, and can be found in standard reference books. The specific volume of a substance is defined as the amount of space (volume) occupied by a unit of mass with common units such as cubic feet per pound (ft^3/lb) or cubic meter per kilogram (m^3/kg) etc. Because specific volume describes how much space a given mass takes up, it is the mathematical inverse of density. In other words, as density increases, specific volume decreases, and vice versa.

An important density term that finds applications with particles (see later chapters) is the bulk density. For particles, a distinction should always be made between true density and bulk density. The true density is the actual density of a discrete particle or solid, whereas the bulk density is the density with the void volume between the particles is included in the determination. In lieu of data, the bulk density may be assumed, for purposes of engineering calculations, to be approximately 60% of the true density [3].

The *specific gravity* (SG) is the ratio of the density of a substance to the density of a reference substance at a specific condition:

$$\mathrm{SG} = \frac{\rho}{\rho_{\mathrm{ref}}} \tag{5.9}$$

The reference most commonly used for *solid* and *liquids* is water at its maximum density, which occurs at 4 °C; this reference density is 1.000 g/cm^3, 1,000 kg/m^3, or 62.23 lb/ft^3. Note that, since the specific gravity is a ratio of two densities, it is dimensionless. Therefore, any set of units may be employed for the two densities as long as they are consistent. The specific gravity of *gases* is used only rarely; when it is, air at the same conditions of temperature and pressure as the gas is usually employed as the reference substance.

Another dimensionless quantity related to density is the API (American Petroleum Institute) gravity, which is often used to indicate densities of fuel oils and some liquid hazardous wastes. The relationship between the API scale and the specific gravity is

$$\text{Degree API} = \frac{141.5}{\mathrm{SG}\left(\frac{60}{60}\ {}^\circ\mathrm{F}\right)} - 135.5 \tag{5.10}$$

where $\mathrm{SG}\left(\frac{60}{60}\ {}^\circ\mathrm{F}\right)$ = specific gravity of the liquid at 60 °F using water at 60 °F as the reference.

5.3.5 Viscosity

Viscosity is a property associated with a fluid's resistance to flow; more precisely, this property accounts for energy losses which result from shear stresses occurring between different portions of the fluid that are moving at different velocities. The absolute viscosity (μ) has units of mass per length time; the fundamental unit is the poise, which is defined as 1 g/cm s. This unit is inconveniently large for many practical purposes and viscosities are frequently given in centipoises (0.01 poise), which is abbreviated cP. The viscosity of pure water at 68.6 °F is 1.00 cP. In English units, absolute viscosity is expressed either as pounds (mass) per foot-second (lb/ft s) or pounds per foot-hour (lb/ft hr). The absolute viscosity depends primarily on temperature and to a lesser degree on pressure.

The kinematic viscosity (ν) is the absolute viscosity divided by the density of the fluid and is useful in certain fluid flow problems; the units for this quantity are length squared per time, e.g., square foot per second (ft^2/s) or square meters per hour (m^2/hr). A kinematic viscosity of 1 cm^2/s

is called a stoke. For pure water at 70 °F, $\nu = 0.983$ cS (centistokes). Because fluid viscosity changes rapidly with temperature, a numerical value of viscosity has no significance unless the temperature is specified.

Liquid viscosity is usually measured by the amount of time it takes for a given volume of liquid to flow through an orifice. The *Saybolt universal viscometer* is the most widely used device in the United States for the determination of the viscosity of fuel oils and liquid wastes.

The viscosity of air at atmospheric pressure as functions of temperature is presented in Table 5.1. Viscosities of other substances are available in the literature [4].

5.3.6 Heat Capacity

The heat capacity of a substance is defined as the quantity of heat required to raise the temperature of that substance by 1 °. The term specific heat is frequently used in place of specific heat capacity. This is not strictly correct, because specific heat has been traditionally defined as the ratio of the heat capacity of a substance to the heat capacity of water. However, since the specific heat of water is approximately 1 cal/g °C or 1 Btu/lb °F, the term specific heat is often interchanged with heat capacity. For gases, the addition of heat to cause the 1 ° temperature rise may be accomplished either at constant pressure or at constant volume. Since the amounts of heat necessary are different for the two cases, subscripts are used to identify which heat capacity is being used c_p for constant pressure and c_v for constant volume. For liquids and solids, this distinction does not have to be made since there is little difference between the two. Values of heat capacities are available in the literature [1, 2, 4].

Heat capacities are often used on a molar basis instead of a mass basis, in which case the units become cal/gmol °C or Btu/lbmol °F. To distinguish between the two, uppercase letters $\left(C_p, C_v\right)$ are used in this text to represent the molar-based heat capacities, and lowercase letters $\left(c_p, c_v\right)$ are used for the mass-based heat capacities.

Heat capacities are functions of both the temperature and pressure, although the effect of pressure is generally small and is neglected in almost all engineering calculations. The effect of temperature on C_P has been describe by

$$C_P = \alpha + \beta T + \gamma T^2 \tag{5.11}$$

$$C_P = a + bT + cT^{-2} \tag{5.12}$$

where α, β, γ, a, b, and c are constant specific to a particular chemical.

Table 5.1 Viscosity of air at 1 atmosphere.

T, °C	Viscosity, micropoise (μP)[a]
0	170.8
18	182.7
40	190.4
54	195.8
74	210.2
229	263.8

1P = 100 cP = 10^6 μP; 1 cP = 6.72×10^{-4} lb/ft · s.

5.3.7 Reynolds Number

The Reynolds number (Re) is a dimensionless number that indicates whether a fluid flowing is in the laminar or turbulent regime. Laminar flow is characteristic of fluids flowing slow enough so that there are no eddies (whirlpools) or macroscopic mixing of different portions of the fluid. (Note: In any fluid, there is always molecular mixing due to the thermal activity of the molecules; this is distinct from macroscopic mixing due to the swirling motion of different portions of the fluid.) In laminar flow, a fluid can be imagined to flow like a deck of cards, with adjacent layers sliding past one another. Turbulent flow is characterized by eddies and macroscopic currents. In practice, moving gases are generally in the turbulent region. For flow in a pipe, a Reynolds number above 2,100 is an indication of turbulent flow [5].

The Reynolds number is dependent on the fluid velocity, density, viscosity, and some characteristic length of the system or conduit; for pipes, the characteristic length is the inside diameter, and for particles, it is the particle diameter:

$$\mathrm{Re} = \frac{dV\rho}{\mu} = \frac{D\nu}{v} \tag{5.13}$$

where Re = Reynolds number (dimensionless)
D = inside diameter of the pipe (ft)
V = fluid velocity (ft/s)
ρ = fluid density $(\mathrm{lb/ft^3})$
ν = fluid viscosity (lb/ft · s)
v = fluid kinematic viscosity $(\mathrm{ft^2/\,s})$

A consistent set of units (other than the above) may be used with Equation (5.13).

5.3.8 pH

An important chemical property of an aqueous solution is its pH. The pH measures the acidity or basicity of the solution. The pH scale provides a measure of the concentration of hydrogen ions (H^+) and hydroxide ions (OH^-). The balance between these ions determines if the solution is acid, basic, or neutral. At ordinary temperatures (mostly like to be 25 °C), this concentration is

$$C_{\mathrm{H^+}} = C_{\mathrm{OH^-}} = 10^{-7}\mathrm{g \cdot ion/L} \tag{5.14}$$

where $C_{\mathrm{H^+}}$ = hydrogen ion concentration
$C_{\mathrm{OH^-}}$ = hydroxyl ion concentration.

The unit g ion stands for g ion, which represents an Avogadro number of ions. In all aqueous solutions, whether neutral, basic, or acidic, a chemical equilibrium or balance is established between these two concentrations, so that

$$K_{\mathrm{eq}} = (C_{\mathrm{H^+}})(C_{\mathrm{OH^-}}) = 10^{-14} \tag{5.15}$$

where K_{eq} = equilibrium constant.

The numerical value for K_{eq} given in Equation (5.15) holds for room temperature and only when the concentrations are expressed in gram-ion per liter (g ion/L). In acid solutions, $C_{\mathrm{H^+}}$ is $> C_{\mathrm{OH^-}}$;

in basic solutions, predominates. The OH^- is a direct measure of the hydrogen ion concentration and is defined by

$$pH = -\log C_{H^+} \tag{5.16}$$

Thus, an acidic solution is characterized by a pH below 7 (the lower the pH, the higher the acidity)—a basic solution, by a pH above 7—and, a neutral solution by a pH of 7.

5.3.9 Boiling and Freezing Points

The boiling point is the temperature at which the vapor pressure of a liquid is equal to its surrounding pressure. Since the vapor pressure remains constant during boiling, so does the temperature. It is important to note that the boiling point is not a fixed value for a substance; instead, it depends strongly on the surrounding pressure. As the external pressure increases, a higher temperature is required for the liquid to boil. The higher the system pressure, the higher the temperature must be to induce boiling meaning that the boiling point increases. The boiling point is specific to each individual substance and is a strong function of pressure. The freezing point is the temperature at which the liquid and solid states of a substance coexist in equilibrium at a given pressure. At this point, the rate at which the substance leaves the solid state equals the rate at which it leaves the liquid state.

5.3.10 Vapor Pressure

This section concludes with a discussion of vapor pressure (denoted p'), which is an important property of liquids, and, to a much lesser extent, of solids. If a liquid is allowed to evaporate in a confined space, the pressure in the vapor space increases as the amount of vapor increases. If there is sufficient liquid present, a point is eventually reached at which the pressure in the vapor space is exactly equal to the pressure exerted by the liquid at its own surface. At this point, a dynamic equilibrium exists in which vaporization and condensation take place at equal rates and the pressure in the vapor space remains constant. The pressure exerted at equilibrium is called the vapor pressure of the liquid. The magnitude of this pressure for a given liquid depends on the temperature, but not on the amount of liquid present. Solids, like liquids, also exert vapor pressure. The process where a solid changes directly into a vapor without first becoming a liquid is sublimation; this phenomenon is only noticeable for solids that have a significant vapor pressure at a given temperature; if a solid's vapor pressure is low, the sublimation happens so slowly that it is hardly observed.

5.4 Ideal Gas Law

Observations based on physical experimentation often can be synthesized into simple mathematical equations called laws. These laws are never perfect and hence are only an approximate representation of reality. The ideal gas law (IGL) was derived from experiments in which the effects of pressure and temperature on gaseous volumes were measured over moderate temperature and pressure ranges. This law works well in the pressure and temperature ranges that were used in taking the data; extrapolations outside of the ranges have been found to work well in some cases and poorly in others. As a general rule, this law works best when the molecules of the gas are far apart, i.e., when the pressure is low, and the temperature is high. Under these conditions, the gas is said to behave ideally, and its behavior is a close approximation to the so-called perfect or ideal gas, a hypothetical entity that obeys the ideal gas law perfectly. For engineering calculations, the ideal gas law is almost always assumed to be valid since it generally works well (usually within a few percent of the correct result) up to the highest pressures and down to the lowest temperatures used in air pollution applications.

The two precursors of the ideal gas law were Boyle's and Charles' laws [1, 2, 4]. Boyle found that the volume of a given mass of gas is inversely proportional to the absolute pressure if the temperature is kept constant:

$$P_1 V_1 = P_2 V_2 \tag{5.17}$$

where V_1 = volume of gas at absolute pressure P_1 and temperature T_1
V_2 = volume of gas at absolute pressure P_2 and temperature T_2.

Charles found that the volume of a given mass of gas varies directly with the absolute temperature at constant pressure:

$$\frac{V_1}{T_1} = \frac{V_2}{T_2} \tag{5.18}$$

where V_1 = volume of gas at absolute pressure P_1 and temperature T_1
V_2 = volume of gas at absolute pressure P_2 and temperature T_2.

Boyle's and Charles' laws may be combined into a single equation in which neither temperature nor pressure need be held constant:

$$\frac{P_1 V_1}{T_1} = \frac{P_2 V_2}{T_2} \tag{5.19}$$

For Equation (5.19) to hold, the mass of gas must be constant as the conditions change from (P_1, T_1) to (P_2, T_2). This equation indicates that for a given mass of specific gas, PV/T has a constant value. Since, at the same temperature and pressure, volume and mass must be directly proportional, this statement may be extended to:

$$\frac{PV}{mT} = C \tag{5.20}$$

where m = mass of a specific gas
C = constant that depends on the gas.

Moreover, experiments with different gases showed that Equation (5.20) could be expressed in a far more generalized form. If the number of moles (n) is used in placed of the mass (m), the constant is the same for all gases:

$$\frac{PV}{nT} = R \tag{5.21}$$

where R = universal gas constant.

Equation (5.21) is called the ideal gas law. Numerically, the value of R depends on the units used for P, V, T, and n (see Table 5.2). In this text, all gases are assumed to approximate ideal gas behavior. As is generally the case in engineering practice, the ideal gas law is assumed to be valid for all problems to follow later.

Table 5.2 Value of *R* in various units.

R	Temperature scale	Units of *V*	Units of *n*	Units of *P*
10.73	°R	ft^3	lbmol	Psia
0.7302	°R	ft^3	lbmol	atm
21.85	°R	ft^3	lbmol	in Hg
555	°R	ft^3	lbmol	m Hg
297	°R	ft^3	lbmol	in H_2O
0.7398	°R	ft^3	lbmol	bar
1545	°R	ft^3	lbmol	psfa
62.361	K	L	gmol	mm Hg
0.08205	K	L	gmol	atm
0.08314	K	L	gmol	bar
8314	K	L	gmol	Pa
8.314	K	m^3	gmol	Pa
82.057	K	cm^3	gmol	atm

Other useful forms of the ideal gas law are shown in Equations (5.22) and (5.23). Equation (5.22) applies to gas flow rather than to gas confined in a container:

$$Pq = \dot{n}RT \tag{5.22}$$

where q = gas volumetric flow rate (ft^3/hr)
P = absolute pressure (psia)
$\dot{n}$ = molar flow rate (lbmol/hr)
T = absolute temperature (°R)
R = 10.73 psia ft^3/lbmol · °R.

Equation (5.23) combines *n* and *V* from Equation (5.21) to express the law in terms of density:

$$P(\text{MW}) = \rho RT \tag{5.23}$$

where MW = Molecular weight of gas (lb/lbmol)
ρ = density of gas(lb/ft^3).

Volumetric flow rates are often given not at the actual conditions of pressure and temperature but at arbitrarily chosen standard conditions (STP, standard temperature and pressure). To distinguish between flow rates based on the two conditions, the letters *a* and *s* are often used as part of the unit. The units acfm and scfm stand for actual cubic feet per minute and standard cubic feet per minute, respectively. The ideal gas law can be used to convert from standard to actual conditions, but since there are many standard conditions in use, the STP being employed must be known or specified. Standard conditions most often used are shown in Table 5.3.

The reader is cautioned on the incorrect use of acfm and/or scfm. The use of standard conditions is a convenience; when predicting the performance of or designing equipment, the actual

Table 5.3 Value of *R* in various units.

System	Temperature	Pressure	Molar volume
SI	273 K	101.3 kPa	22.4 m^3/Kmol
Universal scientific	0 °C	760 mm Hg	22.4 L/Kmol
Natural gas industry	60 °F	14.7 psia	379 ft^3/lbmol
American engineering	32 °F	1 atm	359 ft^3/lbmol
Hazardous waste	60 °F	1 atm	379 ft^3/lbmol
Incinerator industry	70 °F	1 atm	387 ft^3/lbmol

conditions must be employed. Designs based on standard conditions can lead to disastrous results, with the unit usually under-designed.

5.4.1 Partial Pressure and Partial Volume

In engineering practice, mixtures of gases are more often encountered than single or pure gases. The ideal gas law is based on the number of molecules present in the gas volume; the kind of molecules is not a significant factor, only the number. This law applies equally well to mixtures and pure gases alike. Dalton and Amagat both applied the ideal gas law to mixtures of gases. Since pressure is caused by gas molecules colliding with the walls of the container, it seems reasonable that the total pressure of a gas mixture is made up of pressure contributions due to each of the component gases. These pressure contributions are called partial pressures. Dalton defined the partial pressure of a component as the pressure that would be exerted if the same mass of the component gas occupied the same total volume alone at the same temperature as the mixture. The sum of these partial pressures equals the total pressure:

$$P = p_a + p_b + p_c + \dots + p_n = \sum_{i=1}^{n} p_i \tag{5.24}$$

where P = total pressure

n = number of components

p_i = partial pressure of component i.

Equation (5.24) is known as Dalton's law. Applying the ideal gas law to one component (a) only,

$$p_a V = n_a RT \tag{5.25}$$

where n_a = number of moles of component a. Eliminating R, T, and V between Equations (5.21) and (5.25) yields

$$\frac{p_a}{P} = \frac{n_a}{n} = y_a$$

or

$$p_a = y_a P \tag{5.26}$$

where y_a = mole fraction of component a.

Amagat's law is similar to Dalton's law. Instead of considering the total pressure to be made up of partial pressures where each component occupies the total container volume, Amagat considered the total volume to be made up of partial volumes in which each component is at the total pressure. The definition of the *partial volume* is therefore the volume occupied by a component gas alone at the same temperature and pressure as the mixture:

$$V = V_a + V_b + V_c + \ldots + V_n = \sum_{i=1}^{n} V_i \tag{5.27}$$

Applying Equation (5.21), as before yields

$$\frac{V_a}{V} = \frac{n_a}{n} = y_a \tag{5.28}$$

or

$$V_a = y_a V \tag{5.29}$$

where V_a = particular volume of component a.

It is common in the air pollution control industry to describe low concentrations of pollutants in gaseous mixtures in parts per million by volume (ppmv). Since partial volumes are proportional to mole fractions, it is only necessary to multiply the mole fraction of the pollutant by 1 million to obtain the concentration in parts per million. (For liquids and solids, parts per million [ppm] is also used to express concentration, although it is usually on a mass [ppmw] basis rather than a volume basis. The terms ppmv and ppmw are sometimes used to distinguish between the volume and mass bases, respectively.)

5.5 Phase Equilibrium

In air pollution control equipment applications (particularly in absorption), the most important equilibrium phase relationship is that between liquid and vapor. Henry's law theoretically describes liquid–gas behavior and under certain conditions is applicable in practice.

Raoult's law states that the partial pressure of a component over a solution is the product of the vapor pressure of that component and the mole fraction of that component. Raoult's law is expressed as

$$p_A = p'_A x_a \tag{5.30}$$

where p_A = partial pressure of A, Pa

p'_A = vapor pressure of A, Pa

x_A = mole fraction of A in solution.

Raoult's law applies over the entire concentration range from 0 to 1.0, but is applicable only if the solution behaves in an ideal manner. Unfortunately, few solutions behave ideally. A special case of Raoult's law is Henry's law.

Henry's law is an empirical relationship used for representing data on many systems:

$$p_i = H_i x_i \tag{5.31}$$

where H = Henry's law constant for component i (in units pressure).

If the gas behaves ideally, Equation (5.31) may be written as

$$y_i = m_i x_i \tag{5.32}$$

where m_i = constant (dimensionless).

This is a more convenient form of Henry's law to work with. The constant (or) has been determined experimentally for a large number of compounds and is usually valid at low concentrations [4, 6, 7].

Another phase equilibrium application involves the psychrometric or humidity chart. A humidity chart is used to determine the properties of moist air and to calculate moisture content in air. The ordinate of the chart is the absolute humidity $\mathcal{H}$, which is defined as the mass of water vapor per mass of bone-dry air. (Some charts base the ordinate on moles instead of mass.) On the basis of this definition, Equation (5.33) gives $\mathcal{H}$ in terms of moles and also in terms of partial pressure:

$$\mathcal{H} = \frac{18n_{H_2O}}{29\left(n - n_{H_2O}\right)} = \frac{18p_{H_2O}}{29(P - p_{H_2O})} \tag{5.33}$$

where n_{H_2O} = number of moles of water vapor

p_{H_2O} = partial pressure of water vapor

n = total number of moles in gas

P = total system pressure.

The wet-bulb temperature (T_{WB}) is another measure of humidity; it is the temperature at which a thermometer with a wet wick wrapped around the bulb stabilizes. As water evaporates from the wick to the ambient air, the bulb is cooled; the rate of cooling depends on how humid the air is. No evaporation occurs if the air is saturated with water; hence, T_{WB} and T_{DB} (dry bulb) are the same. Furthermore, the lower the humidity, the greater the difference between these two temperatures.

The term enthalpy is a measure of the energy content of the mixture and is defined later in this chapter. Steam tables, charts, and diagrams may also be required in air pollution control equipment calculations. Abbreviated and more comprehensive information is available in the literature [1, 2, 4].

5.6 Conservation Laws

5.6.1 Conservation of Mass

The conservation law for mass can be applied to any process or system. The general form of this law is given in Equation (5.34):

$$\text{mass}_{\text{in}} - \text{mass}_{\text{out}} + \text{mass}_{\text{generated}} = \text{mass}_{\text{accumulated}} \tag{5.34}$$

Equation (5.34) may be applied to the total mass involved or to a particular species, on either a mole or mass basis. The conservation law for mass can be applied to steady-state or unsteady-state processes and to batch or continuous systems. To isolate a system for study, one separates it from the surroundings by a boundary or envelope. This boundary may be real (e.g., the walls of a vessel) or imaginary. Mass crossing the boundary and entering the system is part of the mass-in term in Equation (5.34), while mass leaving the system is part of the mass-out term. Equation (5.34) may be written for any compound whose quantity is not changed by chemical reaction, or for any chemical element, regardless of whether it has participated in a chemical reaction. It may be written for one piece of equipment, around several pieces of equipment, or around an entire process. It may be applied to a single piece of equipment, to several pieces of equipment, or to an entire process. The equation can be used to calculate an unknown quantity directly, to check the validity of experimental data, or to express one or more of the independent relationships among the unknown quantities in a particular problem.

A steady-state process is one in which there is no change in conditions (temperature, pressure, etc.) or rates of flow with time at any given point in the system. The accumulation term in Equation (5.34) is then zero. If there is no chemical reaction, the generation term is also zero. All other processes are classified as unsteady-state. In a batch process, the container holds the product or products. In a continuous process, reactants are fed in an unending flow to a piece of equipment or to several pieces in series, and products are continuously removed from one or more points. A continuous process may or may not be steady state. As indicated previously, Equation (5.34) may be applied to the total mass of each stream (referred to as an overall or total material balance) or to the individual components of the streams (referred to as componential or component material balance). Often, the primary task in preparing a material balance is to develop the quantitative relationships among the streams [1].

5.6.2 Conservation of Energy

A presentation of the conservation law for energy would be incomplete without a brief review of some introductory thermodynamic principles. Thermodynamics is defined as the science that deals with the relationships between the various forms of energy.

A system may possess energy as a result of its temperature, velocity, position, molecular structure, surface, and so on. The energies corresponding to these conditions are internal, kinetic, potential, chemical, surface, and so on, respectively. Engineering thermodynamics is founded on three basic laws. Energy, like mass, is conserved. Application of the conservation law for energy gives rise to the first law of thermodynamics. (The reader is directed to the literature for details on the second and third law.) This law, in steady-state form for batch processes, is presented here (potential, kinetic, and other energy effects are not considered):

$$\Delta U = Q + W \tag{5.35}$$

or

$$\Delta H = Q + W_S \tag{5.36}$$

where Q = heat energy transferred across the system boundary to the system
W = work energy transferred across the system boundary to the system
W_S = mechanical work energy transferred across the system boundary to the system
U = internal energy of the system
H = enthalpy of the system
ΔU = change in internal energy during the process
ΔH = change in enthalpy during the process.

The internal energy and enthalpy in Equations (5.35) and (5.36), as well as in the other equations in this discussion, may be on a mass or a mole basis, or they may represent the total internal energy and enthalpy of the entire system. Most air pollution equipment, as well as industrial facilities, operate in a steady-state flow mode. If no significant mechanical or shaft work is added or withdrawn from the system, Equation (5.36) reduces to

$$Q = \Delta H \tag{5.37}$$

If a unit or system is operated adiabatically, where $Q = 0$, Equation (5.37) then becomes

$$\Delta H = 0 \tag{5.38}$$

Although the topics of material and energy balances are covered separately in this subsection, it should be emphasized that this segregation does not occur in reality; one must often work with both energy and material balances simultaneously [1].

5.7 Thermochemistry and Kinetics

Consider now the energy effects associated with a chemical reaction. To introduce this subject, the reader is reminded that engineers and applied scientists are rarely concerned with the magnitude or amount of energy in a system; their primary concern is with changes in the amount of energy. It has been found in measuring energy changes for systems that the enthalpy is the most convenient term to work with. There are many different types of enthalpy effects; these include sensible heat, latent heat, and heat of reaction.

The heat of reaction is defined as the enthalpy change of a system undergoing chemical reaction. If the reactants and products are at the same temperature and in their standard states, the heat of reaction is termed the standard heat of reaction. For engineering purposes, the standard state of a chemical may be taken as the pure chemical at 1 atm pressure. Heat of reaction data for many reactions are available in the literature [2, 8].

5.7.1 Chemical Reaction Equilibrium

With regard to chemical reactions, two important questions are of concern to the engineer:

1. How far will the reaction go?
2. How fast will the reaction go?

Chemical thermodynamics provides the answer to the first question; however, it tells nothing about the second. Reaction rates fall within the domain of chemical kinetics and are treated later in this section. Both equilibrium and kinetic effects must be considered in an overall engineering analysis of a chemical reaction.

Chemical reaction equilibrium calculations are structured around another thermodynamic term called the free energy. This so-called free energy (G) is a property that cannot be defined easily without some further basic grounding in thermodynamics. We make no such attempt, and the interested reader is directed to the literature. Free energy has the same units as enthalpy and internal energy, and may be on a mole or total mass basis [8–10].

5.7.2 Chemical Kinetics

Chemical kinetics involves the study of reaction rates and the variables that affect these rates. It is a topic that is critical for the analysis of reacting systems, as in a combustion device or incinerator. The topic here is treated from an engineering point of view, that is, in terms of physically measurable quantities [9, 10].

The rate of a chemical reaction can be described in any of several different ways. The most commonly used definition involves the time rate of change in the amount of one of the components participating in the reaction. This rate is usually based on some arbitrary factor related to the reacting system size or geometry, such as volume, mass, or interfacial area. The definition shown in Equation (5.39), which applies to homogeneous reactions, is convenient from an engineering point of view:

$$R_A = (1/V)\ (\mathrm{d}n_A/\mathrm{d}t) \tag{5.39}$$

where R_A = heat energy transferred across the system boundary to the system
n_A = mechanical work energy transferred across the system boundary to the system
V = work energy transferred across the system boundary to the system
t = internal energy of the system

If the volume term is constant, one may write Equation (5.39) as

$$R_A = (\mathrm{d}C_A/\mathrm{d}t) \tag{5.40}$$

where C_A is the molar concentration of A at time t.

Based on Equation (5.40), the reaction rate is positive if species A is being formed (since C_A increases with time), and negative if A is reacting (since C_A decreases with time). The rate is zero if the system is at chemical equilibrium.

An equation expressing the rate in terms of measurable and/or desirable quantities may now be developed. Based on experimental evidence, the rate of reaction is almost always a function of the concentration of the components present in the reaction mixture, the temperature, the pressure, and catalyst variables. In equation form,

$$R_A = R_A(C_i, P, T,\ \text{catalyst variables}) \tag{5.41}$$

Equation (5.41) may be condensed into

$$R_A = (\pm)k_A f(C_i) \tag{5.42}$$

where k_A incorporates all the variables in Equation (5.41) other than the concentration variable. The ($\pm$) notation is included to indicate whether component A is being consumed ($-$) or produced ($+$). The k_A may be regarded as a constant of proportionality; it is termed the reaction velocity constant. Although the "constant" is independent of concentration, it is a function of other variables. This term is very definitely influenced by temperature and catalyst activity. For most applications, it is assumed that k is solely a function of temperature. Thus,

$$R_A = (\pm)k_A(T)f(C_i) \tag{5.43}$$

The effect of temperature on k is generally represented by the Arrhenius equation, and the value of k is obtained from experimental data.

Author Biography

Ernesto Cabezas is a final-year Chemical Engineering student at Universidad San Francisco de Quito (USFQ). His strengths include data analysis and investigative skills. He's passionate about innovation and knowledge transfer. As he approaches graduation, he is eager to apply his expertise as a qualified Chemical Engineer and contribute professionally to future projects of this nature.

References

1 J. Reynolds, *Material and Energy Balances*, A Theodore Tutorial, Theodore Tutorials, East Williston, NY, 1992 (Originally published by USEPA/APTI, RTP, NC).
2 L. Theodore and J. Reynolds, *Thermodynamics*, A Theodore Tutorial, Theodore Tutorials, East Williston, NY, 1992 (Originally published by USEPA/APTI, RTP, NC).
3 L. Theodore, *Personal Notes*, East Williston, NY, 1991.
4 P. Abulencia and L. Theodore, *Fluid Flow for the Practicing Chemical Engineer*, John Wiley & Sons, Hoboken, NJ, 2022.
5 L. Theodore (Section editor); D. Green and M. Southard, editors, *Perry's Chemical Engineers' Handbook*, McGraw-Hill, New York City, NY, 2019.
6 L. Theodore and J. Bardon, *Mass Transfer Operations*, A Theodore Tutorial, Theodore Tutorials, East Williston, NY, 1995 (Originally published by USEPA/APTI, RTP, NC).
7 L. Theodore and F. Ricci, *Mass Transfer Operations for the Practicing Engineer*, John Wiley & Sons, Hoboken, NJ, 2010.
8 L. Theodore, F. Ricci and T. VanVleet, *Thermodynamics for the Practicing Engineer*, John Wiley & Sons, Hoboken, NJ, 2009.
9 L. Theodore, *Chemical Reactor Analysis and Applications for the Practicing Engineer*, John Wiley & Sons, Hoboken, NJ, 2012.
10 L. Theodore, *Chemical Reaction Kinetics*, A Theodore Tutorial, Theodore Tutorials, East Williston, NY, 1992 (Originally published by USEPA/APTI, RTP, NC).

6

Absorbers

6.1 Introduction

In the absorption process, a soluble component of a gas mixture is dissolved into a relatively nonvolatile liquid. In addition to simply being dissolved, the gas mixture may react chemically with the liquid. This technique is common and profitable as a step in some petroleum and petrochemical operations in which a gas has a relatively high concentration of solvent vapor. From an air pollution standpoint, absorption has been used primarily to control inorganic compounds, rather than organic vapors, because low concentrations of organic vapors tend to require long contact time and large quantities of absorbent. The economics of the method under these circumstances are often unfavorable unless the absorbent can be regenerated or the solution can be used as a process make-up stream. Otherwise, the absorbent stream can present an additional disposal problem. In emission control applications, absorption is best used in conjunction with other control techniques such as combustion incineration or adsorption, as required, to achieve the prescribed degree of emissions removal.

Absorption of hydrocarbons in the petroleum and petrochemical industries is an important manufacturing step. Products often absorbed in a product-recovery step in industrial processes include light hydrocarbons (e.g., acetylene, butadiene), acetic acid, methanol, ethanol, propanol, chloroform, formaldehyde, formic acid, amines, and ketones. Actually, this type of absorption must be considered as a preliminary step in air pollution control as it does not usually remove as much of the material as would be required for emission control. Absorption has also been employed in scrubbers to remove condensable vapors and particulate matter from the following sources: asphalt batch plants, ceramic tile manufacture, coffee roasters, chromium-plating units, petroleum coker units, fish-meal systems, pipe coating, smoke generators, smoke houses, and varnish and resin cookers.

The ideal absorbent should meet six requirements:

1. The gas should be quite soluble to enhance the rate of absorption and to decrease the quantity of absorbent required; solvents that are chemically similar to the solute generally provide good solubility.
2. The solvent should be relatively nonvolatile.
3. The solvent should be noncorrosive, if possible, to reduce equipment costs.
4. The solvent should be inexpensive and readily available.
5. The solvent should have low viscosity to increase absorption and reduce flooding.
6. The solvent should be low in toxicity, nonflammable, and chemically stable, and have a low freezing point.

The common absorbents for gaseous pollutant are water, nonvolatile hydrocarbon oils, and aqueous solutions (e.g., solutions of oxidizing gents, sodium carbonate, or sodium hydroxide).

6.2 Description of Absorber Equipment

Gas absorption equipment is designed to provide thorough contact between the gas and the liquid solvent to permit interphase diffusion of the materials. This contact is provided by several types of equipment, namely bubble-plate columns, jet scrubbers, packed towers, spray towers, and venturi scrubbers. Bubble-plate columns employ stepwise contact by means of a number of plates or trays arranged so that the gas is dispersed through a layer of liquid on each plate, as shown in Figure 6.1. Each plate is more or less a separate stage, and the number of plates required depends upon the

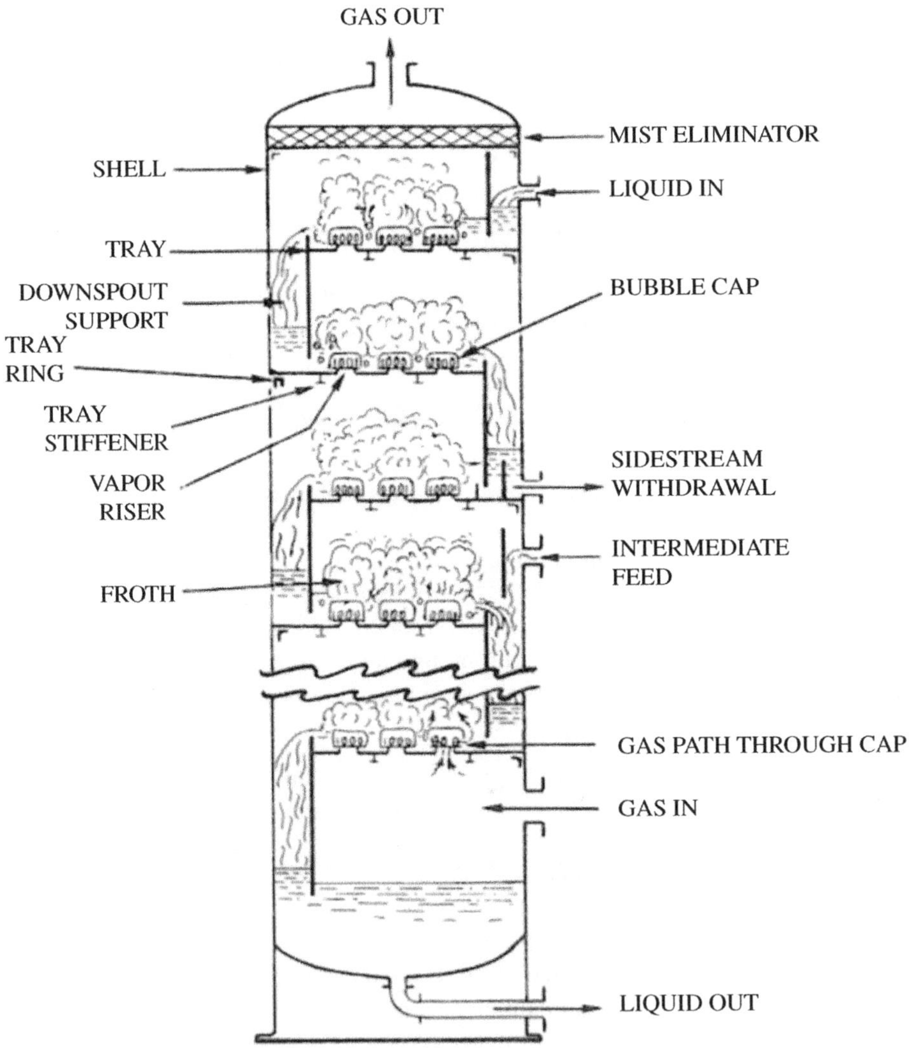

Figure 6.1 Schematic of a bubble-cap tray tower.

difficulty of mass transfer operations and the degree of separation desired. Jet scrubbers are basically spray nozzles. Packed towers are filled with a packing material having a large surface-to-volume ratio; the packing is wetted by the absorbent to provide a large surface area of liquid film for continuous contacting of the gases (Figure 6.2).

The packing should provide a large surface area and should give enough void space when packed to permit good liquid flow. The material should not break easily in handling and should be light in weight; it should be chemically inert enough to prevent deterioration. Manufactured packing has various shapes, as shown in Figure 6.3, and Raschig rings, consisting of hollow cylinders having a diameter equal to the length, are the most common type of packing (see Figure 6.3). Spray towers dispense the liquid in the form of a spray and pass the gas through this spray. Venturi scrubbers, contact the gas and the absorbent in the throat of a venturi nozzle (see Figure 6.4). The gas–liquid mixture then enters an entrainment separator tangentially, and centrifugal force separates the liquid droplets from the gas [1, 2].

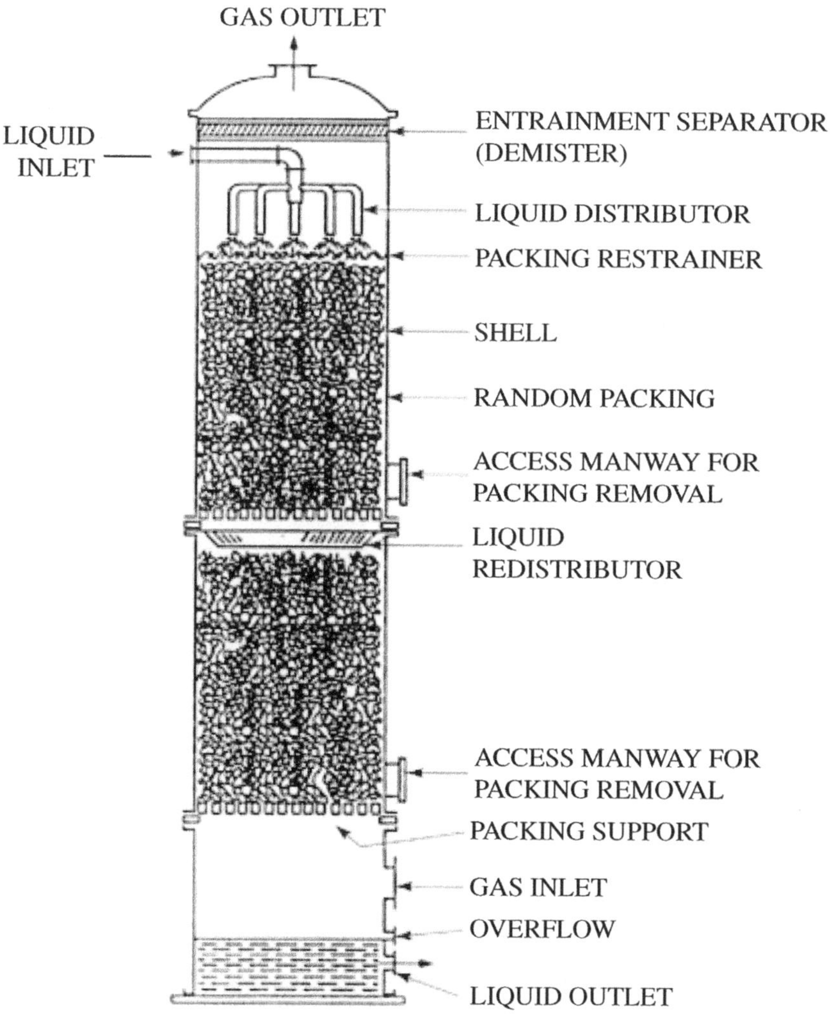

Figure 6.2 Packed tower.

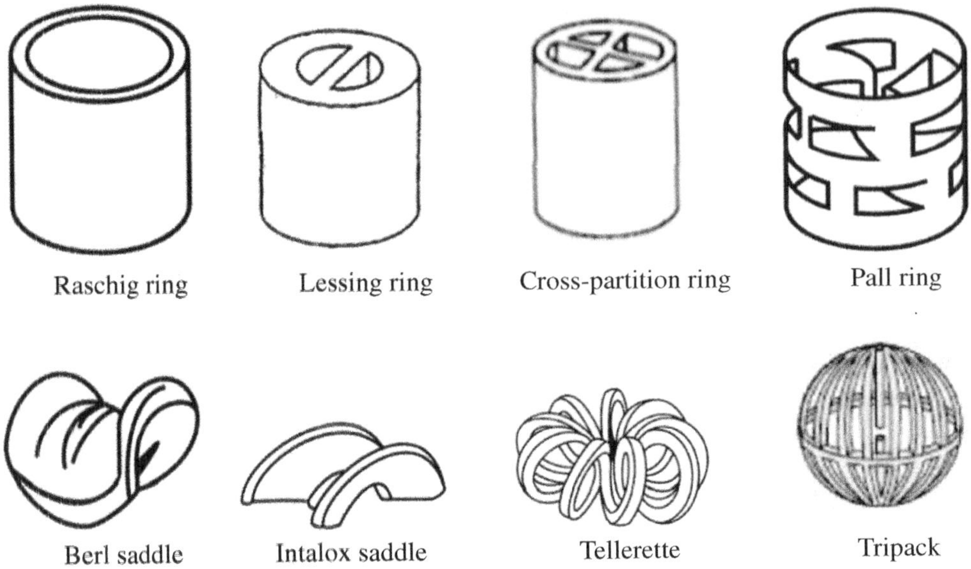

Figure 6.3 Common tower packing materials.

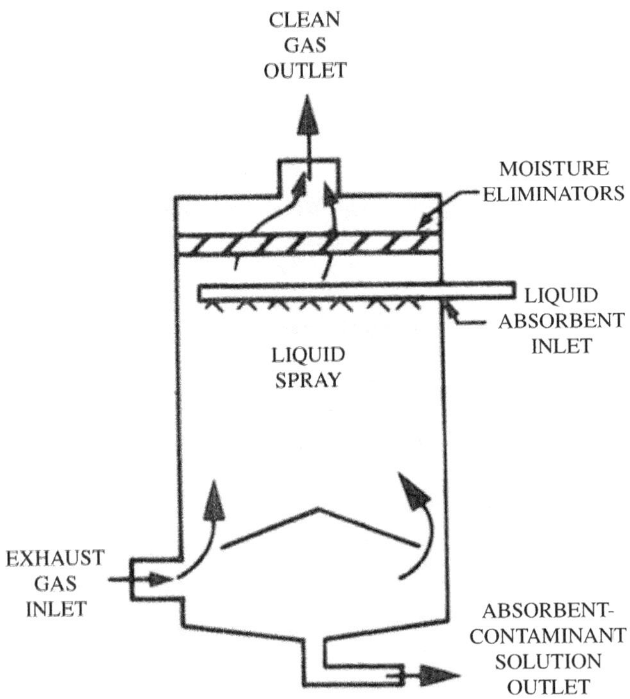

Figure 6.4 Spray tower.

Packed and spray towers are most often used because they introduce relatively lower pressure losses than the bubble-plate columns. Low pressure losses are important because large volumes of exhaust gases with relatively low concentrations of contaminants are treated in many air pollution control installations. Spray chambers have the advantage of being able to handle exhaust gases

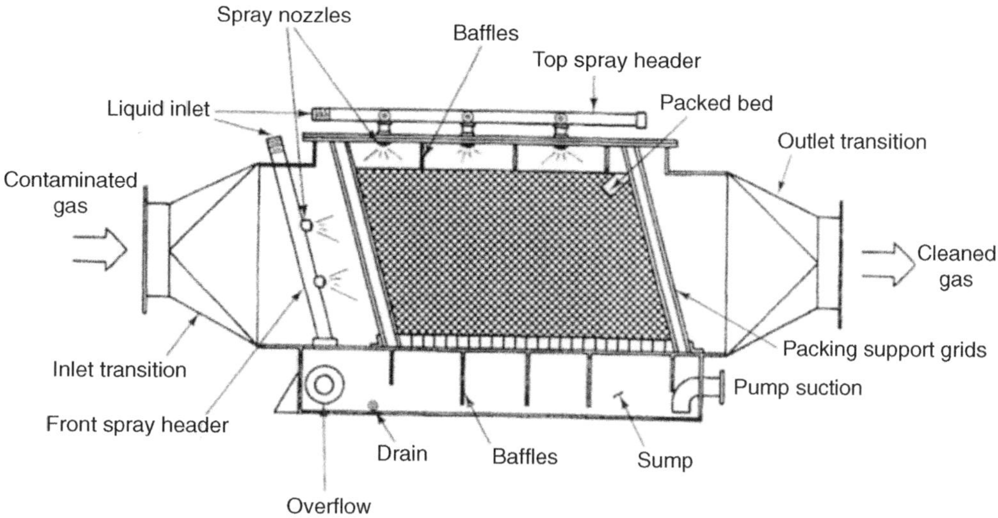

Figure 6.5 Cross-flow design.

containing particulate matter without plugging. The spray chamber is, however, the least effective of the various types of absorption equipment. Since very fine droplets of liquid are necessary for good contact, spray nozzles operated with high pressure drops are required. These fine droplets tend to be entrained in the gas and must be separated.

The cross-flow design (Figure 6.5) is the most unique and has become a commonly used mode for the removal of highly or moderately soluble gases, or where rapid chemical reactions will occur. In this mode, the airstream flows horizontally through the packing while the liquid flows vertically downward. The cross-flow mode has some attractive features over the countercurrent design, while affording lower pressure drop. Lower pressure drop results in the cross-flow where the liquid is at right angles to the air flow resulting in less pressure drop when compared to the countercurrent mode, where the gas and liquid flow in opposite directions opposite each other. In addition, the cross-flow has greater solids-handling capacity without causing fouling. This can be achieved by increasing the liquid flow over the first section of packing to flush off any collected solids to prevent buildup, and by adding a front cocurrent spray to prevent blinding at the front packing support plate. Care must also be taken to prevent gas from bypassing the packed bed. This bypassing occurs primarily in the liquid distribution zone and in the sump. By the proper use of internal baffles, this can be eliminated.

6.3 Process Considerations

The equilibrium of interest in gas absorption operations is that between a relatively nonvolatile absorbing liquid (solvent) and solute gas (usually the pollutant). As described earlier, the solute is ordinarily removed from a relatively large amount of a carrier gas that does not dissolve in the absorbing liquid. The equilibrium relationship of importance is a plot (or data) of x, the mole fraction of solute in the liquid, against y^* (sometimes denoted simply as y), the mole fraction in the vapor in equilibrium with x. For cases that follow Henry's law, Henry's law constant m can be defined by the equation:

$$y^* = y = mx \tag{6.1}$$

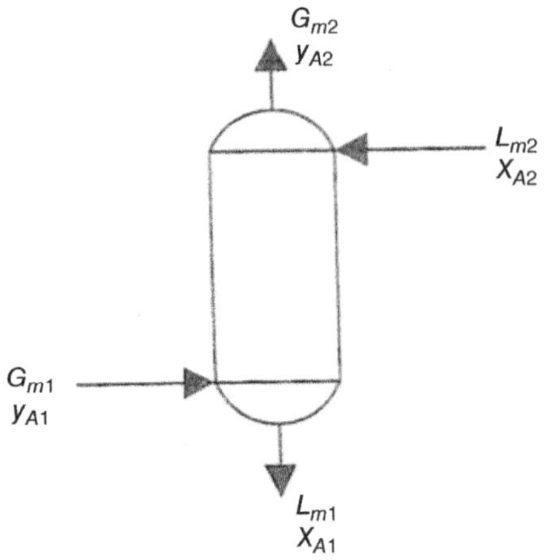

Figure 6.6 Material (mole) balance for the absorption of component *A* in an absorption column.

The usual operating data to be determined or estimated for isothermal systems are the liquid rate(s) and the terminal concentrations or mole fractions. An operating line that describes operating conditions in the column is obtained by a mole balance around the column (as shown in Figure 6.6).

$$\text{Total moles in} = \text{Total moles out} \qquad G_{m1} + L_{m2} = G_{m2} + L_{m1} \tag{6.2}$$

The terms *G* and *L* represent the molar flow rates of the gas and liquid, respectively.

For component *A*, the mole (or mass) balance becomes

$$G_{m1}y_{A1} + L_{m2}x_{A2} = G_{m2}y_{A2} + L_{m1}x_{A1} \tag{6.3}$$

Assuming $G_{m1} = G_{m2}$ and L_{m1} and L_{m2} (reasonable for most air pollution control applications where contaminant concentrations are usually extremely small), one may write Equation 6.3 as

$$G_m y_{A1} + L_m x_{A2} = G_m y_{A2} + L_m x_{A1} \tag{6.4}$$

Rearranging the componential mole balance in Equation 6.4 leads to

$$\frac{L_m}{G_m} = \frac{y_{A1} - y_{A2}}{x_{A1} - x_{A2}} \tag{6.5}$$

This is the equation of a straight line, known as the operating line, since it describes operating conditions in the column. It has a slope L_m/G_m and passes through the points (x_{A1}, y_{A1}), and (x_{A2}, y_{A2}) on *x* and *y* coordinates as indicated in Figure 6.7.

In the design of most absorption columns, the quantity of gas to be treated G_m, the terminal concentrations y_{A1} and y_{A2}, and the composition of the entering liquid x_{A2} are ordinarily fixed by process requirements; however, the quantity of liquid solvent to be used is subject to some choice. Should this quantity already be specified, the operating line in the figure is fixed. If the quantity of solvent is unknown, the operating line is consequently unknown. However, this can be obtained through setting the determination of the minimum liquid-to-gas ratio, a topic to be discussed shortly [3, 4].

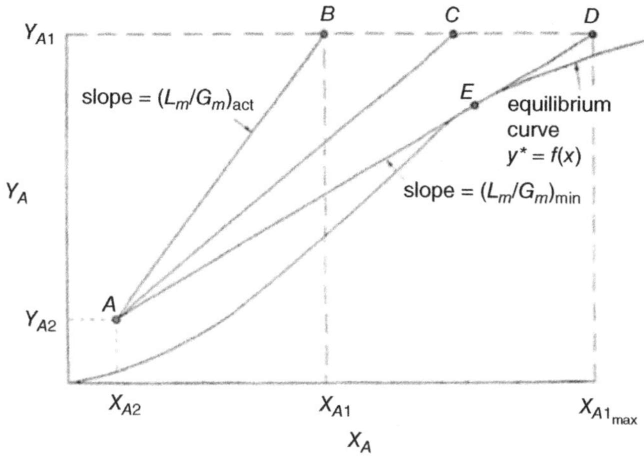

Figure 6.7 Operating and equilibrium lines.

With reference to Figure 6.7, the operating line must pass through point A and must terminate at the ordinate y_{A1}. If such a quantity of liquid is used to determine the operating line AB, the existing liquid will have the composition x_{A1}. If less liquid is used, the exit liquid composition will clearly be greater, as at point C, but since the driving forces (displacement of the operating line from the equilibrium line) for mass transfer are less, the absorption is more difficult. The time of contact between gas and liquid must then be greater, and the absorber must be correspondingly taller.

The minimum liquid that can be used corresponds to the operating line AD, which has the greatest slope for any line touching the equilibrium curve and is tangent to the curve at E. At point E, the diffusional concentration difference driving force is zero; the required contact time for the concentration change desired is infinite and an infinitely tall column is required. This then represents the limiting or minimum liquid-to-gas ratio.

The importance of the minimum liquid-to-gas ratio lies in the fact that column design and operation is frequently specified as some factor of the minimum liquid–gas ratio. For example, a typical situation frequently encountered is that the slope of the actual operating line, $(L_m/G_m)_{act}$, is 1.5 times the minimum, $(L_m/G_m)_{min}$.

Once all the streams entering and leaving the column and their constituents are identified, flow rates calculated, and operating conditions determined, the physical dimensions of the column can be calculated. The column must be of sufficient diameter to accommodate the gas and liquid, and of sufficient height to ensure that the required amount of mass is transferred.

6.3.1 Packed Column

Regarding the diameter, consider a packed column operating at a given liquid rate while the gas rate is gradually increased. After a certain point, the gas rate is so high that the resistive force (drag) on the liquid is sufficient to keep the liquid from flowing freely down the column. Liquid begins to accumulate and tends to block the entire cross-section for flow (so-called flooding). This, of course, increases both the pressure drop and prevents the packing from mixing the gas and liquid effectively, and ultimately some liquid is even carried back up the column. This undesirable condition, known as flooding, occurs fairly abruptly, and the superficial gas velocity (the velocity if the column is empty) at which it occurs is called the flooding velocity. The calculation of column diameter is often based on flooding considerations; the usual operating range is approximately 50–75% of the flooding rate.

One of the more commonly used correlations that has withstood the test of time is U.S. Stoneware's generalized pressure drop correlation, as presented in Figure 6.8. The procedure to determine the column diameter is as follows:

1. Calculate the abscissa, $(L/G)\,(\rho_G/\rho_L)^{0.5}$
2. Proceed to the flooding line and read the ordinate (design parameter)
   The following units must be employed in using Figure 6.8:

   $$L, G = \text{lb/ft}^2 \cdot \text{s}$$

   $\rho_G, \rho_L = \text{lb/ft}^3$ (ρ is often employed for ρ_G)
   F = packing factor, dimensionless
   ϕ = ratio of water density to liquid density, dimensionless
   $g_c = 32.2\ \text{ft} \cdot \text{lb/lb}_f \cdot \text{s}^2$
   μ_L = viscosity of liquid, cP

   Note that the flooding curve can also be approximately represented in equation form by [5]:

   $$Y = -3.85 - 1.06X - 0.119X^2$$

   where e^Y = vertical coordinate
   $X = \ln(\text{horizontal coordinate})$.
3. Solve the ordinate term for G at flooding.

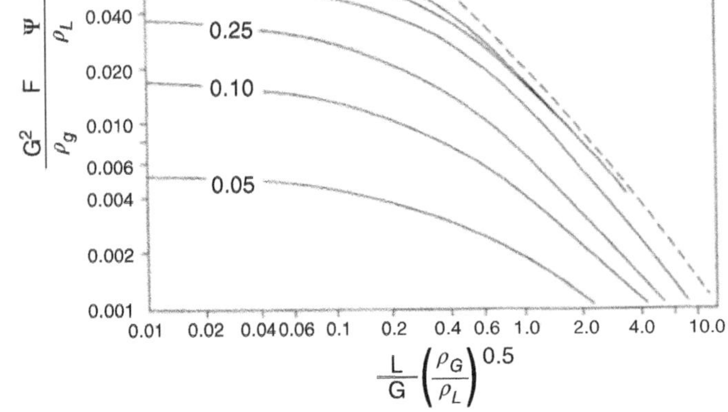

Figure 6.8 Generalized pressure drop correlation to estimate column diameter.

4. Calculate the column cross-sectional area S for the fraction f of flooding velocity chosen for operation by the equation:

$$S = \frac{\dot{m}}{fG} \tag{6.6}$$

where $\dot{m}$ = lb/s of gas, S-area, ft^2

5. The diameter of the column is then determined by

$$D = 1.13(S)^{0.5} \tag{6.7}$$

Note that the proper units as designated in the correlation must be used as the plot is not dimensionless. The flooding rate should also be evaluated using (total) flows of the phases at the bottom of the column, where they are at their highest value. The pressure drop may be evaluated directly from Figure 6.8 using a revised ordinate that contains the actual, not flooding, value of G.

The column height is given by

$$Z = N_{OG} H_{OG} \tag{6.8}$$

where N_{OG} is the number of overall transfer units, dimensionless; H_{OG} is the height of overall transfer units, ft; and, Z is the height of packing, ft. In most air pollution design practice, the number of transfer units (N_{OG}) is obtained experimentally or calculated using any of the methods to be explained later in this section. The height of a transfer unit (H_{OG}) is also usually determined experimentally for the system under consideration. Information on many different systems using various types of packings has been compiled by the manufacturers of gas absorption equipment and should be consulted prior to design. The data are usually in the form of graphs depicting, for a specific system and packing, the H_{OG} vs. the gas rate (lb/hr · ft^2) with the liquid rate (lb/hr · ft^2) as a parameter. The packing height (Z) is then simply the product of the H_{OG} and the N_{OG}. Although

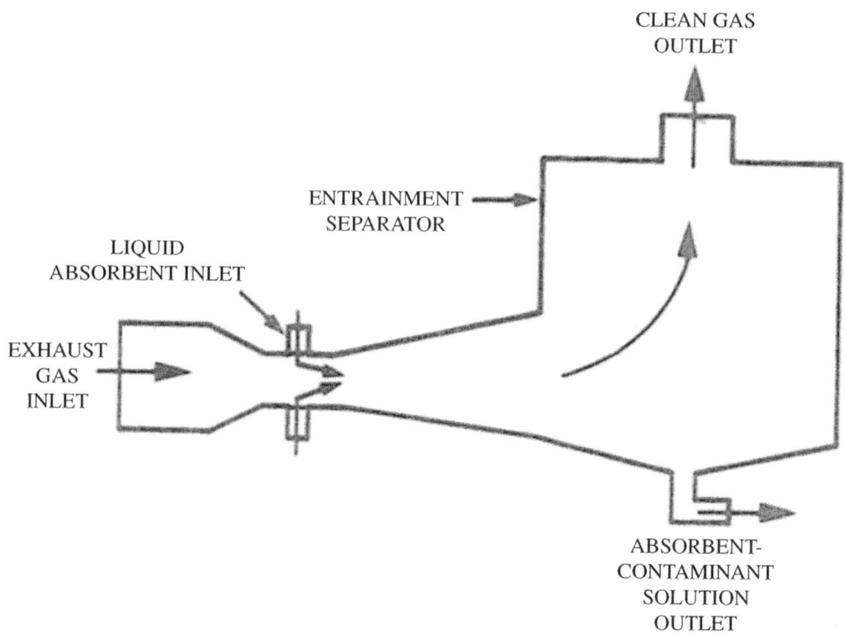

Figure 6.9 Venturi scrubber.

there are many different approaches to determine the column height, the $H_{OG} - N_{OG}$ approach is the most frequently used at the present time, with H_{OG} usually being obtained from the manufacturer. A more "theoretical" approach is available and a brief discussion on this method follows (Figure 6.9).

In air pollution operations, the pollutant to be absorbed is often very dilute. For this condition

$$N_{OG} = \int \frac{dy}{y - y^*} \tag{6.9}$$

If the operating line and equilibrium line are *both* parallel and straight

$$N_{OG} = \frac{y_1 - y_2}{y - y^*} \tag{6.10}$$

If both the operating line and equilibrium line are linear

$$N_{OG} = \frac{y_1 - y_2}{(y - y^*)_{ln}}; \ (y - y^*)_{ln} = \frac{(y - y^*)_1 - (y - y^*)_2}{\ln\left(\frac{(y - y^*)_1}{(y - y^*)_2}\right)} \tag{6.11}$$

If the operating line and/or equilibrium line is curved, the integral above in Equation 6.9, should be evaluated. However, one can show that if Henry's law applies (operating line and equilibrium line straight), the number of transfer units is given by

$$N_{OG} = \frac{\ln\left(\left(\frac{y_1 - mx_2}{y_2 - mx_2}\right)\left(1 - \frac{1}{A}\right) + \frac{1}{A}\right)}{1 - \frac{1}{A}}; \ A = L/mG \tag{6.12}$$

Where the absorption factor is A (not to be confused with area) and is equal to L/mG, where m is the slope of the equilibrium curve. The solution to this equation can conveniently be found graphically in Figure 6.10. Note also that the flow rates L and G are based on moles in Equation 6.12. However, if $m = 0$, Equation 6.12 reduces to

$$N_{OG} = \ln\left(\frac{y_1}{y_2}\right) \tag{6.13}$$

This equation may be applied if the gas is highly soluble or if the absorbate (pollutant) reacts with the liquid [6].

Qualitatively, the height of a transfer unit is a measure of the height of a contactor required to affect a standard separation, and it is a function of the gas flow rate, the liquid flow rate, the type of packing, and the chemistry of the system. As indicated above, experimental values for N_{OG} are generally available in the literature or from vendors.

Some general rules of thumb in the equipment design of packed columns do exist. They are by no means all-inclusive in that there are other considerations that might have to be taken into account, e.g., allowable pressure drop, possible column height restrictions, etc. The rules must therefore be applied discriminately. For approximation purposes, if the gas rate is greater than about 500 ACFM, a nominal packing size smaller than 1 inch would probably not be practical; similarly, at about 2,000 ACFM, packing sizes smaller than 2 inch would also likely be impractical. The nominal size of the packing should never exceed about 1/20th of the column diameter.

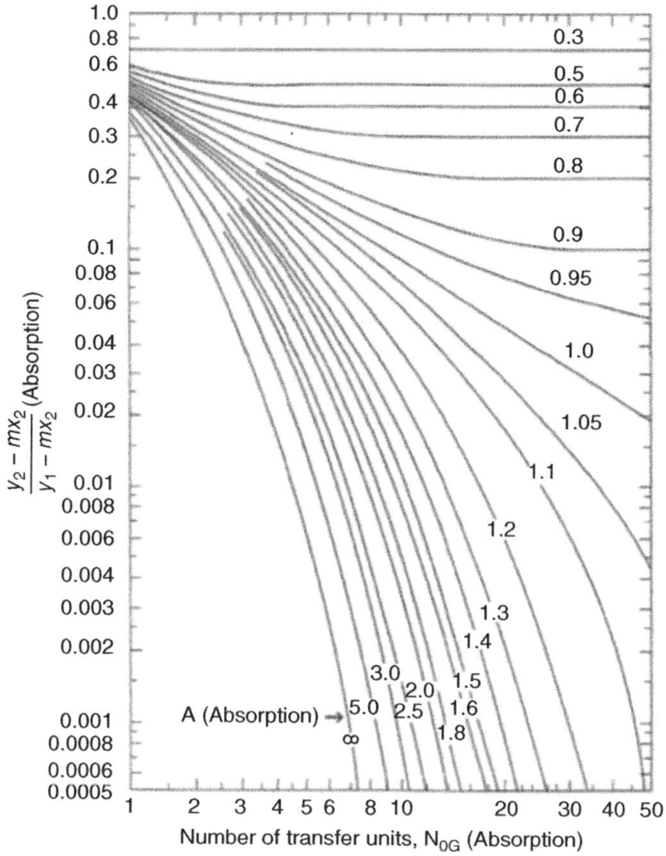

Figure 6.10 N_{OG} for an absorption column with constant absorption factor.

As noted earlier, the usual practice is to design so that the operational gas rate is approximately 75% of the rate that would cause flooding. If possible, column dimensions should be in readily available sizes (i.e., diameters to the nearest half-foot and heights to the nearest foot). If the column can be purchased "off the shelf" as opposed to being specially made, a substantial savings can be realized.

6.3.2 Plate Column

The most important design considerations for plate columns include calculation of the column diameter, type and number of plates to be used (usually sieve plates), actual plate layout and physical design, and plate spacing (which, in turn, determines the column height). To consider each of these to any great extent is beyond the scope of this book. Details are available in any standard chemical engineering unit operations or mass transfer (e.g., distillation) text. The discussion that follows, therefore, will be a relatively concise presentation of the general design techniques that will provide satisfactory results for the purpose of estimation.

The column diameter and consequently its cross-section must be sufficiently large to handle the gas and liquid at velocities that will not cause flooding or excessive entrainment. The superficial gas velocity for a given type of plate at flooding is given by the relation:

$$V_F = C_F \left(\frac{\rho_L - \rho_G}{\rho_G} \right)^{0.5} \tag{6.14}$$

Where V_F is the gas volumetric flow rate through the net column cross-sectional area for gas flow, $ft^3/s \cdot ft^2$; ρ_L and ρ_G are the liquid and gas densities, respectively, lb/ft^3; and C_F is an empirical coefficient that depends on the type of plate and operating conditions.

The net cross-section is the difference between the column cross-section and the area taken up by downcomers. In actual design, some percentage of V_F is usually used: for nonfoaming liquids 80–85% of V_F, and 75% or less for foaming liquids. Of course, the value is subject to a check of entrainment and pressure drop characteristics. The calculation of column diameter based on Equation 6.14 assumes that the gas flow rate is the controlling factor in its determination.

After a plate layout has been assumed, it is necessary to check the plate for its liquid-handling capacity. If the liquid-to-gas ratio is high and the column diameter large, the check will indicate whether the column will show a tendency toward flooding or gas maldistribution on the plate. If this is the case, then the liquid rate is the controlling factor in estimating column diameter and a satisfactory assumption for design purposes is a plate-handling capacity of 30 gal/min of liquid per foot of diameter. However, a well-designed single-pass, cross-flow plate can ordinarily be expected to handle up to 60 gal/min of liquid per foot of diameter without excessive liquid gradient on the plate. It should also be noted that low gas rates can lead to weeping, a condition where the liquid flows down through the holes in the plate rather than across the plate.

The column height is determined from the product of the number of actual plates (theoretical plates divided by the overall plate efficiency) and the plate spacing chosen. The theoretical plate (or "stage," as it is sometimes called) is the theoretical unit of separation in plate column calculations. It is defined as a plate in which two dissimilar phases are brought into intimate contact with each other until equilibrium is reached and then are mechanically or otherwise separated. During the contact, various diffusing components of the mixture redistribute themselves between the phases. In an equilibrium stage, the two phases are well mixed for a time sufficient to allow establishment of equilibrium between the phases leaving the stage. At equilibrium, no further net change of composition of the phases is possible for a given set of operating conditions. The number of theoretical plates can be determined graphically from the operating diagram composed of an operating line and equilibrium curve.

For cases when both the operating line and the equilibrium curve may be considered straight (dilute solutions), the number of theoretical plates may be determined directly without recourse to graphical techniques. This will frequently be the case for relatively dilute gases (as usually encountered in air pollution control) and liquid solutions where, more often than not, Henry's law is usually applicable. Since the quantity of gas absorbed is small, the total flows of liquid and gas entering and leaving the column remain essentially constant. Hence, the operating line will be substantially straight. For such cases, the Kremser–Brown–Souders equation applies for determining the number of theoretical plates N:

$$N = \frac{\log\left(\left(\frac{y_{N+1} - mx_0}{y_1 - mx_0} \right) \left(1 - \frac{1}{A} \right) + \frac{1}{A} \right)}{\log A} \tag{6.15}$$

Note: ln may also be employed in both the numerator and denominator. Here, mx_0 is the gas composition in equilibrium with the entering liquid (m is Henry's law constant = slope of equilibrium curve). If the entering liquid contains no solute gas, then $x_0 = 0$ and Equation 6.15 can be simplified further. The solute concentrations in the gas stream, y_{N+1} a y_1, represent inlet and outlet concentrations, and L and G the total mole rates of liquid and gas flow per unit time per unit column cross-sectional area, respectively. Small variations in L and G may be roughly compensated for by using the geometric average value of each taken at the top and bottom of the column. Equation 6.15 has been plotted in Figure 6.11 for convenience and may be used for the solution to this equation.

The number of actual trays, which is based on the tray efficiency, is determined by the mechanical design used and the conditions of operation. For the case where the equilibrium curve and operating lines are straight, the overall tray efficiency E_0 can be computed, and the number of actual trays determined analytically:

$$E_0 = \frac{\text{equilibrium trays}}{\text{actual trays}} = \frac{\left(\log(1 + E_{MGE})\left(\frac{1}{A} - 1\right)\right)}{\log\left(\frac{1}{A}\right)} \tag{6.16}$$

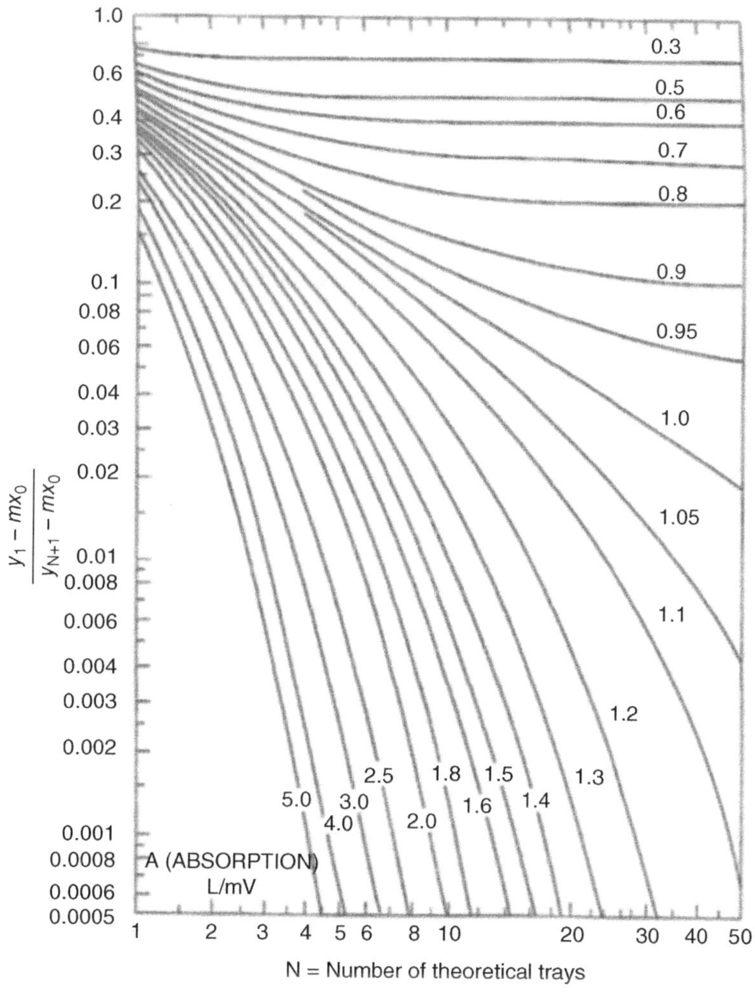

Figure 6.11 KSB plot.

Where E_{MGE} = Murphree efficiency corrected for entrainment (values available in the literature). Empirical data for standard tray designs within standard ranges of liquid and gas rates are available. These data, as shown in Figure 6.12, are accurate for bubble-cap trays and can be used as rough estimates for sieve and valve trays. After the overall efficiency of the tower is determined, the number of actual trays is calculated as:

$$N_{act} = \frac{N}{E_0} \tag{6.17}$$

The general procedure to follow in sizing a plate column is as follows:

1. Calculate the number of theoretical stages N using Figure 6.11 or Equation 6.15.
2. Estimate the efficiency of separation E. This may be determined at the local (across plate), plate (between plates), or overall (across column) level. The overall efficiency E_0 is generally employed.
3. Calculate the actual number of plates:

$$N_{act} = \frac{N}{E_0} \tag{6.17}$$

4. Obtain the height between plates h. This is usually in the 12–36 inch range. Most towers use a 24 inch plate spacing.
5. The tower height Z is then

$$Z = (N_{act})(h) \tag{6.18}$$

6. The diameter may be calculated directly from Equation 6.14.
7. The plate or overall pressure drop is difficult to quantify. It is usually in the 2–6 inch H_2O per plate range for most columns with the lower and upper values applying to small and large diameters, respectively. This pressure drop value represents the approximate height of the liquid on the plates.

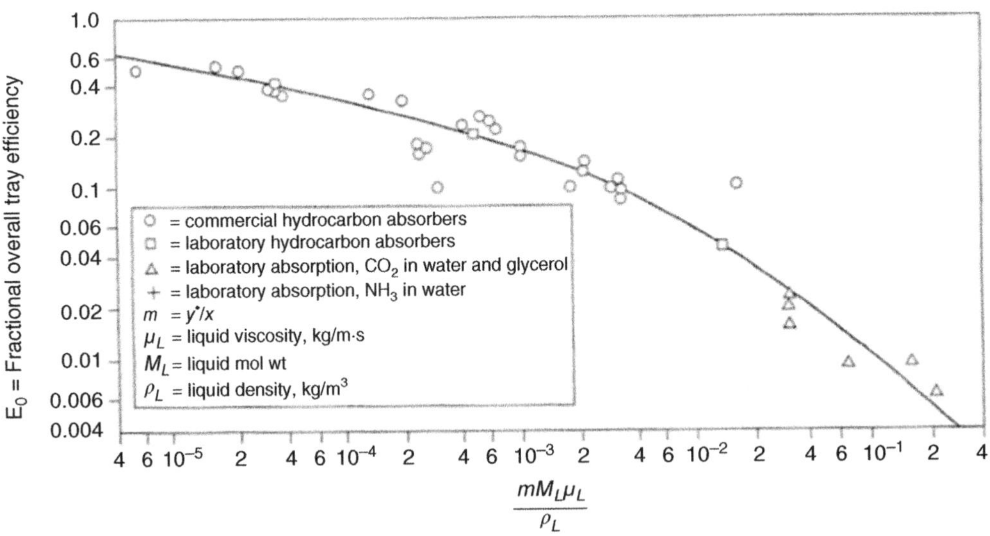

Figure 6.12 Tray efficiency.

6.3.3 Strippers

Absorbers are also used for stripping to regenerate the liquid. During this operation, the pollutant (absorbate) is transferred from the liquid to the gas (the stripping medium). The calculations essentially remain the same for both packed and plate towers. Theodore provides extra details on the strippers in the literature [4, 5].

6.4 Equipment Design Considerations

Basic gas absorption principles are common to all packed scrubbers and wet collectors used for gas absorption, regardless of their operating modes. As noted earlier, gas to be absorbed into the liquid must be soluble or be capable of reacting chemically with the scrubbing liquid. The degree of solubility depends on the type of gas, its concentration (mole fraction or vapor pressure), the system pressure, the type of absorbing liquid, and the liquid temperature. Any liquid has limited amount of gas which it can absorb before it becomes saturated. This is known as equilibrium. If the liquid reaches equilibrium within a packed scrubber, it cannot absorb additional gas. In most packed scrubbers, the liquid is recirculated from a reservoir (sump) to the liquid distributor. This liquid cannot be recycled indefinitely or it will reach saturation. Equilibrium or saturation will be reached first at the gas outlet of the scrubber because it is here that the contaminant has the lowest concentration in the air stream. To prevent equilibrium from occurring and drastically reducing the removal efficiency, fresh liquid is introduced and contaminated liquid is withdrawn from the system continuously, or by introducing chemicals to react with the absorbed gaseous pollutant.

In addition to equilibrium considerations, one must consider how the gas can be absorbed into the liquid most effectively. Since it is the liquid that is absorbing the gas, a means of extending the liquid surface must be provided. This is the premise for the packed absorber itself. By providing a packing medium over which the liquid is spread, a greater area of contact can be achieved. One could interpret, therefore, that the area of gas–liquid contact is the area of the packing. Were this true, the relative efficiency of two packings would be proportional to the ratio of their areas. Yet, ½ inch Raschig rings have twice the surface area than 1 inch Raschig rings but are only 70% more efficient. Therefore, there must be more than the area of the packing relating to performance. This additional factor, liquid surface renewal, has a greater effect with the use of thermoplastic packing than with ceramic packings.

Liquid surface renewal is related to the number of agglomeration points (sometimes referred to as interstitial holdup points). At an agglomeration point, liquid that has flowed down a section of packing will be momentarily held at this point until the weight of the liquid exceeds the surface tension forces and then flow will continue. Liquid flowing down an adjacent section of packing can also flow into this retention zone and mix with the resident liquid. By falling onto the next packing section, the droplet breaks up and presents fresh liquid surface for gas contact. This retention (agglomeration) and continued flow (dispersion) mechanism causes a surface renewal. This agglomeration–dispersion mechanism is also one of the reasons for the effectiveness of packed columns; it is not the surface area of the packing alone, especially where thermoplastic packings are considered [7, 8].

The choice of type, size, and material of construction of the packing medium therefore becomes extremely important. The specific packing that is chosen will affect the initial size and subsequent cost of the equipment. It will also relate to pressure drop, type of support for the packing, type of liquid distribution, fouling, and packing replacement. Packings are available in three basic materials of construction: metal, ceramic, or thermoplastics. Metal packings are rarely used for gas absorption because most gaseous pollutants are so highly corrosive that the type of exotic metal required becomes cost prohibitive. Ceramic materials are used, but in limited applications, primarily where high system temperatures exist. Ceramic packings are very brittle, and extreme care must be taken in filling the absorber to prevent or reduce breakage. This is often accomplished by filling the column with water and allowing the packing to slowly settle into position, or by physically stacking the packing in a dry column. During operation, ceramics can erode or may be subjected to thermal shock, causing packing wear or breakage. Occasionally, the columns should be inspected and additional packing added, if required. Generally, because of the rough surfaces of ceramic packings, fouling or plugging will occur more readily than with thermoplastic packings. If fouling is observed, periodic scheduled maintenance is recommended to prevent eventual total blinding of the packing media. The packing can usually be cleaned by water flushing, chemical cleaning, or steam cleaning, depending on the deposited material. Other factors that should be considered in choosing ceramic materials are their higher pressure drop and greater weight compared to thermoplastic materials.

Thermoplastic materials of construction for packings are the most widely used for gas absorption in air pollution control. Care must be taken to choose the proper thermoplastic material for the corrosive environment and temperature.

Thermoplastic packings are available in polypropylene, polyethylene, polyvinyl chloride (PVC), CPVC, Noryl[†], Kynar[‡], Udel[§] polysulfone, and Tefzel[ɸ], to name the most common. If the proper material is chosen, it can be virtually inert to a corrosive environment. The smooth surface of plastic packings reduces fouling since any solids that may be deposited do not cling, as they would on a rough surface, and are more easily washed away. Lightweight plastic allows a more open support plate to be used, reducing pressure drop and solids deposition. This lighter weight also reduces the shell cost and is less expensive to install and support. Plastic packings are easily installed by simply dumping the packing into a dry column. However, when installing the packing, care should be taken to prevent bridging the packing.

† Noryl is a registered trademark of General Electric Co.
‡ Kynar is a registered trademark of Pennwalt Corp.
§ Udel is a registered trademark of Union Carbide Corp.
ɸ Tefzel is a registered trademark of E. I. du Pont de Nemours and Co.

If bridging occurs, this area will be void of packing, causing what is known as channeling. This will reduce the efficiency of the absorber. Channeling is a condition providing nonuniform cross-sectional distribution of a liquid or a gas flowing through the packing media. Most plastic packings will settle slightly during the first two weeks of operation. It is recommended that the absorber packing height be checked after this initial 2-week period, and if it has settled, that additional packing be added to bring the packing level to its proper height.

Although the type and size of packing are important in designing the physical absorber size and packing depth to achieve a specified removal efficiency of the gaseous pollutants, an equally important factor is the method of liquid distribution over the packing. If the liquid is not distributed evenly across the packing, or if portions of the packing are not wetted, channeling will occur and reduce performance. Again, bear in mind that it is the liquid that absorbs the gases, not the packing. Also, packing that is not wetted will tend to plug much faster because there is no liquid continually washing away any solids deposition.

There are four basic types of liquid distributors: spray, weir, perforated pipe, and open splash plate. The spray type using spray nozzles provides the most effective means of evenly distributing the liquid over the packing. It also provides an area above the packed section for additional contact between the air stream and the spray droplets for gas absorption. A properly designed spray distributor should provide overlapping sprays, and the top of the packing should be at a specified distance below the sprays for optimum effectiveness. If the packing level is too high, some areas near the tower wall will not be wetted; conversely, if the packing level is too low, too great a portion of the liquid is sprayed on the vessel wall and becomes ineffective. Full cone-type spray nozzles are suggested and should be as nonclogging as possible. Spray nozzle plugging is the major disadvantage; however, with the proper nozzle selection and by installing a strainer in the recycle piping which is capable of removing solids smaller than the nozzle orifice size, nozzle plugging can be drastically reduced, if not eliminated.

Weir-type distributors are also used frequently, primarily to eliminate the plugging associated with spray types. Liquid is fed into the open weir boxes, which contain V notches from which the liquid overflows onto the packing. These V notches (points of liquid overflow) are typically spaced at 6 inch increments throughout the absorber cross-sectional area. With weir distributors the distance between the packing and the point of liquid feed is not as critical. This is normally 4 to 12 inches below the weir box. When these liquid streams contact the packing, they splatter and spread out over the packing surfaces. The spreading of the liquid occurs quickly on rough-surfaced packing such as ceramic material. This occurs less rapidly on plastic packing having extended surfaces, and the liquid may not spread out evenly on plastic packings having no extended surfaces,, if at all, prior to leaving the packed section. Uneven liquid distribution will also occur if the weir box is not level. This liquid channeling will drastically reduce tower performance.

Perforated pipe distributors are not recommended. Because of the extremely small hole size and number of holes required to provide even distribution, plugging becomes a major operational problem. These holes (orifices) can be as small as 1/16 inch and can be very difficult to clean. If the holes are made larger and spaced less frequently to reduce clogging, the liquid distribution becomes uneven [8].

Open-splash-type distributors are used very infrequently. This type is simply an open pipe with a plate suspended below it to provide a means of distributing the liquid over a larger area. They are used when the liquid being introduced is known to have large quantities of suspended solids and filtration is not practical, such as the use of river water where silt would be present periodically. Should this type of distributor be required, it is suggested that greater liquid flows be used to compensate for uneven liquid distribution [8].

Another major consideration in absorption equipment is the entrainment separation section (sometimes referred to as the mist eliminator). In most packed absorbers, the entrainment separator is incorporated as an integral part of the absorber. It is placed in the gas outlet above the liquid distributor in the countercurrent absorber, or externally in the cocurrent unit or after the wetted packed section in the cross-flow mode. Some manufacturers provide this as a dry section integral with the wetted bed in the cross-flow mode. The purpose of an entrainment separator is sometimes misunderstood. Its specific function is to remove entrained liquid particles which the air stream captures as it passes through the packed section and liquid distributor. It is not designed for the specific purpose of removing liquid mists entering the inlet air stream. The entrained liquid particles are generally large in size and can be removed easily by impingement or interception mechanisms, and in excess of 99.9% by weight. Three types of entrainment separators are used: mesh pads, packing media, and chevron baffles. Because of their fine filamentous construction, mesh pads provide the most effective removal. However, their fine filamentous wire construction can cause rapid plugging and require frequent cleaning.It is because of this potential plugging that a shallow bed of the more open packing media is often used. Although not quite as efficient, it does provide a much larger open area, reducing the frequency of cleaning if plugging or scaling occurs. Where large plugging or scaling potentials exist, chevron-type separators are used. These are less effective than mesh or packing, but do provide maximum protection against solids plugging.

The material of construction of the shell is significant as well. If the proper material is not chosen, repair and maintenance costs can be extremely high because of corrosion. The material chosen for the absorber shell should be dictated by the types of gaseous pollutants present in the airstream, the liquid used for absorption, the operating temperature, and the reaction products formed in solution (if a reaction is required for removal). Typical gaseous pollutants include HCl, HF, Cl_2, SO_2, H_2S, NH_3, NO_x, amines, alcohols, and organics. Often, caustic or acid solutions are used, rather than water, to react with the absorbed gases to form soluble salts. As can be seen, most of these are extremely corrosive and require special care in selecting the proper materials of construction. Absorbers are constructed of stainless steel, lined steel, exotic metals, PVC, and fiberglass-reinforced plastic (FRP). In most applications, economics dictate the use of either lined steel or fiberglass. If fiberglass is chosen, care must be taken to select the proper resin material since not all resins have equal protection against corrosion. If lined steel is selected, care must be exercised not only in selecting the appropriate lining material, but also on the surface preparation of the steel to which the lining is to be applied. In addition, it is recommended that steel shells be externally coated or protected against atmospheric corrosion.

As can be seen, gas absorbers within themselves are very simple devices with no moving parts. With the selection of proper materials of construction to protect against corrosion, there is very little maintenance required other than to guard against scaling or plugging.

It becomes apparent that the absorber cannot function without equipment to provide a means of motivating the air a liquid stream to and from the device. In order to provide an operational system for a typical gas absorption process, one must also consider the operation and maintenance of the duct, fan, pumps, piping, and valves, but also chemical feed system, liquid reservoir (sump), and instrumentation.

The motivating force for the liquid is a pump, which in most cases is used for recycling the liquid from the absorber sump to the liquid distributor. This is done to conserve liquid consumption and reduce treatment of the contaminated liquid. This pump must be carefully chosen because it provides the means by which the absorber functions. Care must also be taken in choosing the proper material of construction not only against corrosion, but also against abrasion if solids are present. Normally, a base-mounted centrifugal pump with a coupling connection to the motor is used. A pump with a packed stuffing box is used in many cases for easy maintenance; however, it should be recognized that leakage will and should occur. This drippage can be a nuisance and an external corrosion problem. It can be eliminated by the use of a properly chosen mechanical seal or by piping away this leakage. On pumps that handle solids, it is generally recommended that a continuous freshwater flush be provided to prevent scoring of the shaft. To reduce impeller wear and maintenance where solids are present, a 1,750- or 1,150-rpm pump is recommended. If the recycle pump is to be used outdoors in a cold climate where winter freezing problems may exist, it is recommended that a drain be provided in the casing or, if this cannot be provided, that the pump be thermally protected against the liquid freezing in the bottom of the casing when the pump is not operating.

Many problems associated with pumps are not necessarily due to the pump itself but to the piping connections to and from the pump and the physical placement of the pump with respect to the absorber sump. The pump should be placed so that the pump suction line is completely filled with liquid (flooded suction) and the liquid level in the sump at a level above the pump suction to provide the Net Positive Suction Head Required (NPSHR) for the pump. In addition, the pump suction connection in the absorber should be designed to prevent cavitation or vortexing from occurring, especially in shallow sump reservoirs. Cavitation causes air to be drawn into the pump with the liquid and will cause the pump to operate erratically. This will produce low liquid pumping rates and excessive wear. The recycle piping should be of a proper material of construction to withstand the corrosive environment, temperature, and pressure conditions. Typical materials used are PVC, polypropylene, fiberglass (FRP), lined steel, and stainless steel. The suction-line piping should be at least one pipe size larger than the pump suction connection on the recycle pump. It is also suggested that a large coarse screen or filter be used on the suction side of the pump to prevent broken pieces of packing, rags, or other large foreign material from entering the pump and damaging the impeller or casing. This screen or filter should be large enough to prevent restriction of flow to the pump. The discharge-line piping should be at least one pipe size larger than the pump discharge size. A throttling valve should be provided in the discharge line of the pump to allow for adjustment of the proper liquid flow. Where spray-type liquid distributors are used in the absorber, it is recommended that a strainer or filter be provided in the pump discharge piping. The strainer should be designed to filter out particles equal to or smaller than the spray nozzle orifice or particle-free passage size. This will reduce or prevent plugging of the spray nozzles. Normally, a shutoff valve is installed in both the suction and discharge piping to allow for isolation of the pump for maintenance. In cold climates, consideration must also be given for insulation of piping systems that are not self-draining when the system is shut down.

Another essential piece of equipment is the fan or blower to circulate the air through the absorber and duct system. Most absorber systems locate the fan to induce the draft (vacuum) on the system, to prevent out-leakage of the contaminated air. There are occasions, however, where a forced-draft system, using the fan on the inlet side of the absorber, is necessary, and more practical, because of the nature of the corrosion environment. In the selection of the fan material of construction, the location of the fan with respect to the absorber is an important factor to consider, as well as the maintenance required for operational reliability. On systems where hot unsaturated airstreams are entering the absorber, with low or little particulate solids present, it may be advantageous to provide a forced-draft system since, at high temperatures with corrosive gases, corrosion is not as large a problem as saturated airstreams containing corrosives, and the fan material of construction could be of lower-grade alloy materials. If, however, the inlet air is at low temperature and highly saturated with water vapor, or contains a high, or relatively high, solids loading, it is suggested that the fan be located after the absorber to reduce solids buildup on the fan blades. It should be recognized that systems using induced-draft fans after the absorber will still be exposed to highly corrosive gases because no absorber is 100% effective. In addition, one should be cognizant of the fact that the airstream coming from any absorber is saturated with water vapor. This in turn will cause a certain amount of condensation to take place in the fan housing at the walls, which will absorb any remaining corrosive pollutants and become very corrosive. Therefore, particular attention should be given to the material of construction of the fan and housing. Materials generally used are PVC, fiberglass (FRP), or lined steel housings with exotic-alloy wheels and shafts. In certain cases, lined impellers are used, provided that the fan tip speed is low enough to prevent erosion of the lining. Corrosion and erosion of the fan wheel and blades is a major consideration, and a fan should be selected at minimal rpm and tip speed to reduce the abrasion that may occur and reduce maintenance. Corrosion of the fan blades can be caused not only from solids abrasion, but also from water droplets or condensate, which can be very abrasive.

Fans used on absorber systems are normally the centrifugal type and should be equipped with access openings and drains for inspection of the wheel. The drains on the fans should be U-trapped or barometrically sealed to allow any condensate formed in the fan housing to drain away. In designing the static pressure for the fan, system designers usually add safety factors to the estimated system pressure drop. Although this has advantages in overdesign of the fan, it can cause problems resulting in more air being delivered to the system than is designed. This increased capacity will be detrimental to the absorber performance and also require more horsepower, and thus increased energy consumption for the system. To prevent this from happening, many system designers employ a damper in the duct ahead of the fan for volume adjustment. On fans with adjustable sheaves or with belt drives, if large increases in capacity are encountered, it may be advantageous in reducing horsepower consumption simply to change the sheave to lower the fan speed to the desired system capacity. All blowers or fans in a system should be located for ease of accessibility for maintenance, and all drives should be adequately protected for safety reasons. The fans or blowers should be adequately isolated by use of flexible connectors to the duct on the inlet and outlet side to prevent vibration from being transmitted to the duct system. Fans can be equipped with vibration switches to shut down the fan automatically should vibration be excessive. This is suggested on systems where solids or excessive buildup may be expected. The system ductwork should be designed of an appropriate material of construction against the corrosive environment. If the system is handling solids or any amount of particulate matter that may be deposited in the duct, it is recommended that periodic cleanout ports be designed into the ductwork for inspection and/or cleaning. The duct should also contain a slope in its installation, so that any condensate formed will run to a low point and can be drained off so that no pools are created in the ductwork. In extremely corrosive applications, the duct is generally constructed of materials such as PVC, polypropylene, fiberglass (FRP), or lined steel.

Other auxiliary equipment includes:

1. Chemical feed systems
2. Liquid-level controls
3. Freeze protection (insulation and heat tracing)
4. Instrumentation

More information on the operation and maintenance of systems can be found in Richardson and Theodore [9].

6.5 Applications

Three illustrative examples complement the presentation for this chapter concerned with absorbers.

6.5.1 Absorber Design with No Data

You have been requested to qualitatively outline how one can size (diameter, height) a packed tower to achieve a given degree of separation without any information on the physical and chemical properties of a gas to be cleaned.

Approximately 1,600 ACFM gas stream is to be treated in a packed tower containing ceramic packing. The gas stream contains 100 ppm of a toxic pollutant that is to be reduced to 1 ppm. Estimate the tower's cross-sectional area, diameter, height, pressure drop, and packing size.

Solution: To calculate the height, one needs both the height of a gas transfer unit H_{OG} and the number of gas transfer units N_{OG}. Since equilibrium data are not available, assume that m (slope of equilibrium curve) approaches zero. This is not an unreasonable assumption for most solvents that preferentially absorb (or react with) the solute. For this condition (see Section 6.3),

$$N_{OG} = \ln\left(\frac{y_1}{y_2}\right) \tag{6.13}$$

Where y_1 and y_2 represent inlet and outlet concentrations, respectively. Since it is reasonable to assume the scrubbing medium to be water or a solvent that effectively has the physical and chemical properties of water, H_{OG} can be assigned values usually encountered for water systems. These are given in Table 6.1.

For plastic packing, the liquid and gas flow rates are both typically in the range of 1,500–2,000 lb/(hr · ft^2) of cross-sectional area. (Note: The "flow rates" are more properly termed "fluxes"; a flux is a flow rate per unit cross-sectional area with units of lb/hr · ft^2 in this problem.) However, it is common practice to use the term flow rate. For ceramic packing, the range of flow rates is 500–1,000 lb/hr · ft^2. For difficult-to-absorb gases, the gas flow rate is usually lower and the liquid flow rate higher. Superficial gas velocities (velocity of the gas if the column is empty) are in the 3–6 ft/s range. The height Z is then calculated from

$$Z = H_{OG}\, N_{OG}\, S_f \tag{6.19}$$

Where S_f is a safety factor, the value of which can range from 1.25 to 1.5. Pressure drops can vary from 0.15 to 0.40 inch H_2O/ft packing. Packing size increases with increasing tower diameter. Note: This problem and design procedure were originally developed by one of the authors in 1985 and later published in 1988. The reader is left with the exercise of verifying the chart in Table 6.2 for plastic packing.

Key information for ceramic packing is provided in Table 6.3. The equation for the cross-sectional area of the tower S in terms of the gas volumetric flow rate q in ACFS is

$$S(\text{ft}^2) = q(\text{ACFS})/4$$

An equation to estimate the tower packing pressure drop ΔP in terms of Z is

$$\Delta P(\text{inH}_2O) = (0.2)Z;\ Z = \text{ft}$$

Table 6.1 Packing diameter vs. H_{OG}.

Packing diameter, in	Plastic packing H_{OG}, ft	Ceramic packing H_{OG}, ft
1.0	1.0	2.0
1.5	1.25	2.5
2.0	1.5	3.0
3.0	2.25	4.5
3.5	2.75	5.5

Table 6.2 Packing height, Z(ft), as a function of efficiency and plastic packing size.

Removal efficiency, %	Plastic packing size, in				
	1.0	**1.5**	**2.0**	**3.0**	**3.5**
63.2	1.0	1.25	1.5	2.25	2.75
77.7	1.5	1.9	2.25	3.4	4.1
86.5	2.0	2.5	3.0	4.5	5.5
90	2.3	3.0	3.45	5.25	6.25
95	3.0	3.75	4.5	6.75	8.2
98	3.9	4.9	5.9	8.8	10.75
99	4.6	5.75	6.9	10.7	12.7
99.5	5.3	6.6	8.0	11.9	14.6
99.9	6.9	8.6	10.4	15.5	19.0
99.99	9.2	11.5	13.8	20.7	25.3

Table 6.3 Packing height, Z(ft), as a function of efficiency and ceramic packing size.

Removal efficiency, %	Ceramic packing size, in				
	1.0	**1.5**	**2.0**	**3.0**	**3.5**
63.2	2.0	2.5	3.0	4.5	5.5
77.7	3.0	3.7	4.5	6.75	8.25
86.5	4.0	5.0	6.0	9.0	11.0
90	4.6	5.75	6.9	10.4	12.7
95	6.0	7.5	9.0	13.5	16.5
98	7.8	9.8	11.7	17.6	21.5
99	9.2	11.5	13.8	20.7	25.3
99.5	10.6	13.25	15.9	23.8	29.1
99.9	13.8	17.25	20.7	31.1	38.0
99.99	18.4	23.0	27.6	41.4	50.7

The following packing size(s) is (are) recommended:

- For $D \approx 3$ ft, use 1-inch packing
- For $D < 3$ ft, use < 1-inch packing
- For $D > 3$ ft, use > 1-inch packing

As a rule, recommended packing size increases with increasing diameter. For the problem at hand,

$$S = \frac{1{,}600}{4} = 400\ \text{ft}^2$$

The diameter D is

$$D = \left(\frac{4S}{\pi}\right)^{0.5}$$

$$= \left[\frac{(4)(400)}{\pi}\right]^{0.5}$$

$$= 22.6 \text{ ft}$$

For a tower this large, the 3.5 inch packing should be used. Additional information is available from Theodore [10].

6.5.2 Tower Height and Diameter

A packed column is used to absorb a toxic pollutant from a gas stream. From the data given below, calculate the height of packing and column diameter. The unit operates at 50% of the flooding gas mass velocity, the actual liquid flow rate is 40% more than the minimum, and 95% of the pollutant is to be collected. Employ the generalized correlation provided earlier in Figure 6.4 to estimate the column diameter.

- Gas mass flow rate = 3,500 lb/hr
- Pollutant concentration in inlet gas stream = 1.1 mol%
- Scrubbing liquid = pure water
- Packing type = 1 inch Raschig rings; packing factor F = 160
- H_{OG} of the column = 2.5 ft
  Henry's law constant m = 0.98
  Density of gas (air) = 0.075 lb/ft3
  Density of water = 62.4 lb/ft3
  Viscosity of water =1.8 cP

Solution: To calculate the number of overall gas transfer units N_{OG}, first calculate the equilibrium outlet concentration x_1^* at $y_1/m = 0.011$:

$$x_1^* = \frac{y_1}{m}$$

$$= 0.011/0.98$$

$$= 0.0112$$

Determine y_2 for 95% removal:

$$y_2 = \frac{(0.05)y_1}{(1 - y_1) + (0.05)y_1}$$

$$y_2 = \frac{(0.05)(0.011)}{(1 - 0.011) + (0.05)(0.011)}$$

$$= 5.56 \times 10^{-4}$$

The minimum ratio of molar liquid flow rate to molar gas flow rate, $(L_m/G_m)_{\min}$, is determined by a material balance:

$$\left(\frac{L_m}{G_m}\right)_{\min} = \frac{y_1 - y_2}{x_1^* - x_2}$$
$$= \frac{0.011 - 5.56 \times 10^{-4}}{0.0112 - 0}$$
$$= 0.933$$

The actual ratio of molar liquid flow rate to molar gas flow rate L_m/G_m is

$$\frac{L_m}{G_m} = 1.40\left(\frac{L_m}{G_m}\right)_{min}$$
$$= (1.40)(0.933)$$
$$= 1.306$$

In addition,

$$\frac{mG_m}{L_m} = \frac{0.98}{1.306} = 0.7504$$

The absorption factor A is defined (see Equation 6.12) as

$$A = \frac{L_m}{mG_m} = \frac{1}{0.7504} = 1.333$$

The value of $(y_1 - mx_2)(y_2 - mx_2)$ is

$$\frac{y_1 - mx_2}{y_2 - mx_2} = \frac{(0.011 - (0.98)(0))}{5.56 \times 10^{-4} - (0.98)(0)} = 19.78$$

N_{OG} is then calculated from Colburn's equation (or read from Colburn's chart—Figure 6.10).

$$N_{OG} = \frac{\ln\left[\frac{y_1 - mx_2}{y_2 - mx_2}\left(1 - \left(\frac{1}{A}\right)\right) + (1 + A)\right]}{1 - \left(\frac{1}{A}\right)}$$

$$= \frac{\ln\left[19.78\left(1 - \left(\frac{1}{1.333}\right)\right) + (1 + 1.333)\right]}{1 - \left(\frac{1}{1.333}\right)}$$

$$= 6.96$$

The packing height Z is then

$$Z = N_{OG} H_{OG}$$
$$= (6.96)(2.5)$$
$$= 17.4 \text{ ft}$$

To determine the diameter of the packed column, the ordinate of Figure 6.4 is first calculated:

$$\left(\frac{L}{G}\right)\left(\frac{\rho_G}{\rho_L}\right)^{0.5} = \left(\frac{L_m}{G_m}\right)\left(\frac{18}{29}\right)\left(\frac{\rho_G}{\rho_L}\right)^{0.5} = 1.306\left(\frac{18}{29}\right)\left(\frac{0.075}{62.4}\right)^{0.5} = 0.0281$$

The value of the abscissa at the flooding line is determined from the same figure:

$$\frac{G^2 F \psi \mu_L^{0.2}}{\rho_L \rho g_c} = 0.21;\ \rho = \rho_G$$

The flooding gas mass velocity G_f in lb/(ft^{2}s) is

$$G_f = \left(\frac{0.21\rho_L \rho g_c}{G^2 F \psi \mu_L^{0.2}}\right)^{\frac{1}{2}} = \left(\frac{((0.21)(62.4)(0.075)(32.2))}{(160)(1)(1.8)^{1.8}}\right)^{\frac{1}{2}}$$

$$= 0.419\ lb/(ft^2 \cdot s)$$

6.5.3 Absorber Failure to Meet Performance Guarantee

Engineers designed an absorber unit to operate with a maximum discharge concentration of 50 ppm. Once the unit was installed and running, the unit operated with a discharge of 60 ppm. Rather than purchase a new unit, what options are available to bring the unit into compliance with the specified design concentration?

Solution: This is obviously an open-ended question. One may employ any one or a combination of the following suggestions [11].

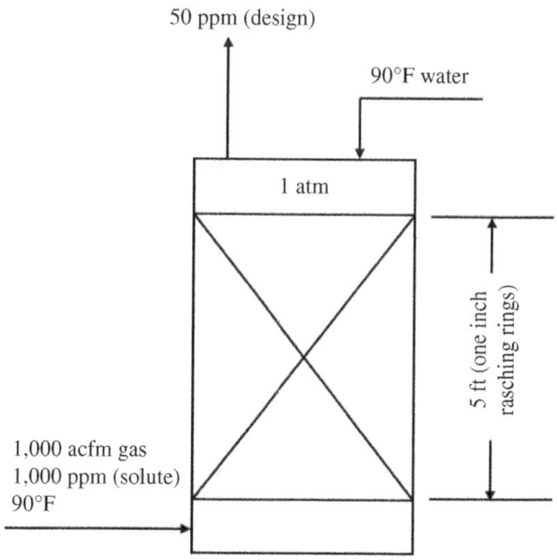

1. Use a smaller packing size. A different size may produce a better packing factor.
2. Use a different type of packing. A different packing type may also produce a better packing factor.
3. Make sure there is no channeling inside the column. The packing should be randomly distributed and there should be no open spots where water will accumulate.

4. Increase the water flow entering the column. In effect, the slope of the operating line will increase' and provide a greater driving force for mass transfer. Thus, there will be a greater mass transfer between the liquid and gas phase and the efficiency of the column will increase.
5. Use a liquid other than water to scrub the gas. Another liquid may cause better mass transfer between the gas and liquid phase.
6. Check to see if flooding is occurring. If the column is near the flooding point, there will not be much mass transfer. The water will be pushed up the column by the gas rather than the water coming down the column.
7. Lower the initial concentration in the carrier gas (if possible).
8. Increase the height of the bed. This will provide more surface area for mixing and more time in the absorber.
9. Increase the pressure. This will reduce the slope of the operating line' and lead to a greater driving force.
10. Decrease the temperature. This will reduce the slope of the operating line and lead to a greater driving force.
11. Increase the pressure and decrease the temperature.
12. Additional packing height could be added at the top of the column. This would increase the amount of the solute removed from the gaseous phase.
13. Add sprays at the bottom of the column.
14. Modify the process producing the problem. Since the details of this process are not included in the problem statement, no specific recommendations can be mentioned in this discussion.
15. Finally, before considering changes to the system, one should undertake a thorough inspection of the unit and surrounding components. The emission monitoring system should be recalibrated. All valves, fittings, and pipes should be checked for plugging or leaks. The liquid distributor should be checked to make sure it is functioning properly. The distribution of the packing should be inspected to make sure it is as uniform as possible. Any problems encountered during this inspection should be corrected immediately. Following the maintenance check, the performance of the unit should be reevaluated.

Any other suggestions?

References

1 L. Theodore and R. Allen, *Pollution Prevention*, A Theodore Tutorial, East Williston, NY, 1994 (originally published by the ES EPA/APRI, RTP, NC).

2 L. Theodore and R. Bardon, *Mass Transfer Operations*, A Theodore Tutorial, East Williston, NY, 1994 (originally published by the ES EPA/APRI, RTP, NC).

3 L. Theodore and F. Ricci, *Mass Transfer Operations for the Practicing Engineer*, John Wiley & Sons, Hoboken, NJ, 2010.

4 L. Theodore, *Air Pollution Control Equipment Calculations*, John Wiley & Sons, Hoboken, NJ, 2008.

5 L. Theodore, *Chemical Engineering: The Essential Reference*, McGraw Hill, New York City, NY, 2012.

6 L. Theodore, *Personal Notes*, East Williston, NY, 1976.

7 Ceilcote Co., Inc. *Tellerette Manual*, Berea, OH, 1970.

8 J. MacDonald, *Personal Notes*, Berea, OH, 1974.

9 J. Richardson and L. Theodore, *Optimizing Air Pollution Equipment Performance: Operation and Maintenance*, John Wiley & Sons, Hoboken, NJ, 2025.

10 L. Theodore, *Engineering Calculations Sizing Packed Tower Absorbers Without Data*, CEP, 18–19, 2005.

11 L. Theodore, *Personal Notes*, East Williston, NY, 1986.

Suggested Reading

1 G. Holmes, R. Singh, and L. Theodore, *Handbook of Environmental Management and Technology*, Wiley-Interscience, New York City, NY, 1993.

2 J Mycock, J. McKenna, and L. Theodore. *Handbook for Air Pollution Control Engineering and Technology*, CRC Press/Taylor & Francis Group, Boca Raton, FL, 1995.

3 L. Theodore and R. Bardon, *Mass Transfer Operations*, A Theodore Tutorial, East Williston, NY, 1994 (originally published by the ES EPA/APRI, RTP, NC).

4 L. Theodore, *Air Pollution Control Equipment Calculations*, John Wiley & Sons, Hoboken, NJ, 2008.

5 L. Theodore, *Chemical Engineering: The Essential Reference*, McGraw Hill, New York City, NY, 2012.

6 L. Theodore (section editor), D. Green, and M. Southard, *Perry's Chemical Engineering Handbook*, 9th edition, McGraw Hill, New York City, NY, 2019.

7 J. Richardson and L. Theodore, *Optimizing Air Pollution Equipment Performance: Operation and Maintenance*, John Wiley & Sons, Hoboken, NJ, 2025.

7

Adsorbers

Note: The bulk of the writing for this chapter on adsorbers has been drawn from material developed by one of the authors that appeared in several earlier government publications [1–5].

7.1 Introduction

Adsorption is a process wherein molecules of a fluid or gas contact and adhere to the surface of a solid. This phenomenon is believed to occur naturally in nature, regardless of the particular solids and gases involved. The forces present in this intermolecular activity have been identified as being of two major types: van der Waals forces and chemical or ionic forces. The bonding force playing the major role in the adsorption process identifies either physical adsorption, in which the van der Waals molecular interaction force predominates, or chemisorption, where chemical reaction forces provide the predominant intermolecular bonds. Thus, if in the process of adsorption, the individuality of the adsorbed molecule (adsorbate) and of the solid surface (adsorbent) is preserved, this is physical adsorption. If a chemical reaction occurs and the nature of the adsorbate or the adsorbent is altered through a valence electron exchange representing chemical bonds, this is chemisorption.

Both physical adsorption and chemical adsorption are exothermic reactions, and both liberate heat during the adsorption phase. The heat of physical adsorption, with its somewhat weak physical bonds, is of the same order of magnitude as the heat of liquefaction (or vaporization); the heat of chemisorption, possessing stronger bonding forces, is in the same order of magnitude as its associated chemical reaction. Regeneration therefore requires a corresponding equivalent amount of energy to free the gas molecule from the solid–gas bonding force. The strong bonding forces characterized in chemisorption make this particular method undesirable for use in regenerative applications because of the excessive amount of energy required for desorption. However, regeneration of adsorbents possessing predominately physical bonds can be accomplished quite easily utilizing the heat energy available in steam, hot gas, or electricity. The solid–gas interface may also be broken by lowering the vapor pressure above the adsorbent, providing the liquefied gas molecule that is attached to the solid surface a sufficient amount of energy to break the weak van der Waals forces and again move into the gaseous state.

Although it is probable that all solids adsorb gases to some extent, adsorption as a rule is not very pronounced unless an adsorbent possesses a large surface area for a given mass. For this reason, such adsorbents as charcoals, activated alumina, silica gel, and molecular sieves are particularly effective as adsorbing agents. These substances have a very porous structure and with their large, exposed surfaces can take up appreciable volumes of various gases. The first three of these adsorbents are amorphous in nature with a nonuniform internal structure. Molecular sieves, however, are crystalline and possess an internal structure of regularly spaced cavities with interconnecting pores of a definite size. Adsorbent details are provided below.

7.1.1 Charcoal

Charcoal is an inefficient form of activated carbon. The activation process involves a high-temperature distillation of hydrocarbon impurities from a charcoal and leads to exposure of a larger free surface for possible adsorption of the gaseous molecule on the solid surface.

A large majority of the total surface area of an individual adsorbent is in the molecular-sized pores. Obviously, the surface of the carbon cannot be "wetted" by the vapor if the gas molecule is too large to penetrate the pore structure. Logically, then, an adsorbent in industrial applications must be selected that will present to the particular gas molecule the largest surface area within the molecular-sized pore structure of the granule [6].

Each individual activated carbon granule offers and abundance of surface area of the exposed gas molecule. Gaseous molecules travel into the interior of the where they are adsorbed and held fast by van der Waals forces until they are liberated by a sufficient amount of energy. Activated carbon with micropore structures ranging from 15 to 20 angstroms (Å) in diameter, coupled with a larger system of macropores of approximately 1,000 Å provides an effective adsorbent for many organic compounds used in industry today, and has found extensive use in many air pollution control applications [7].

7.1.2 Activated Alumina

Activated alumina (hydrated aluminum oxide) is produced by special heat treatment which is precipitated from native aluminas or bauxite. It is available in either granule or pellet form and is used principally for drying of a wet gas stream.

7.1.3 Silica Gel

The manufacture of silica gel consists of the neutralization of sodium silicate by mixing with dilute mineral acid, washing the gel formed to rid it of salts produced during the neutralization reaction, followed by drying, roasting, and grading processes. The name "gel" arises from the jellylike form of the material present during one stage of its production. It is generally used in granular form, although bead forms are also available. Silica gel also finds primary use in gas drying, although it has other applications (e.g., gas desulfurization and purification).

7.1.4 Molecular Sieves

Molecular sieves are dehydrated zeolites (i.e., aluminosilicates) in which atoms are arranged in a definite pattern. The fundamental building block of the molecular sieve is a tetrahedron of four oxygen anions surrounding a smaller silicon or aluminum cation. The cations serve to make up the positive charge deficit in the alumina tetrahedra. Each of the four oxygen anions is shared with another silica or alumina tetrahedron to extend the crystal lattice in three dimensions. The resulting crystal is unusual in that it is honeycombed with relatively large cavities, each cavity connected with six adjacent ones through apertures or pores. These pores serve as the adsorption surface within a molecular sieve. Molecular sieves will effectively remove acetylene from olefins, and ethylene or propylene from saturated hydrocarbons [8].

7.1.5 Applications of Adsorbents

Although these above four adsorbents are the most commonly used, others are available. These adsorbents and others are listed with their applications in Table 7.1 [9]. Industrial adsorbents are usually capable of adsorbing both organic and inorganic gases. However, their preferential adsorption characteristics and other physical properties make each one more or less specific for a particular application. For organic vapor adsorption, activated carbon has superior qualities, possessing hydrocarbon selective properties and high adsorption capacity for such materials. The three other common adsorbents discussed above (activated alumina, silica gel, and molecular sieves) are both hydrophilic (affinity for water molecules) and organic-adsorbing by nature. The hydrophilic properties of the adsorbent create a considerable drawback when considering solvent recovery by steam regeneration since the water molecule also attaches itself to the solid surface by adsorption and is not easily removed by conventional desorption methods. A list of gases and vapors selectively adsorbed by activated carbon is provided in Table 7.2.

7.1.6 Equilibrium Considerations

Most available data on adsorption systems are determined at equilibrium conditions. Adsorption equilibrium is the set of conditions at which the number of molecules arriving on the surface of the adsorbent equals the number of molecules that are leaving. The adsorbent bed is then said to be "saturated with vapors" and can remove no more vapors from the exhaust stream. Thus, equilibrium determines the maximum amount of vapor that may be adsorbed at a given set of

Table 7.1 Common adsorbents and their applications.

Adsorbent	Application
Activated carbon[a]	Solvent recovery, elimination of odors, and purification of gases
Alumina	Drying of gases, air, and liquids
Bauxite	Treatment of petroleum fractions; drying of gases and liquids
Bone char	Decolorizing sugar solutions
Decolorizing carbons	Decolorizing oils, fats, and waxes; deodorizing of domestic water
Fuller's earth	Refining of lube oils and vegetable and animal oils, fats, and waxes
Magnesia	Treatment of gasoline and solvents; removal of metallic impurities from caustic solutions
Molecular sieves	Selective removal of contaminants from hydrocarbons
Silica gel	Drying and purification of gases
Strontium sulfate	Removal of iron from caustic solutions

a. A list of gases and vapors selectively adsorbed by activated carbon is provided in Table 8.2.

Table 7.2 Gases and vapors selected by activated carbon.

Acetaldehyde	Acetone
Benzene	Isobutane
Normal butane	Normal butene
Butyne	Carbon dioxide
Carbon disulfide	Carbon tetrachloride
Carbonyl sulfide	Chloroform
Cumene	Cyclohexane
Cyclohexanone	Cyclopentadiene
Dichloroethane	Dichloroethylene
Dimethyl formamide	Ethane
Ethanol	Ethyl acetate
Ethyl chloride	Ethyl mercaptan
Ethylene	Ethylene oxide
Freon 12	Heptane
Normal hexane	Hexanol
Hydrogen cyanide	Hydrogen sulfide
Isopentane	Isoprene (methyl butadiene 1,3)
Isovaleric acid	Simulated kerosene ($C_{14}H_{30}$)
Methane	Methyl ethyl ketone
Methyl mercaptan	Mineral spirits
Neopentane	Normal pentane
Perchloroethylene	Propane
Propylene	Propyl mercaptan
Pyridine	Tetrahydrofuran
Tetrahydrothiamine	Toluene
Trichloroethylene	Vinyl chloride
Metaxylene	

operating conditions. Although a number of variables affect adsorption, the two most important ones in determining equilibrium for a given system are temperature and pressure. Three types of equilibrium graphs are used to describe adsorption systems: isotherm at constant temperature, isobar at constant pressure, and isostere at constant amount of vapors adsorbed.

The most common and useful adsorption equilibrium data is the adsorption isotherm. The isotherm is a plot of the adsorbent capacity vs. the partial pressure, P, of the adsorbate at a constant temperature, T. Adsorbent capacity is usually given in weight percent expressed as grams of adsorbate per 100 g of adsorbent. Figure 7.1 shows a typical example of an adsorption isotherm for carbon tetrachloride on activated carbon. Graphs of this type are used to estimate the size of adsorption systems, as described later. Attempts have been made to develop generalized equations that can predict adsorption equilibrium from physical data. This is very difficult because adsorption isotherms take many shapes depending on the forces involved. Isotherms may be concave

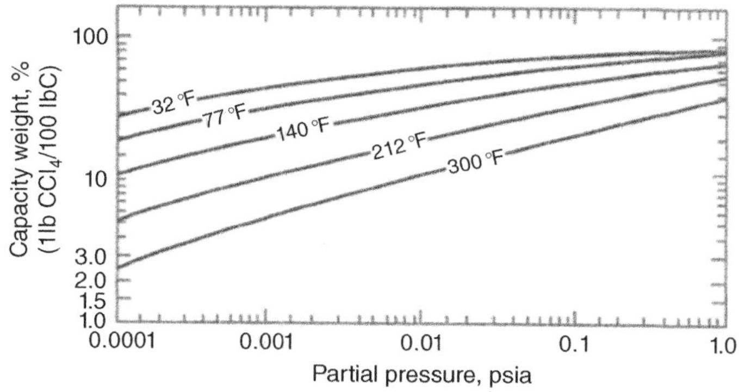

Figure 7.1 Adsorption isotherms for carbon tetrachloride on activated carbon.

upward, concave downward, or "S"-shaped. To date, most of the theories agree with data only for specific adsorbate systems and are valid over limited concentration ranges.

Two additional adsorption equilibrium relationships are the isostere and the isobar. The isostere is a plot of the natural log of the partial pressure p vs. 1/T at a constant amount of vapor adsorbed. Adsorption isostere lines are usually straight for most adsorbate–adsorbent systems. The isostere is important in that the slope of the isostere (approximately) corresponds to the heat of adsorption. The isobar is a plot of the amount of vapors adsorbed vs. T at a constant partial pressure. However, in the design (see Section 7.3) of an air pollution control equipment system, the adsorption isotherm is by far the most commonly used equilibrium relationship.

7.2 Description of Adsorber Equipment

Adsorbers can have fixed, moving, or fluidized beds. They can be set vertically or horizontally. A single fixed-bed unit is satisfactory if process downtime is available for regeneration of the adsorbent. For example, an adsorber for a spray-paint booth that is in use only 6 hours a day can be designed to extract the total emission for this period and to be desorbed after the operating period.

A typical adsorption process is shown diagrammatically in Figure 7.2. One adsorber handles the vapor-laden stream while the other is undergoing regeneration. When the first, or onstream, adsorbent bed approaches the break-through point, the second adsorber, which has been regenerated, is placed onstream. This procedure ensures that the vapor is removed from the air continuously. Caution must be exercised to ensure that an adsorber is adequately cooled after desorption before it is placed onstream again. Contact of organic vapors with a hot carbon bed may promote decomposition or partial oxidation and thereby result in the discharge of odorous or irritating gases to the atmosphere.

Adsorbers may be classified as regenerative or nonregenerative. Regenerative systems are used when the adsorbent is to be reactivated by desorption and the desorbed vapors recovered for reuse when the adsorbent is to be replaced with fresh material, the displaced material usually being returned to the vendor for regeneration. Nonregenerative adsorption is often employed in air conditioning systems for large buildings and hospitals.

The most common adsorption system is the fixed bed adsorber. These systems consist of two or more adsorber beds operating on a timed adsorbing/desorbing cycle. One or more beds are adsorbing vapors, while the other bed(s) is being regenerated. If particulate matter or liquid droplets are

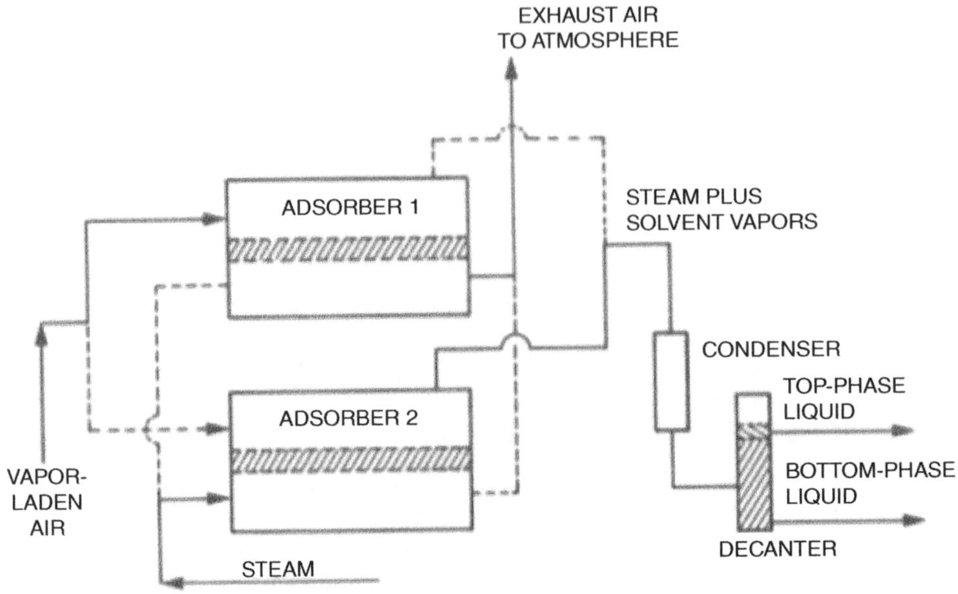

Figure 7.2 Diagrammatic sketch of two-unit fixed-bed adsorber.

present in the vapor-laden airstream, this stream is sent to pretreatment to remove them. If the temperature of the inlet vapor stream is high (much above 120 °F), cooling may also be required. Since all adsorption processes are exothermic, cooling coils in the carbon bed itself may also be needed to prevent excessive heat buildup. Carbon bed depth is usually limited to a maximum of 4 ft, and the vapor velocity through the adsorber is held below 100 ft/min to prevent an excessive pressure drop.

In another application, the adsorbent is injected into the gas stream, mixes with that stream, and adsorbs the contaminants. Usually, at some point after the adsorbent is injected into the gas stream, the gas stream passes through a particulate's removal device (often a fabric filter, see Chapter 15). The particulates removal system separates out the adsorbent, and any other particulates, from the gas stream. When a fabric filter is used, the adsorbents form part of the layer on the filter bags and continue to remove contaminants until the filter is cleaned.

There are stationary-bed solvent recovery systems that employ vertical cylindrical beds wherein the solvent-laden air flows axially down through the bed. This particular design is advertised for use in recovery of solvents used in degreasing and dry cleaning, although equally well suited for recovery of solvents from other industries. Solvents mentioned are trichloroethylene, tetrachloroethylene, toluene, Freon TF, and dichloromethane. Regeneration is accomplished with steam flowing upward through the bed and, since the above-mentioned solvents are immiscible in water, decantation is used to separate the condensed solvent from the steam condensate. The valves are of disk type, opened and closed by air-driven pistons. Water is used as coolant in the condenser. Steam, electric power to drive the blower, and cooling water are three operating-cost items. The cost of steam and cooling water increases with frequency of regeneration.

Dual adsorber systems can also be operated with both beds on-stream simultaneously, especially when solvent concentrations are low. In this situation, regeneration is less frequent than a full work shift and may be accomplished during offwork hours. Operation in parallel almost doubles the air-handling capacity of the adsorption unit and may be an advantage in terms of operating cost.

Moving-bed systems are also available. The rotary component of the system consists of four coaxial cylinders. The outer cylinder is impervious to gas flow except at the slots near the end. They serve as impure air inlet ports where they are shown uncovered and as steam-vapor outlet ports as shown at the lower end of the rotary. The carbon bed is retained between two cylinders made of screen or perforated metal and also segmented by partitions placed radially between the two cylinders. The inner cylinder is again impervious to gas flow except at the slots near the end of the rotary. These slots serve as outlet ports for the purified air and as inlet ports for the regenerating steam. On each rotation of the rotary, each segment of the bed undergoes adsorption and regeneration. The desorbed solvent can then be separated from the steam by decantation or distillation [10].

Because of the continuous regeneration capability of the rotary bed, more efficient utilization of the carbon is possible than with stationary-bed systems. In most solvent recovery operations, the adsorption zone and the saturated bed behind it are idle but add to the bulk of the system and increase airflow resistance through the bed. In deep beds of 12–36 inches, as required in the stationary-bed systems, a large portion of the bed is idle at any one time [10]. By continuous regeneration, the regeneration time for each segment of the bed is shortened, and thereby shorter bed lengths can be used. This leads to two advantages: a more compact system and reduced pressure drop. The disadvantages are those associated with wear on moving parts and maintaining seals in contact with moving parts. The use of shorter beds also decreases the steam utilization efficiency.

In a fluidized-bed solvent recovery system, carbon is recirculated continuously through the adsorption–regeneration cycle. The spent carbon, saturated with solvent, is elevated to the surge bin and then passed down into the regeneration bed, where it is contacted to an upward flow of steam. The regenerated carbon is then metered into the adsorber, where the carbon traverses beds while fluidized with the upward flow of the vapor-laden air. An air velocity of approximately 240 ft/min is required to cause fluidization of the bed. In both the regeneration and adsorption phases, the carbon is moved countercurrent to the gas or vapor [11].

The countercurrent movement increases the efficiency of regeneration, i.e., the steam/solvent ratio is less than for a stationary bed under otherwise comparable conditions. In addition to the beneficial effects of the countercurrent movement, the bed length can be increased to further improve steam utilization. The countercurrent movement also increases the effective use of the carbon; more solvent can be recovered with less carbon than with stationary- or rotary-bed systems. By adjustments or balance of the carbon and solvent input rates, the total carbon in the adsorber can be made part of the adsorption zone reaching saturation in the lowest bed just before it is discharged into an elevator. Very little of the carbon is then idle; hence, maximum utilization is made of the carbon.

The fact that the carbon has reached saturation when delivered to the regeneration phase is another factor in the reduced steam requirement; in this respect, the fluidized-bed system is most favorable. In rotary-bed operations, the carbon moved into the regeneration phase has reached the lowest state of saturation in the three systems.

When large air volumes are treated and available space for the installation is at a premium, the smaller size and lower initial cost are definite advantages over the stationary-bed system. A serious disadvantage is that of high attrition losses of the carbon caused by the fluidization of the beds. Because of the attrition, filtration of the effluent airstream may be required. Influent airstream need not be highly filtered because plugging of the fluidized beds with particulate matter is minimal.

7.3 Process Design Considerations

Two or more adsorbers are required for continuous adsorption, one or more being in operation while another is regenerated. Fluidized-bed and moving-bed equipment may be used for large-scale continuous operation.

The movement of vapors through an adsorbent bed is often referred to as a dynamic process. The term dynamic refers to motion, both in the movement of gas through the adsorbent bed and the change in vapor concentration(s) as it moves through the bed. There are a variety of configurations in which the contaminant air stream and adsorbent are brought into contact. The most common configuration is to pass the airstream down through a fixed volume or bed of adsorbent. Figure 7.3 illustrates how adsorption (mass transfer) occurs as vapors pass down through the bed.

The gas stream containing the pollutant at an initial concentration C_0 is passed down through a (deep) bed of adsorbent material that is free of any contaminant. Most of the contaminant is readily adsorbed by the top portion of the bed. The small amount of contaminant that is left is easily adsorbed in the remaining section of the bed. The effluent from the bottom of the bed is essentially pollutant-free, denoted at C_1.

After a period of time, the top layer of the adsorbent bed becomes saturated with contaminant. The majority of adsorption (approximately 95%) now occurs in a narrow portion of the bed directly below this saturation section. The narrow zone of adsorption is referred to as the mass transfer zone (MTZ). As additional contaminant vapors pass through the bed, the saturated section of the bed becomes larger and the MTZ moves further down the length of the adsorbent. The actual length of the MTZ remains fairly constant as it travels through the adsorbent bed. Additional absorption occurs as the vapors pass through the "unused" portion of the bed. The effluent concentration at C_2 is essentially still zero, since there is still an unsaturated section of the bed.

Finally, when the lower portion of the MTZ reaches the bottom of the bed, the concentration of contaminant in the effluent suddenly begins to rise. This is referred to as the breakthrough point

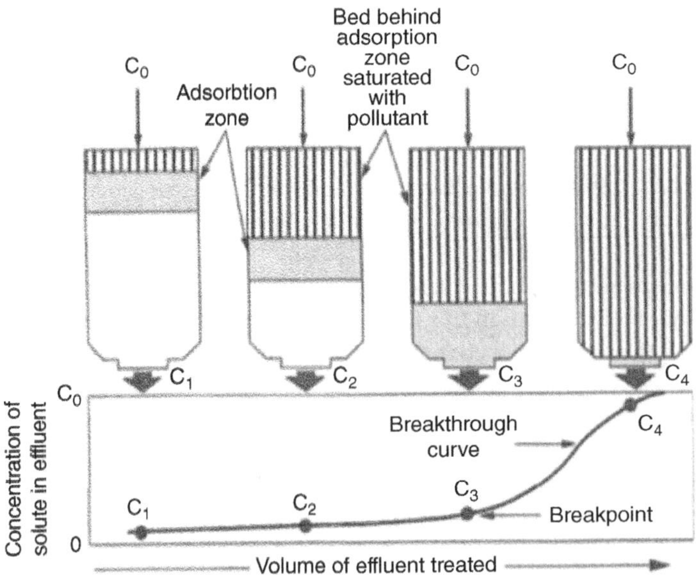

Figure 7.3 Adsorption wavefront.

(or breakpoint)—where untreated vapors are being exhausted from the adsorber. If the contaminated air stream is not switched to a fresh bed, the concentration of contaminant in the outlet will quickly rise until it approaches the initial concentration, illustrated at point C_4. It should be noted that in most air pollution control systems even trace amounts of contaminants in the effluent stream are undesirable.

To achieve continuous operation, adsorbers must be either replaced or cycled from adsorption to desorption before breakthrough occurs. In desorption or regeneration, the contaminant vapors are removed from the used bed in preparation for the next cycle. Most commercial adsorption systems are the regenerable type.

In regard to regenerable adsorption systems, three important terms are used to describe the equilibrium capacity (CAP) of the adsorbent bed. Other capacity terms are measured in mass of vapor per mass of adsorbent. First, the breakthrough capacity (BRK) is defined as the capacity of the bed at which unreacted vapors begin to be exhausted. The saturation capacity (SAT) is the maximum amount of vapor that can be adsorbed per unit mass of carbon (this is the capacity read from the adsorption isotherm) and thus is equal to CAP. The working capacity or working charge (WC) is the actual amount adsorbed in the adsorber. Thus, the working capacity is a certain fraction of the saturation capacity. Working capacities can range from 0.1 to 0.7 times the saturation capacity. (Note: A smaller capacity increases the amount of carbon required.) This fraction is set by the designer for individual systems by often balancing the cost of carbon and adsorber operation vs. preventing breakthrough while allowing for an adequate cycle time.

Another factor in determining the working capacity is that it is uneconomical to desorb all the vapors from the adsorber bed. The small amount of residual vapors left on the bed is referred to as the HEEL. This HEEL accounts for a large portion of the difference between the saturation capacity and the working capacity. In some cases, the working capacity can be estimated by assuming it to be equal to the saturation capacity minus the HEEL.

The following equations and procedures may be used to estimate some of the terms described above for an adsorber bed of height Z:

$$\text{BRK} = [(0.5)(\text{CAP})(\text{MTZ}) + (\text{CAP})(\text{Z} - \text{MTZ})]/\text{Z} \tag{7.1}$$

$$\text{WC} = [(0.5)(\text{CAP})(\text{MTZ})/\text{Z}] + [(\text{CAP})(\text{Z} - \text{MTZ})]/\text{Z} - \text{HEEL} \tag{7.2}$$

As discussed, the working charge can sometimes be taken to be some fraction of the saturated (equilibrium) capacity of the adsorbent (CAP):

$$\text{WC} = f(\text{CAP});\ 0 < f < 1.0 \tag{7.3}$$

For multicomponent adsorption, the maximum working charge may be calculated from the following:

$$\text{WC} = 1.0/\sum_{i=1}^{n}\left(\frac{w_i}{CAP_i}\right) \tag{7.4}$$

where n = number of components

w_i = mass fraction of i in n components (not including carrier gas)

CAP_i = equilibrium capacity of component i

For a two-component (A, B) system, Equation (7.4) reduces to

$$\text{WC} = \frac{(\text{CAPA})(\text{CAPB})}{(\text{w}_\text{A})(\text{CAP}_\text{B}) + (\text{w}_\text{B})(\text{CAP}_\text{A})} \tag{7.5}$$

Note once again that the acronym SAT has also been used to represent the equilibrium capacity, i.e., CAP.

After determining the service time and/or working charge necessary for a particular application, there are other factors to be considered. These include:

1. The adsorbent particle size
2. The physical adsorbent bed depth
3. The gas velocity
4. The temperature of the inlet gas stream to the adsorbent
5. The contaminant concentration to be adsorbed
6. The contaminant concentration not to be adsorbed, including moisture
7. The removal efficiency
8. Possible decomposition or polymerization of the contaminant on the adsorbent
9. The frequency of operation(s)
10. Regeneration conditions
11. The system pressure
12. The system pressure drop

The design of fixed-bed adsorption systems also requires the capability of estimating pressure drop through the bed. Ergun derived a correlation to estimate the pressure drop for the flow of a fluid through a bed of packed solids when it alone fills the void in the bed. This correlation is given by the relationship (S. Ergun, Chem. Eng. Progress, 1952) [11]:

$$\frac{\Delta P g_c d_p \varepsilon^3}{(Z)(2)\rho v^2(1-\varepsilon)} = \frac{75(1-\varepsilon)}{R\varepsilon} + 0.875 \tag{7.6}$$

where ΔP = pressure drop of gas, lb_f/ft^2

Z = depth of packing, ft

g_c = conversion constant, 4.18 (10^8) ft lb/lb_f hr^2

d_p = effective particle diameter (the diameter of a sphere of the same surface/volume ratio as the packing in place), ft

ε = fractional void volume in a dry-packed bed, ft^3 voids/ft^3 packed volume

ρ = gas density, lb/ft^3

v = gas velocity, ft/hr

Re = $d_p v\rho/\mu$, Reynolds Number

μ = gas viscosity, lb/ft hr

Pressure drop information on different mesh carbon sizes is presented in Figure 7.4 (EPA chart). These data are also often used in pressure drop calculations.

Adsorbent regeneration is an integral part of the design procedure. Periodic replacement or regeneration of the adsorbent bed is mandatory in order to maintain continuous operation.

When the adsorbate concentration is high and/or the cycle time is short (less than 12 hours), replacement of the adsorbent is not feasible and in situ regeneration is required. Regeneration is accomplished by reversing the adsorption process, usually by increasing the temperature or decreasing the pressure. Commercially, four methods are used in regeneration:

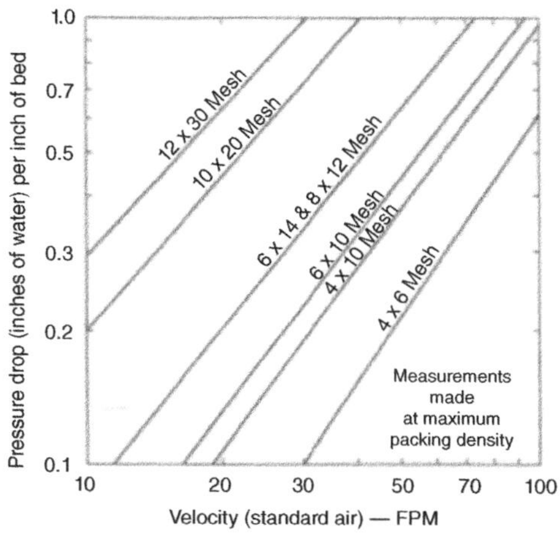

Figure 7.4 Activated carbon pressure drop curves.

1. Thermal Swing. The bed is heated so that the adsorption capacity is reduced to a lower level. The adsorbate leaves the surface of the carbon and is removed from the vessel by a stream of purge gas. Cooling must be provided before the subsequent adsorption cycle begins. Steam regeneration is an example of thermal swing regeneration.
2. Pressure Swing. The pressure is lowered at a constant temperature to reduce the adsorbent capacity.
3. Inert Purge Gas Stripping. The stripping action is caused by an inert gas that reduces the partial pressure or the contaminant in the gas phase, reversing the concentration gradient. Molecules migrate from the surface into the gas stream.
4. Displacement Cycle. The adsorbates are displaced by some preferentially adsorbed material. This method is usually a last resort for situations in which the adsorbate is both valuable and is heat-sensitive, and for which pressure swing regeneration is ineffective.

Because it is simple and relatively inexpensive, steam stripping is one of the more commonly used desorption techniques. Several additional advantages to using steam for desorption are:

1. At high pressures, the steam temperature (>212°F, 100°C) is high enough to desorb most solvents of interest without damaging the carbon or the desorbed vapors.
2. Steam readily condenses in the adsorber bed releasing its (the steam's) heat of condensation, aiding in desorption.
3. Many organic compounds can easily be separated and recovered from the effluent stream by condensation or distillation [3].
4. Residual moisture in the bed is easily removed by a stream of cool dry air (either pure or process effluent).
5. Steam is a more concentrated source of heat than hot air so it is very effective in raising the temperature of the adsorber bed very quickly.

The amount of steam required for regeneration depends on the adsorbate loading of the bed. The longer a carbon bed is steamed, the more adsorbate will be desorbed, thus decreasing the HEEL. It is usually not cost-effective to try to desorb all of the adsorbed vapors from the bed. Acceptable working capacities can be achieved by using less steam and leaving a small portion of adsorbate in the bed.

During the initial heating period, no vapors are desorbed. This arises because a fixed amount of steam is first required to raise the temperature of the cold bed to the desorption temperature. After this initial period, a substantial amount of adsorbate vapor is released until a plateau is reached. The plateau represents the optimum steam requirement, usually in the range of 0.25–0.35 lb of steam/lb of carbon, or approximately 2–3 lb steam/lb adsorbed vapor. In these systems, steam is usually supplied at or slightly above atmospheric pressure.

A simplified overall design procedure for a system adsorbing an organic that consists of two horizontal units (one on/one off) that are regenerated with steam is provided in the literature.

7.4 Equipment Design Considerations

The equipment necessary to constitute a complete adsorption system must be designed and sized adequately to achieve optimum emission reduction efficiencies and maximum solvent reclamation and recovery. The sizing and overall design of the adsorption system depends on the properties and characteristics of both the inlet gas stream and the adsorbent. The design of stationary-bed activated-carbon systems must include consideration of the following operating parameters.

1. Gas stream
   a. Adsorbate concentration
   b. Temperature
   c. Temperature rise during adsorption
   d. Pressure
   e. Flow rate
   f. Presence of adsorbent contaminant material
   g. Gas density at operating temperature and pressure
   h. Gas viscosity at operating temperature and pressure
2. Adsorbent
   a. Adsorption capacity as used on-stream
   b. Temperature rise during adsorption
   c. Isothermal or adiabatic operation
   d. Adsorbent life in presence of particulate matter
   e. Possibility of catalytic effects causing adverse chemical reaction in the gas stream and of formation of solid polymerizates on the adsorbent bed, with consequent deterioration

The adsorbent is generally used in a fixed bed through which the contaminated air is passed. Depending on the concentration and market value, the contaminant is either recovered or discarded when the loading of the adsorbent requires regeneration. The evaluation of the dynamic capacity of an adsorbent bed is essential when determining adsorption rates.

Equipment factors are considered in more detail in the subsections below:

7.4.1 The Adsorbent Particle Size

The shape and size of the adsorbent particle affects several parameters (in particular the pressure drop and diffusion rate). The pressure drop is lowest when the particles are uniform in size and spherical. The pressure decrease will vary with Reynolds number. The mass transfer rate increases inversely with diameter: $d^{3/2}$ and the internal adsorption rate inversely with d^2. With all other parameters considered equal, the pressure drop increases with smaller particle size, yet provides a more efficient adsorptive environment.

7.4.2 The Physical Adsorbent Bed Depth

The actual adsorbent bed depth affects the adsorption mass transfer in two ways: (1) the bed depth is sized to be deeper than the transfer zone, which is unsaturated; and (2) any multiple of minimum bed depth yields a greater than proportional increase in capacity.

7.4.3 The Gas Velocity

The velocity of gas stream through the adsorption bed is determined by the "crushing velocity" of the adsorbent. Crushing-velocity data are available from manufacturers of adsorbents. The length of the MTZ is directly proportional to the velocity; usually, the higher the velocity, the longer the unsaturated zone. The effect of gas velocity on MTZ is presented in Figure 7.5.

7.4.4 The Temperature of the Inlet Gas Stream and the Adsorbent

Temperature yields an inversely proportional effect on adsorption capacity, such as when temperature increases, the adsorption capacity decreases. The adsorption process is exothermic. As the adsorption activity moves through the bed, a temperature front follows and heat is transferred to the gas stream. When the gas leaves the area of adsorption activity, the heat exchange reverses (gas will transfer heat to the bed). Beginning with two assumptions, the temperature differential during the adiabatic operation of the adsorber can be determined as follows:

$$\Delta T = \frac{6.1}{\left(\frac{C_p}{C_i}\right) \times 10^5 + 0.51\left(\frac{C_p}{C_{st}}\right)} \tag{7.7}$$

where ΔT = temperature rise, °F

C_{st} = saturation capacity of bed at T + ΔT, °F

C_i = inlet concentration, ppm

C_p = heat capacity of adsorbent, Btu/ft³/°F

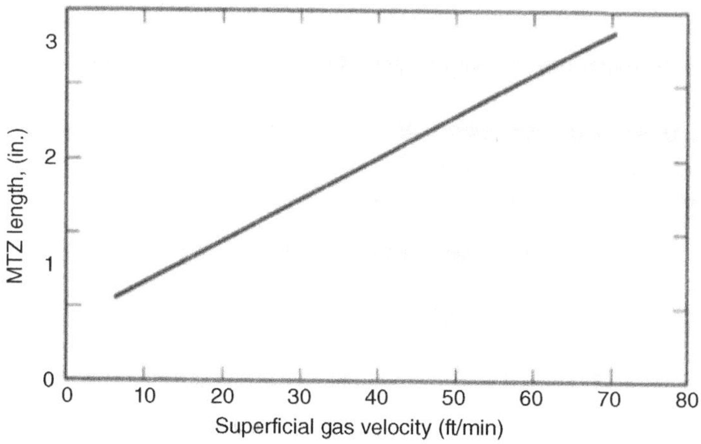

Figure 7.5 Effect of gas velocity on length of MTZ.

1. Thermal equilibrium exists between the gas and the bed.
2. The effluent-gas-stream temperature is similar to the bed temperature.

7.4.5 The Adsorbate Concentration

There is direct correlation between adsorbate concentration and adsorption capacity. The MTZ length is inversely proportional to the adsorbate concentration. Considering these two observations only, a deeper or longer bed will remove a lower-concentration contaminant and a higher concentration of the same contaminant with equal efficiency. In considering recovery, the value of the adsorbate and the concentration are primary factors. In handling combustible gas, it is mandatory that the concentration of the adsorbate be well below the lower explosive limit (LEL).

All gases present will be adsorbed on the adsorbent surface to some degree. The competition between the gases for available surface area lowers the adsorption capacity for the preferred adsorbate to be removed. Under ambient conditions, air is adsorbed only slightly (10–20 ml STP/g) on commercial adsorbents, but moisture and CO_2 (carbon dioxide) adsorb more significantly. Even though activated carbon is less sensitive than other adsorbents to moisture, its adsorption capacity is limited considerably when compared to adsorption from a dry air stream.

7.4.6 The Removal Efficiency

Varying percentages of removal are required with different applications. Of course, deeper beds are required to achieve complete removal than for a partial removal efficiency of 60–80%.

7.4.7 Adsorbate Decomposition and Polymerization

Decomposition and polymerization of solvents can occur when in contact with adsorbents. The decomposed product takes on properties different from those of the original substance. These different properties include corrosivity and the ability to be adsorbed at a lower capacity. Polymerization can significantly lower the adsorption capacity and render it nonregenerable by conventional methods. When adsorption of acetylene on activated carbon is accomplished at higher temperatures, polymerization occurs.

7.4.8 Regeneration Conditions

When deciding whether to employ a regenerative system, several factors should be considered. The principal one is that of economics. It is important to establish if recovery of the adsorbate will be cost-effective or if regeneration of the adsorbent is the prime consideration. If solvent recovery is the main objective, the design should be based on past experimental data to establish the ratio of sorbent fluid to recoverable adsorbent at different working capacities. A well-designed system will have steam consumption in the range of 1–4 lb of steam per pound of recovered solvent.

7.4.9 Installation Procedures

After control equipment design has been completed, the installation becomes an important factor in the future convenience of operating and maintenance procedures on the air pollution control system.

Improper installation may contribute to excessive energy consumption and unnecessary losses of volatile organic materials to the environment and to the workplace. This may result in deterioration of the air resource and the water quality and may present excessive employee exposure to toxic and hazardous chemicals. Because of these consequences, everything possible must be done to ensure correct operation and ease of maintenance of all air pollution control equipment.

7.5 Applications

Three solved design problems are presented below.

7.5.1 Working Capacity

An air stream containing organic vapors is to be cleaned with an adsorber using activated carbon as the adsorbent. The organic vapor concentrations in the air stream are provided in Table 7.3. Calculate the "theoretical" working capacity and the "actual" capacity given a HEEL of 0.025 (fractional basic) and a packing factor of 0.03 (negative).

Solution: The working capacity is calculated from Equation (7.4):

$$\mathrm{WC} = 1.0/\sum_{i=1}^{n}\left(\frac{w_i}{\mathrm{CAP}_i}\right)$$

Substituting, one obtains

$$\mathrm{WC} = \frac{1.0}{\left(\frac{0.67}{0.39}\right) + \left(\frac{0.05}{0.08}\right) + \left(\frac{0.25}{0.4}\right) + \left(\frac{0.02}{0.11}\right) + \left(\frac{0.01}{0.16}\right)}$$

$$= \frac{1.0}{1.718 + 0.625 + 0.625 + 0.182 + 0.063} = 0.3113 = 31.13\%$$

This represents the "theoretical" or maximum working charge. The actual WC, including the HEEL effect, is

$$\mathrm{WC} = 0.3113 - 0.025 = 0.2863 = 28.63\%$$

Including the packing factor leads to:

$$\mathrm{WC} = 0.3113 - 0.025 - 0.03 = 0.2563 = 25.63\%$$

7.5.2 Ethyl Acetate Application

A solvent recovery plant is to recover 1,800 lb/hr of ethyl acetate (EA) vapor from a mixture with air at a concentration of 1.8 lb vapor/1,000 ft^3 air at 60°F, 1 atm pressure. The adsorbent will be activated carbon, 4 × 10 mesh (average particle diameter 0.0091 ft) apparent density of individual particles 42 lb/ft^3, and apparent density of the packed bed 28 lb/ft^3. The carbon is capable of adsorbing 0.25 lb vapor/lb carbon including the HEEL up to the breakpoint.

Table 7.3 Equilibrium data for organic vapors.

Components	Mass fraction, w_i (solute-free basis)	Equilibrium capacity, lb/lb
Methane	0.67	0.39
Toluene	0.05	0.08
Propane	0.25	0.40
Diphenyl	0.02	0.11
Benzyl alcohol	0.01	0.16

Determine the amount of carbon required and choose suitable dimensions for the carbon beds on what you consider a good design. Estimate the pressure drop by using the EPA chart (Figure 7.4). Using this pressure drop, calculate the horsepower requirement for the system.

Solution: In some respects, this is an open-ended problem since design values need to be set. Select a 1 × 1 system, operating 2 hours on and 2 hours off. Thus, select as a basis 2 hours of operation.

$$\begin{aligned} \text{MEA} &= (2)(1{,}800) \\ &= 3{,}600 \text{ lb} \\ \text{MAC} &= 3{,}600/0.25 \\ &= 14{,}400 \text{ lb; } 28{,}800 \text{ total for two units} \\ \text{VAC} &= 14{,}400/28 \\ &= 514.3 \text{ ft}^3 \end{aligned}$$

The volumetric flow rate is

$$\begin{aligned} q &= [(1{,}800)/(1.8/1{,}000)](1/60) \\ &= 16{,}700 \text{ acfm} \end{aligned}$$

A horizontal unit should definitely be used. Select

$$v = 80 \text{ ft/min}$$

so that

$$\begin{aligned} A &= 16{,}700/80 \\ &= 209 \text{ ft}^3 = (L)(D) \\ H &= Z = 514.3/209 \\ &= 2.5 \text{ ft} \end{aligned}$$

If $L/D = 2$, then D = 10.2 ft and L = 20.4 ft. If $L/D = 3$, then D = 8.3 ft, L = 25.0 ft. Select D = 9.5 ft so that

For this design,

$$\begin{aligned} L &= A/D \\ L &= 209/9.5 \\ &= 22 \text{ ft} \\ L/D &= 22/9.5. = 2.3 \\ H/D &= 2.5/9.5 = 0.263 \end{aligned}$$

Select a 4 × 10 mesh carbon. From Figure 7.4, one obtains
$\Delta P/L = 0.72$ in H_2O/ft bed
Thus,

$$\begin{aligned}\Delta P &= (0.72)(12)(2.5)\\ &= 21.6 \text{ in } H_2O\end{aligned}$$

Finally (for 70% efficiency)

$$\begin{aligned}\text{HP} &= (q)(\Delta P)/E_f\\ &= (16{,}700)(21.6)(5.21)/(60)(550)(0.7)\\ &= 81.2\end{aligned}$$

7.5.3 Industrial Plant

A carbon adsorption system is being used to control odors emitting from a drum-filling operation. The material being drummed is a high-purity grade of pyridine (C_5H_5N) that has a human odor detection level of 100 ppm. It has been reported that an odor has been detected from outside the drumming area when the equipment is in service, i.e., when drums are being filled. Discussions with operating personnel have indicated that the adsorption system is the source of the odor. You are requested to determine if the adsorption equipment/emission is the source of the odor or if the equipment is capable of containing/controlling the pyridine emission. Design and actual operating data are provided below.

1. The adsorption units are twin horizontal units with face dimensions for flow of 5 × 12 ft. Each unit contains new 4 × 6 mesh activated carbon B that was installed one month ago. The measured bed height is 12 inches.
2. The carbon manufacturer maintains that the breakthrough capacity of the carbon is 0.49 lb pyridine/lb carbon and that the carbon has a bulk density of 25 lb/ft^3.
3. Laboratory tests performed by plant personnel indicate that the carbon contains a HEEL of approximately 0.03 lb pyridine/lb carbon when regenerated with 4.0 lb of steam/lb pyridine at 10 psig.
4. The ventilation blower for the drum filling station has a flow of 5,000 acfm at 25 °C and 14.7 psia and contains a pyridine concentration of 2,000 ppm (plant hygienist data).
5. The drum filling operation operates on a 24 hours/day basis and the adsorption units are operated on an 8-hour adsorption, 5-hour regeneration cycle, with 3 hours for cooling and standby. The steam used during the 5-hour regeneration cycle was determined to be 2,725 lb (mass flowmeter).
6. The adsorption unit was designed based on a pressure drop through the bed following the relationship

$$\text{AP} = 0.37Z\left(\frac{v}{100}\right)^{1.56}$$

   where Z is the bed depth in inches, v is the velocity in ft/min, and the pressure drop is in inches of water. The measured operational pressure drop is 3.3 inches of water.
7. The fractional fan efficiency is 0.58.

To evaluate the adsorber's performance, determine the following:

1. The mass of pyridine to be captured in the adsorption period
2. The working capacity of the carbon B
3. The mass and volume of carbon that should be used in each unit
4. The required bed height
5. The design pressure drop through the bed using the required bed height for full capture
6. The horsepower requirement for this process
7. The required steam to regenerate the bed to the HEEL level of 0.03 lb pyridine/lb carbon

Solution: Calculate the mole fraction of pyridine (p) in the gas stream.

$$y_p = 2{,}000/10^6 = 0.0020$$

Calculate the volumetric flow rate of P in acfm.

$$q_p = y_p q = (0.0020)(5{,}000) = 10.0 \text{ acfm}$$

Determine the density of the p vapor at the operating conditions.

$$\rho_p = \frac{P(\text{MW})}{RT} = \frac{(14.7)(79)}{(10.73)(537)} = 0.2015 \text{ lb/ft}^3$$

The mass of p collected during the adsorption period is then

$$m_p = (10)(0.2015)(8)(60) = 967.2 \text{ lb}$$

Estimate the working capacity (WC) of carbon B for this system.

$$\begin{aligned} \text{WC} &= \text{BC} - \text{HEEL} \\ &= 0.49 - 0.03 \\ &= 0.46 \text{ lbp/lb carbonB} \end{aligned}$$

Calculate the mass of carbon that should be used for each unit.

$$M_{AC} = m_p/\text{WC} = 967.2/0.46 = 2{,}103 \text{ lb carbon B}$$

The volume of activated carbon V_{AC} is

$$\begin{aligned} V_{AC} &= M_{AC}/\rho_B \\ &= 2{,}103/25 = 84.1 \text{ ft}^3 \end{aligned}$$

The cross-sectional area A of the carbon that is presently available for flow is

$$A = (5)(12) = 60 \text{ ft}^2$$

Since the bed height is 12 inches or 1.0 ft, the actual volume currently employed is 60 ft^3. Because this is below the required 84 ft^3, the odor problem is present; the equipment is not capable of controlling the emission to eliminate the odor.

The "required" height of the adsorbent in the unit is

$$\begin{aligned} Z &= V_{AC}/A \\ &= 84.1/60 \\ &= 1.40 \text{ ft} \\ &= 16.8 \text{ inches} \\ &\approx 17.0 \text{ inches} \end{aligned}$$

Estimate the pressure drop across the required adsorbent in inches of H_2O using the relationship provided.

$$\Delta P = 0.37Z\left(\frac{v}{100}\right)^{1.56} = 0.37Z\left(\frac{\frac{q}{A}}{100}\right)^{1.56}$$

$$= (0.37)(17)\left(\frac{\frac{5{,}000}{60}}{100}\right)^{1.56}$$

$$= 4.73 \text{ in } H_2O$$

The total pressure drop across the bed in lb_f/ft^2 is then

$$\Delta P_{total} = (4.73)(5.2) = 24.6 \ lb_f/ft^2$$

While the HP requirement is

$$HP = \frac{(24.6)(5{,}000)}{(60)(550)(0.58)} = 6.43$$

Finally, the steam requirement for regeneration is
msteam = (4.0)(967.2)= 3,869 lb steam during regeneration
The actual operating steam rate (2,725 lb for 5 hours) is also below the required value.

References

1 L. Theodore, *Air Pollution Control Equipment Calculations*, John Wiley & Sons, Hoboken, NJ, 2008.

2 L. Theodore and R. Allen, Pollution Prevention, A Theodore Tutorial, East Williston, NY, 1994 (originally published by the ES EPA/APRI, RTP, NC).

3 L. Theodore, *Chemical Engineering: The Essential Reference*, McGraw-Hill, New York City, NY, 2014.

4 J. Richardson and L. Theodore, *Optimizing Air Pollution Equipment Performance: Operation and Maintenance*, John Wiley & Sons, Hoboken, NJ, 2025.

5 L. Theodore, *Air Pollution Control Equipment: Gases*, Theodore Tutorials, East Williston, NY, 1976 (originally published by the ES EPA/APRI, RTP, NC).

6 L. Theodore, *Air Pollution Control Equipment: Particulates*, Theodore Tutorials, East Williston, NY, 1976 (originally published by the ES EPA/APRI, RTP, NC).

7 L. Theodore (section editor), D. Green, and M. Southard, *Perry's Chemical Engineering Handbook*, 9th edition, McGraw Hill, New York City, NY, 2019.

8 A.J. Buonicore and L. Theodore, *Industrial Control Equipment for Gaseous Pollutants*, Volume I, CRC Press, Inc., Boca Raton, Florida, 1975.

9 OAQPS, *Air Pollution Engineering Manual*, 2nd edition, EPA, 1973.

10 MSA RESEARCH CORP., *Package Sorption Device System Study*, EPA, 1973.

11 L. Theodore, *Adsorber Sizing Made Easy*, CEP, New York City, NY, Marel, 2005.

Suggested Readings

1 W. Strauss. *Industrial Gas Cleaning*, New York City, NY, Pergamon Press, 1966.
2 *Package Sorption System Study*. NTIS, report NO. EPA-R2-73-202, Washington DC, 1973.
3 J. A. Danielson, *Air Pollution Engineering Manual*, EPA OAP Publication No. AP-40, Government Printing Office, Washington, D.C.
4 A. C. Stern, *Fundamentals of Air Pollution*, Academic Press, New York City, NY, 1974.
5 L. Theodore and A. J. Buonicore, *Industrial Control Equipment for Gaseous Pollutants*. Vols I and II, CRC Press, West Palm Beach, 1975.
6 L. Theodore, *Air Pollution Control Equipment Calculations*, John Wiley & Sons, Hoboken, NJ, 2008.
7 L. Theodore, *Chemical Engineering: The Essential Reference*, McGraw-Hill, New York City, NY, 2014.
8 J. Richardson and L. Theodore, *Optimizing Air Pollution Equipment Performance: Operation and Maintenance*, John Wiley & Sons, Hoboken, NJ, 2025.

8

Incinerators/Combustion Equipment

8.1 Introduction

Combustion control equipment can be divided into three types: flares, thermal incinerators, and catalytic reactors. In many industrial operations, and particularly in chemical plants and petroleum refineries, large volumes of combustible waste gases are produced. These gases result from undetected leaks in the operating equipment, from upset conditions in the normal operation of a plant where gases must be vented to avoid dangerously high pressures in operating equipment, from plant start-ups, and from emergency shut-downs. Large quantities of gases may also result from "off-spec" product or excess product which cannot be sold. Flows and compositions are typically intermittent, with flow rates during major upsets of several million cubic feet per hour.

The preferred control method for excess gases and vapors is to recover them in a blowdown recovery system. However, large quantities of gas, especially those produced during upset and emergency conditions, are difficult to contain and reprocess. In the past, all waste gases were vented directly into the atmosphere. However, widespread venting caused safety and environmental problems. In practice, therefore, it is now customary to collect such gases in a closed flare system and to burn these gases as they are discharged and burned at ground level, usually with shielding for the flame.

Combustion equipment can be used to control the emissions of gaseous or small-particulate air pollutants which are either combustible or which thermally decompose at high temperatures. Three rapid oxidation methods are used to destroy combustible contaminants:

1. Thermal incineration
2. Catalytic incineration
3. Flares (direct combustion)

The thermal and flare methods are characterized by the presence of a flame during combustion. Catalytic incinerators serve the same purpose but utilize a metallic catalyst to promote the required rapid reaction; this unit operates at lower operating temperatures and with higher reaction rates.

Combustion is a chemical process occurring from the rapid combination of oxygen with various elements or compounds resulting in the release of heat. The process of combustion is also referred to as oxidation or incineration. Most fuel used for combustion, along with waste, are composed essentially of carbon and hydrogen, but can include other elements such as sulfur, nitrogen, and chlorine. Although combustion seems to be a very simple process that is well understood, in reality it is not. The exact manner in which a fuel or waste is oxidized occurs in a series of complex, free radical chain reactions. For example, methane undergoes the most simple combustion reaction of a hydrocarbon ($CH_4 + O_2 \rightarrow CO_2 + H_2O$) and that process contains over 300 intermediate free radical reactions [1]. The precise set of reactions by which combustion occurs is termed the *mechanism of combustion*. By analyzing the mechanism of combustion, the rate at which the reaction proceeds, and the variables affecting the rate can be estimated. For most combustion devices, the rate of reaction proceeds extremely fast compared to the mechanical operation of the device. Maintaining efficient and complete combustion is somewhat of an art rather than a science [2].

The main uses of fume incinerators are for the control of odor and smoke with some applications in hydrocarbon and carbon monoxide (CO) emission control. Generally, the gas stream that the incinerator is applied to contains a low concentration of combustible emissions. This comes about because of the safety requirement to keep the concentration of combustible gases within the fume generating equipment at some percentage below the lower flammability limit. If the gas stream, for whatever reason, is of a high enough concentration of combustibles, it can be used directly as a fuel.

The design of all types of fume incinerators must take into consideration the "three Ts" of combustion: time, temperature, and turbulence. These three variables will govern the speed and completeness of the combustion reaction. For complete combustion to occur, the oxygen must come into intimate contact with the combustible molecule at a sufficient temperature, and for a sufficient length of time in order that the reaction be completed. Incomplete reactions may result in the generation of aldehydes, organic acids, carbon, and carbon monoxide.

The greatest variation among different afterburner designs is in how well they achieve the goal of raising all the fume to the required temperature for the required combustion residence time. Most cases of poor performance are due to nonuniform temperatures and flows, which allow some of the pollutants to escape without adequate treatment [3].

Time and temperature requirements will be discussed together since they are, to some degree, interrelated. A higher temperature allows the use of a shorter residence time, and vice versa. Temperature is the expensive variable, with time a direct function of the combustion chamber volume. For example, the residence times for a particular application could be doubled for only a 20 to 30% increase in total capital costs. However, variation of the temperature is directly related to fuel usage, which is the major component of daily operating costs.

8.2 Description of Combustion Equipment

Thermal incinerators are the most widely used method to control the release of hydrocarbon fumes, especially smoke and solvents from coating processes. Typically, thermal incinerators are constructed of a steel outer shell lined with refractory material. The purpose of the refractory lining, which is typically 4–8 inches thick, is to protect the steel shell from direct exposure to the effects of high temperatures and corrosive materials, and also to improve the thermal efficiency of the unit by limiting heat losses. The refractory, which will be discussed in detail later in this section, serves as a thermal insulator. It lowers the temperature from greater than 2,000 °F on the inside of the combustion chamber to a temperature of 180 to 400 °F at the outer surface of the steel shell. In this temperature range, the steel shell will continue to retain its strength while also remaining hot enough to prevent the condensation of water vapor on the surface, thereby reducing corrosive attack.

These incinerators are equipped with a burner at one end that fires a fuel, typically natural gas. There is also a fume inlet located near the burner where the gas stream to be oxidized enters the incinerator. The burner may utilize the air in the process waste stream as the combustion air for the fuel, or it may use a separate source of outside air for this purpose. See Figure 8.1.

Refractory material consists of various types of fireclays, usually containing alumina (typically >60%) or silica, and is available in several forms, including bricks and castable or plastic material. Bricks are manufactured in standard sizes (4 ½ inch wide × 4 inch high × 9 inch long is typical) but may be precut at the factory to form any desired shape. They are best used to form regular surfaces. Castable refractory is usually supplied in a dry powder form which is mixed with water and cast in place in the field. Hence, initial cost and shipping charges are less than those for brick, and irregular surfaces are more easily formed. However, castable, as well as plastic refractory requires the use of a suitable anchoring system. Anchors are normally recommended by the refractory manufacturer for a particular material and service condition. Normally, they are made of alloy steel and are welded to the steel surface before lining. This adds considerably to the cost of installing a castable or plastic type of refractory.

Plastic refractory material is essentially a premixed castable with binders added to improve its plastic-handling qualities. Castable or plastic refractory linings are usually homogeneous (i.e., consist of the same-density material throughout). Brick installations, however, are often composed of layers of different-quality bricks. On the hot side, plain firebrick is normally used. Because it is very dense (~150 lb/ft^3), it has a high degree of "refractoriness" and is able to withstand the furnace environment for long periods. One or more layers of "insulating firebrick" may be placed between the firebrick and steel shell. Insulating firebrick is much less dense than firebrick (50 to 80 lb/ft^3) and so has much better insulating properties. Finally, a layer of "insulating block," which is a ceramic material, may also be used between the insulating firebrick and the steel shell. This material has excellent insulating properties but cannot be exposed for long periods to temperatures above 1,000 °F [4].

A refractory lining is almost always installed in the field. Refractory material is heavy and rather brittle and so must be completely supported during shipment if installed at the factory. The cost of such support is usually prohibitive.

A brick lining is installed much as any other brick structure. Skilled masons are required. Bricks are trimmed if necessary and cemented in place with special mortar. Castable refractory may require the use of forms, much as in concrete work, and the material may be either poured or "gunned" into place. Care must be taken in placing castable refractory to assure that no voids are formed.

Blanket-type insulation is made from ceramic fiber materials. They are very lightweight and have excellent thermal insulating properties. In some cases, a 6-inch refractory lining, weighing 100 lb/ft^2, could be replaced by a 2-inch-thick blanket weighing less than 10 lb/ft^2. This will result in a significant weight reduction and a smaller incinerator. Because the blanket material is so lightweight and flexible, it can be shipped installed in the incinerator. This saves in field-installation and shipment costs.

The installation of insulating blankets requires that a stud system first be added to the incinerator shell. The insulation blanket is then impaled on the studs and secured on the stud with a locking washer. The stud-and-washer system is made of high-alloy steel and represents the most expensive part of the installation.

Insulation blanket is not usually suitable for use when fuel oil is used as an auxiliary fuel. However, fuel oil may saturate the blanket during upset conditions. When the burner is relit, the

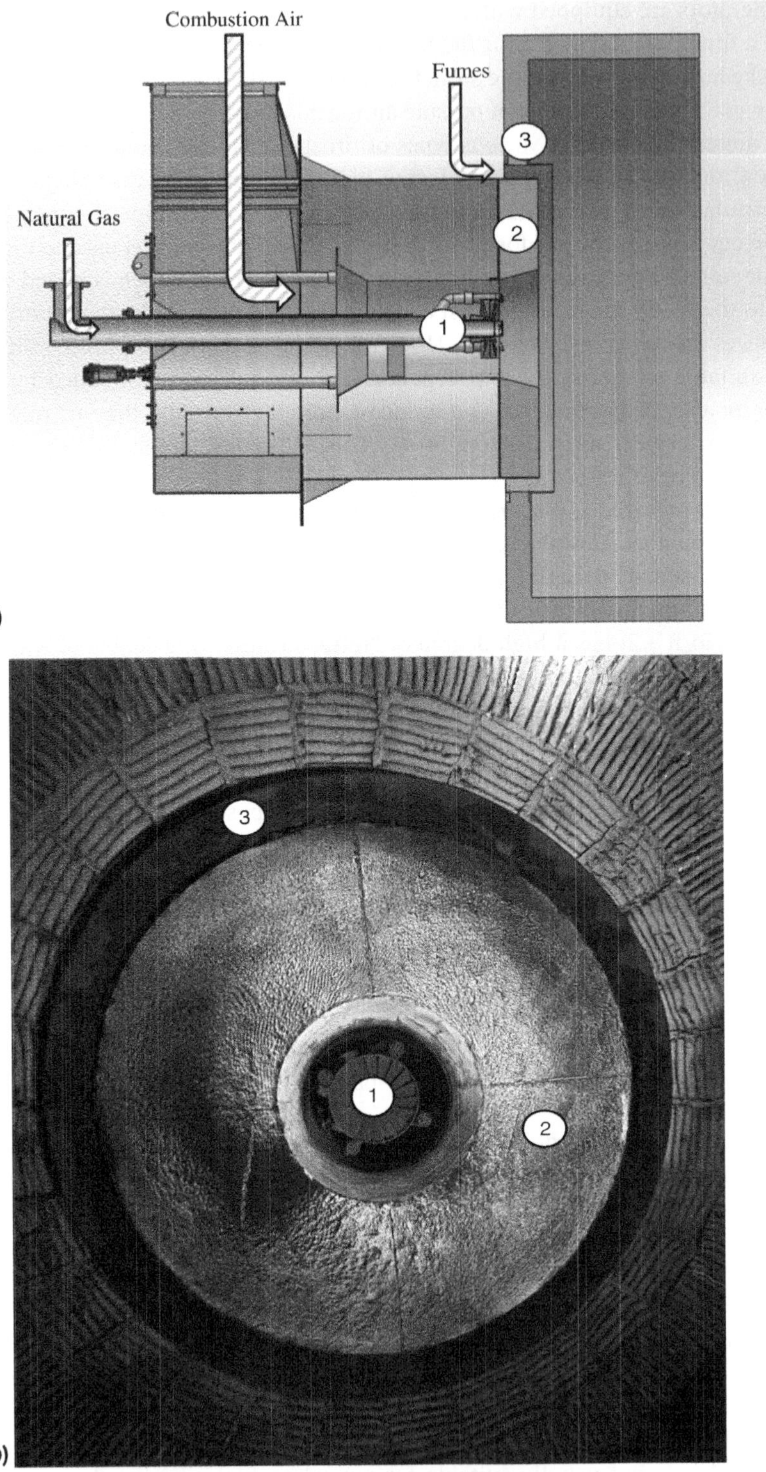

Figure 8.1 (a) Schematic of a thermal oxidizer. (b) Photo from the inside of a thermal oxidizer. 1. Natural gas burner. 2. Refractory. 3. Fume annulus.

fuel oil will ignite, destroy the blanket, and create and uncontrolled flame. Blankets are also very pervious to fumes, which may cause a corrosion problem in particular applications.

The major advantage of insulation blankets is that no warm-up period is required for the incinerator, resulting in a large fuel savings over refractory linings, and increased operational flexibility. Replacement of the insulation blanket does not usually require replacement of the stud-and-washer system and is therefore relatively inexpensive.

8.2.1 Catalytic Incinerators

A catalytic incinerator is an alternative to a thermal incinerator as a means for oxidizing gaseous or oxygenated hydrocarbons to carbon dioxide and water. Contact of waste stream with a catalyst bed allows oxidation reactions to occur rapidly in the temperature range of 700 to 900 °F in contrast to the 1,300 to 1,600 °F typically required of a thermal incinerator. The oxidation reaction that occurs at the surface of the catalyst produces the same products (carbon dioxide and water) and liberates the same heat of combustion as a thermal incinerator would.

The heat required to bring the waste stream up to the required oxidation temperature is usually supplied by a natural gas fueled burner. This type of incinerator is again constructed of a steel outer shell lined with a refractory material. The wall thickness of the refractory for a catalytic incinerator can be less than that of the thermal incinerator simply because the operating temperatures are not as great.

A catalyst bed is located at a distance downstream from the burner sufficient to allow for a uniformly preheated and distributed mixture of combustion products and waste gas stream. The catalyst bed in commercial units is typically a metal mesh mat, ceramic honeycomb, or other ceramic matrix structure with a surface coating of finely divided platinum or other platinum-family metals, such as chromium, vanadium, nickel, or cobalt. This metal coating is the actual catalyst for the oxidation reactions. The matrix serves to support the catalyst on a high geometric surface area and promotes good contact between waste stream and catalyst as seen in Figure 8.2 [5].

8.2.2 Flares

The flare system is used primarily as a safe method for disposing of excess waste gases. All process plants that handle hydrocarbons, hydrogen, ammonia, hydrogen cyanide, or other toxic or dangerous gases are subject to emergency conditions that require the immediate release of large volumes of such gases for protection of plant and personnel. Flares are used for this purpose. In operation, the gas containing the organics is continually fed to and discharged from a stack, with the combustion occurring near the top of the stack and characterized by a flame at the end of the stack.

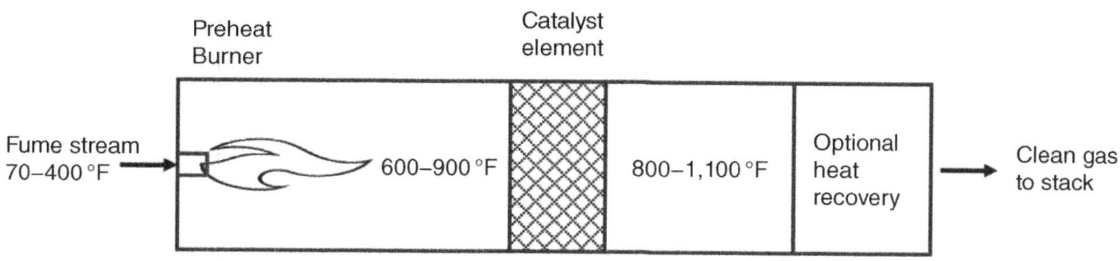

Figure 8.2 Schematic of a catalytic incinerator.

In general, three types of flares are in use today: elevated, ground, and forced draft. The elevated flare is made up of several components as shown schematically in Figure 8.3. The ground of "bilevel" flare is employed where noise, luminosity, or smoke are objectionable. They are, for all practical purposes, enclosed in combustion chambers which produce smokeless burning and conceal the flare flame. However, this type of flare has been outlawed in most states. The forced-draft flare uses air provided by a blower to supply primary air and the turbulence necessary to provide smokeless burning of exhaust gases without the use of a steam injection system as used in the steam-assisted flare [6, 7].

Typical maximum flare capacities are presented in Table 8.1.

Thermal incinerators, catalytic incinerators, and flares may each be used in appropriate situations to control smoke, odor, and combustible fumes. They are not particularly complicated

Figure 8.3 Flare stack schematic.

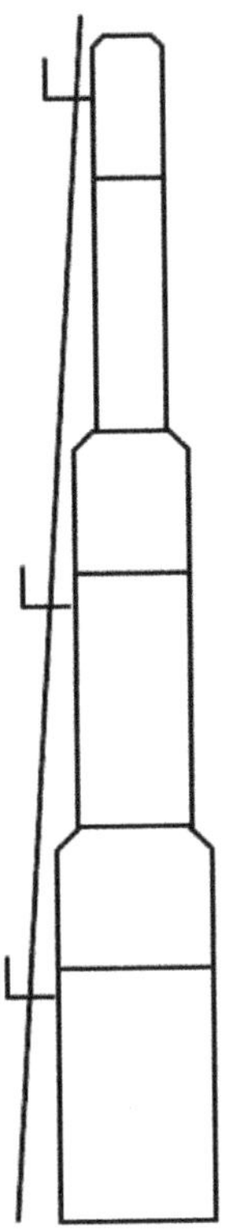

Table 8.1 Maximum flare capacities.

Type	Capacity (1,000 lb/hr)
Ground flare	80–100
Elevated flare	1,000–2,000
Elevated, forced draft	100

devices, but their efficiency depends, to a large degree, upon proper operation, and maintenance. Improper operation and maintenance can lead to unsafe combustion conditions and premature failure of the insulation or catalyst, which can significantly increase the system's lifetime costs. Due to the nature of enclosing large combustion reactions, improper operation can also lead to catastrophic failures resulting in loss of life. It is absolutely essential, therefore, that operation and maintenance personnel be qualified and skilled in their tasks.

8.3 Process Design Considerations

In describing any combustion process, there are numerous terms used to define heat. These terms can be split into two categories: combustion and thermodynamic. Combustion terms are included to aid in standardizing fuel usage calculations; these terms are applied to heat that is produced by combustion methods. Thermodynamic terms apply to all systems. Since the combustion terms are specific examples of the thermodynamic terms, some overlap is involved in defining them.

The following are important terms describing heat thermodynamically:

1. Heat content or enthalpy (H)—the sum total of latent and sensible heat present in a substance (gas, liquid, or solid) minus that contained at an arbitrary set of conditions chosen as the base or zero point. Values for various gases are listed in Table 8.2.
2. Sensible heat (H_s)—heat that when added or removed causes a change in temperature.
3. Latent heat (H_l)—heat given off by a vapor condensing to a liquid or gained by a liquid evaporating to a vapor, without a change in temperature. The latent heat of vaporization of water 212 °F is 970.3 Btu/lb.

Some useful terms describing the heat produced by the combustion of a fuel are:

1. Gross heating value (HV_G)—the total heat obtained from the complete combustion of a fuel at 60 °F. Constant pressure, normally 1 atm, is maintained throughout the entire combustion process. Gross heating values are also referred to as total or higher heating values (HHV).
2. Net heat value (HV_N)—the gross heating value minus the latent heat of vaporization of the water formed by the combustion of the hydrogen in the fuel. For a fuel containing no hydrogen, the net and gross heating values are the same. Net heating values are also referred to as the lower heating values (LHV).
3. Available heat (HA)—the gross quantity of heat released within a combustion chamber minus (1) the sensible heat carried away by the dry flue gases and (2) the latent heat and sensible heat carried away in water vapor contained in the flue gases. The available heat represents the net quantity of heat remaining for useful heating.

Depending on the user, the above terms can also have more than one definition. For example, a laboratory chemist may describe latent heat as the energy used in the chemical combustion of a fuel to carbon dioxide and water; while a boiler operator may describe latent heat as the difference between the gross and net heating values. It is important to ensure all parties on using the same designation for heat requirements, for example, the boiler industry tends to use HHV, while the chemical process industry tends to use LHV.

Another important term used in performing combustion calculations is the specific heat, or more appropriately the heat capacity c_p, of a substance. One set of units for heat capacity is Btu/lb· °F in English units. Heat capacity also depends on temperature.

Table 8.2 Heat contents (enthalpy) of various gases.

	Hear content H; Btu/lb at 1 atm								
Temperature, °F	O_2	N_2	Air	CO	CO_2	SO_2	H_2	CH_4	H_2O
60	0	0	0	0	0	0	0	0	0
100	8.8	9.9	9.6	10	8	5.9	137	21	—
200	30.9	34.8	33.6	34.9	29.3	21.4	474	76.1	—
300	53.3	59.9	57.7	59.9	52	34.5	832	136.4	1,165
400	76.2	85	81.8	85	75.3	54.4	1,182	202.1	1,212
500	99.4	110.3	106	110.6	99.8	71.8	1,532	272.6	1,259
600	123.1	136.1	130.2	136.3	125.1	89.8	1,882	347.8	1,307
700	147.2	161.7	154.5	162.4	149.6	108.2	2,233	427.4	1,355
800	171.7	187.7	178.9	188.7	17.8	127	2,584	511.2	1,404
900	196.6	213.9	203.4	215.6	205.6	146.1	2,935	599.2	1,454
1,000	221.7	240.7	235	242.7	233.6	165.5	3,291	691.1	1,505
1,200	272.5	294.7	228.5	297.8	290.9	205.1	4,007	886.2	1,609
1,400	324.3	350.8	343	354.3	349.7	245.4	4,729	1094.1	1,717
1,600	377.3	407.3	398	407.5	416.3	286.4	5,460	1,313	1,829
1,800	430.7	465	455	465.3	470.9	327.8	6,198	1542.6	—
2,000	484	523.8	513	523.8	532.8	369.1	6,952	—	—
2,200	539.3	583.2	570.7	583.3	596.1	411.1	7,717	—	—
2,400	594.4	642.3	628.5	643	659.2	452.7	8,490	—	—
2,600	649	702.8	687.3	703.2	723.2	495.2	8,272	—	—
2,800	702.8	763.1	746.6	771.3	787.4	557.5	1,0060	—	—
3,000	758.6	824.1	806.3	832.6	852	580	1,0870	—	—
3,200	816.4	885.8	866	894	916.7	622.5	1,1680	—	—
3,400	873.4	947.6	925.9	956	981.6	665	1,2510	—	—
3,600	931	1010.3	986.1	1018.3	1047.3	707.5	1,3330	—	—

An area of concern in incineration deals with calculations involving sensible heat and heat (energy) balances. Some approaches with key concepts and equations are presented below.

1. Compute the heat required to perform a heat balance around the combustion system. From the law of conservation of energy (for steady-state conditions):

   $$\text{heat in} = \text{heat out} \tag{8.1}$$

   One may also write this equation in terms of rate

   $$\text{Rate of heat in} = \text{rate of heat out} \tag{8.2}$$

   As discussed earlier, heat is a relative term that is compared at a reference temperature. The heat content of a substance is arbitrarily taken as zero at a specified reference temperature. In the gas industry (natural gas), the reference temperature employed is normally 60 °F.

2. The heat content can be computed from Equation 8.3 as well as using tables such as Table 8.2.

$$\Delta H = c_p(T - T_0) \quad (8.3)$$

where ΔH = enthalpy, Btu/lb
c_p = heat capacity, Btu/lb·°F (heat capacity over temperature range from T to T_0
T = temperature of the substance, °F
T_o = reference temperature, °F

3. Subtracting the heat going in from the heat leaving the system gives the heat that must be supplied by the fuel. This is referred to as a change in enthalpy or heat content. Using Equation 8.3, the enthalpy going in (T_1) is subtracted from the enthalpy leaving (T_2), giving

$$\Delta H = \left[c_{p2}(T_2 - T_0) - c_{p1}(T_1 - T_0)\right] \quad (8.4)$$

where ΔH = change in enthalpy
c_{p2} = heat capacity over the temperature range T_2 to T_0 c_{p1} = heat capacity over the temperature range T_1 to T_0

4. To simplify this calculation, an average specific heat value ($\bar{c}_p$) can be used over the temperature range involved. This reduces Equation 8.4 to

$$\Delta H = \bar{c}_p(T_2 - T_1) \quad (8.5)$$

where $\bar{c}_p$ = average heat capacity over the temperature range T_1 to T_2.

The heat capacity varies with temperature and composition of the gas stream. Therefore, Equation 8.5 is used to obtain an approximate value. For most incineration systems, the waste gases are considered to be essentially air. For air, the average heat capacity value $\bar{c}_p$ is 0.26 Btu/lb·°F for typical temperature ranges normally encountered in fume incinerators.

5. Equation 8.5 depicts the amount of heat required to raise a set quantity of gas from T_1 to T_2. Of more importance is the heat rate Q that is required. The heat rate is easily determined by multiplying either side of Equation 8.5 by the mass flow rate ($\dot{m}$) of the process gas stream. The heat rate required is given by

$$\dot{Q} = \dot{m}\Delta H = \dot{m}c_p(T_2 - T_1) \quad (8.6)$$

where $\dot{Q}$ = heat rate, Btu/hr
$\dot{m}$ = mass flow rate of gases, lb/hr

Equation 8.6 can therefore be used to compute the heat rate required to raise the gas temperature from T_1 to T_2. These equations are simple heat (enthalpy) balances. They do not account for any heat losses in the system. Heat losses from refractory or ducting are usually accounted for by assuming a fixed percent (or fraction) of the total theoretical heat is lost. For example, if an afterburner is required to supply heat at the rate of 1×10^6 Btu/hr and there is a 10% heat loss from the combustion chamber, the total heat rate would have to be 1.1×10^6 Btu/hr to account for the losses.

The following procedure for calculating fuel requirements and the physical design of an incinerator are available in the literature [8] and are presented below.

1. Employing Equations 8.2–8.6, calculate the heat load required to raise the process shown from its inlet temperature to the operating temperature of the combustion device:

$$Q = \Delta H \quad (8.7)$$

2. Correct the heat load term for any radiant losses (RL):

$$Q = (1 + \text{RL})(\Delta H) \tag{8.8}$$

where RL = fractional basis.

3. Assuming natural gas at a known heating value HV_G as fuel, calculate the available heat at the operating temperature:

$$\text{HA}_T = (\text{HV}_G)\left(\frac{\text{HA}_T}{\text{HV}_G}\right)_{\text{ref}} \tag{8.9}$$

Where the subscript "ref" refers to a reference fuel. The available heat for natural gas with a reference HV_G of 1,059 Btu/scf (assuming stoichiometric air) is given by Theodore [9].

$$(\text{HA}_T)_{\text{ref}} = -0.237(\text{T}) + 981 \tag{8.10}$$

where T = incinerator operating temperature, °F.

4. Calculate the natural gas required (NG):

$$\text{NG} = \text{Q}/\text{HA}_T;\ \text{consistent units} \tag{8.11}$$

5. Determine the volumetric flow rate of both the process gas stream q_p and the flue products of combustion of the natural gas q_c at the operating temperature

$$q_T = q_p + q_c \tag{8.12}$$

A good estimate for q_c is

$$q_c = (11.5)(\text{NG}) \tag{8.13}$$

1. The cross-sectional area, S, of the combustion device is given by

$$S = q_T/v \tag{8.14}$$

where v = average throughput velocity of the hot gases in the combustion device.

2. The diameter of the (tubular) combustion device is given by

$$D = (4S/3.14)^{0.5} \tag{8.15}$$

3. The length can be obtained from information on the residence time t_r or the length-to-diameter ratio L/D. Each is considered below:
   a. Since $t_r = L/v$, then

   $$L = (t_r)(v) \tag{8.16}$$

   b. Typical L/D ratios are in the 2–3 range. For $L/D = 2.5$, then

   $$L = (2.5)(D) \tag{8.17}$$

The reader can refer to the literature for additional details [8].

Variations and unknowns in many of the parameters (particularly those of a kinetic nature) in the describing equation(s) make it nearly impossible to provide a generalized procedure for the design of a catalytic incinerator.

The following information is recommended for the proper design of a flare:

1. Type of material to be flare
2. Average molecular weight, M

3. Percent unsaturation
4. Lower (net) heating values (Btu/scf)
5. Specific heat ratio, $k = c_p/c_v$
6. Mass flow rate at maximum discharge, lb/hr
7. Average vapor temperature, °F
8. Flowing pressure, psig
9. Percent toxic, odorous, or noxious gases

The parameters applicable to the calculation of a proper flare burner diameter include mass flow rate and discharge conditions, the type of flame holder used by the manufacturer, and density and heat-capacity ratios, which serve to define the sonic velocity in the fluid stream. In addition, the pressure drop, rather than flow velocity, is usually the controlling factor in flare design.

The following sequences of steps are involved in the catalytic conversion of reactants to products [9, 10]:

1. Transfer of reactants to and products from the outer catalyst surface
2. Diffusion of reactants and products within the pore of the catalyst
3. Activated adsorption of reactants and the desorption of the products on the active centers of the catalyst
4. Reaction(s) on active centers of the catalyst surface.

At the same time, energy effects arising due to chemical reaction can result in the following:

1. Heat transfer to or from active centers to catalyst particle surface
2. Heat transfer to and from reactants and products within the catalyst particle
3. Heat transfer to and from moving streams in the reactor
4. Heat transfer from one catalyst particle to another within the reactor
5. Heat transfer to or from the walls of the reactor

8.4 Equipment Design Considerations

Thermal incinerators or afterburners can be used over a fairly wide, but low, range of organic vapor concentrations. The concentration of the organics in air must be substantially below the lower flammable level, LFL (or lower explosive limit, LEL). As a rule, a factor of 4 is employed for safety precautions. Reactions are conducted at elevated temperatures in order to insure high chemical reaction rates for the organics. To achieve this temperature, it is sometimes necessary to preheat the feed stream with auxiliary energy. Along with the contaminant-laden gas stream, air and fuel are continuously delivered to the reactor, where the fuel is combusted with air in a firing unit (burner). The burner may utilize the air in the process waste stream as the combustion air for the auxiliary fuel, and/or it may use a separate source of outside air for this purpose. The products of combustion and the unreacted feed stream are mixed completely and enter the reaction zone of the unit. The pollutants in the process gas stream are then reacted at the elevated temperature. The unit requires operating temperatures in the 1,200–1,500 °F range for the combustion of most pollutants. A residence time of 0.2–1.0 sec is recommended in the literature, but this factor is primarily dictated by kinetic considerations. A length to-diameter ratio of 2.0–3.0 is usually employed. The end products are continuously discharged at the outlet of the reactor. The average gas velocity can range from as low as 10 ft/sec to as high as 50 ft/sec. (The velocity increases from inlet to outlet due to the increase in the number of moles of the reacting fluid and to the increase in temperature due to reaction.) These high velocities are required to prevent settling of particulates (if present) and to minimize the dangers of flashback and fire hazards.

The fuel is usually natural gas. The energy liberated by reaction may be directly recovered in the process or indirectly recovered by suitable external heat exchanger [11–14]. This should definitely be included in a design analysis since the energy is the only commodity of value that is usually derived from the combustion process.

Due to the high operating temperatures, the unit must be constructed of metals capable of withstanding this condition. Combustion devices are usually constructed with an outer steel shell that is lined with refractory material. However, refractory material is heavy, with densities as low as 50 lb/ft^3 for lightweight insulating firebrick, and as high as 175 lb/ft^3 for castable refractories. Refractory wall thickness is in the 3–9-inch range. This weight adds considerably to the cost. Because of its light weight, firebrick wall construction is being used in some units. All-metal construction has found limited application. These combustion reactors also vary in shape, from tanks to tubular pipes. The latter type is usually used since it has a high surface-to-volume ratio, which is advantageous for energy recovery and suitable for continuous operation.

Some of the advantages of the thermal combustion reactor are removal of organic gases, removal of submicron particles, simplicity of construction, small space requirements, and low maintenance costs. Some of the disadvantages include high operating costs, fire hazards, and flashback possibilities. Catalytic reactors are an alternative to thermal incinerators. If a solid catalyst is added to the reactor, the reaction is said to be heterogeneous. For simple reactions, the effect of the presence of a catalyst is to

1. Increase the rate of reaction
2. Permit the reaction to occur at a lower temperature
3. Permit the reaction to occur at a more favorable pressure
4. Reduce the reactor volume
5. Increase the yield of a reactant(s) relative to the other(s)

A typical catalyst incinerator for the combustion of organic vapors is presented in Figure 8.4. The gas stream is delivered to the reactor continuously by a fan at a velocity in the 10–30 ft/sec range, but at a lower temperature—usually in the 650–800 °F range—than the thermal unit. A length-to-diameter

Figure 8.4 Photo of catalytic incinerator internals.

ratio less than 0.5 is usually employed. The gases, which may or may not be preheated, pass through the catalyst bed where the reaction occurs. The combustion products, which are again made up of vapor, carbon dioxide, inert products, and unreacted vapors, are continuously discharged from the outlet at a higher temperature. Energy savings can again be affected with heat recovery from the exit stream. Here, the hot discharge gases transfer energy to the incoming feed stream.

Metals in the platinum family are recognized for their ability to promote combustion at low temperatures. Other catalysts include various oxides of copper, chromium, vanadium, nickel, and cobalt. These catalysts are subject to poisoning, particularly from halogens, halogen and sulfur compounds, zinc, arsenic, lead, mercury, and particulates. High temperatures can also reduce catalyst activity. It is therefore important that catalyst surfaces be clean and active to insure optimum performance. Catalysts can usually be regenerated with superheated steam.

Catalysts may be porous pellets, usually cylindrical or spherical in shape, ranging from 1/16 to 1/2 inch in diameter. Small sizes are recommended, but the pressure drop through the reactor increases. Other shapes include honeycombs, ribbons, and wire mesh. Since catalysis is a surface phenomenon, a physical property of these particles is that the internal pore surface can be considered infinitely greater than the outside surface.

Thermal incinerators are generally shipped full assembled. They may, however, be shipped in two sections if too large to be shipped as one unit. Installation involves mounting the unit on a concrete slab, bolting the sections together if required, and connecting the incinerator's utility lines (electric, oil, gas, and air) to plant supply lines. Small or specially designed lightweight units may be roof-mounted. If necessary, additional structural support may be provided. As previously stated, the incinerator's fume inlet is normally connected to the discharge side of the fan, which exhausts the fume-generating equipment. The incinerator's combustion chamber outlet is connected to a stack or to heat-recovery equipment preceding the stack.

The concrete support pad should extend several feet beyond the incinerator on all sides, should be at least 8 inches thick, and should be properly bedded on a gravel drainage bed laid on firm, compact earth.

The incinerator discharge duct must be 100% tight. Discharge gases may be in the range of 1,400 °F and duct leakage could lead to fires. The fan and ducting to the incinerator should also be insulated to prevent loss of heat by the fumes. The fan must be firmly secured.

In areas where below-freezing temperatures are encountered, steam or electric tracer lines are required for air and fuel lines to prevent freezing of entrained water. Control valves must also be protected against freezing.

Installation is generally provided by a contractor engaged by the incinerator owner, with supervision being provided by a representative of the manufacturer.

The installation of catalytic incinerators differs from installation of thermal incinerators only in that the catalytic bed must be inserted in its holder and temperature and/or pressure instruments relative to operating conditions preceding and following the catalyst be connected. The installation procedures presented for thermal incinerators also apply, with little or no modification, to flares.

Another key design consideration with combustion equipment is the selection of the burner technology. In order to destroy waste pollutants, fossil fuels are combusted, and this process inherently produces its own set of pollutants. One of the primary concerns from fossil fuel combustion is nitrogen oxide (NO_x) emissions. NO_x is formed by the oxidation of both atmospheric and fuel-bound nitrogen during the combustion process. NO_x is of environmental concern as it negatively affects the atmosphere by producing acid rain, smog, and upper atmosphere ozone depletion. NO_x formation from a burner is determined by the interaction of chemical and physical processes occurring within the incinerator furnace. If temperature, oxygen concentrations, or nitrogen concentrations can be reduced quickly after combustion, thermal NO_x formation is suppressed. This is a major issue for incinerators because it is desired to maintain elevated temperature and residence time to ensure complete destruction of the fumes. Advancements in burner design have led to significantly reduce thermal NO_x emissions by two orders of magnitude over the last thirty years, but different burner manufacturers approach this in a variety of different ways. Not all techniques are suited for every application, and the burner selection should be given close consideration [15].

8.5 Applications

Three illustrative examples, originally from the USEPA literature, complement this chapter concerned with combustion equipment.

8.5.1 Available Heat Calculation

You have been requested to evaluate the gross heating value of a natural gas of a given composition. To determine the available heat of the natural gas at a given temperature, the rate of auxiliary fuel (natural gas) required to heat a known amount of contaminated air to a given temperature, the dimensions of an afterburner treating the contaminated air stream, and the residence time. Operating and pertinent design data are provided in Table 8.3. The natural gas composition (mole or volume fraction) is provided in Table 8.4.

Molar enthalpy data are provided in Table 8.5.

It is required to heat the contaminated air from 200 °F to 1,200 °F.

Solution: The gross heating value of this natural gas is determined below:

$$
\begin{aligned}
\mathrm{HV}_G &= \sum x_i \mathrm{HV}_{G,i} \\
&= 0.0515(0) + 0.8111(1{,}013) + 0.0967(1{,}792) + 0.0351(2{,}590) \\
&\quad + 0.0056(3{,}370) \\
&= 1{,}105 \frac{\mathrm{BTU}}{scf} \text{ natural gas}
\end{aligned}
$$

Table 8.3 Pertinent design parameters.

Gas velocity	20 ft/sec
Length-to-diameter ratio of the afterburner	2.0
Temperature of dry natural gas	60 °F
Volumetric flow rate of contaminated air	5,000 scfm (60 °F, 1 atm)

Table 8.4 Natural gas composition.

Component	Mole fraction	Gross heating value, Btu/SCF
N_2	0.0515	0
CH_4	0.8111	1,013
C_2H_6	0.0967	1,792
C_3H_8	0.0351	2,590
C_4H_{10}	0.0056	3,370

Table 8.5 Approximate molar enthalpies of combustion gases (Btu/lbmol).

T (°F)	N_2	Air (MW = 28.97)	CO_2	H_2O
32	0	0	0	0
60	195	195	243	224
77	312	313	392	360
100	473	472	598	545
200	1,170	1,170	1,527	1,353
400	2,570	2,576	3,537	3,001
600	3,991	4,010	5,714	4,700
800	5,443	5,479	8,026	6,560
1,000	6,929	6,984	10,447	8,284
1,200	8,452	8,524	12,960	10,176
1,500	10,799	10,895	16,860	13,140
2,000	14,840	14,970	23,630	18,380
3,000	23,280	23,460	37,750	29,780

The balanced chemical combustion equations for each of the four combustible components of the natural gas using 1 scf of natural gas as a basis is

$$0.8111\ CH_4 + 1.6222\ O_2 \rightarrow 0.8111\ CO_2 + 1.6222\ H_2O$$
$$0.0967\ C_2H_6 + 0.3385\ O_2 \rightarrow 0.1934\ CO_2 + 0.2901\ H_2O$$
$$0.0351\ C_3H_8 + 0.1755\ O_2 \rightarrow 0.1053\ CO_2 + 01404\ H_2O$$
$$0.0056\ C_4H_{10} + 0.0364\ O_2 \rightarrow 0.0224\ CO_2 + 0.0280\ H_2O$$

The number of standard cubic feet for each of the following components of combustion may now be determined.

For O_2: 1.6222 + 0.3385 + 0.1755 + 0.0364 = 2.172 scf/scf natural gas
For CO_2: 0.8111 + 0.1934 + 0.1053 + 0.0224 = 1.132 scf/scf natural gas
For H_2O: 1.6222 + 0.2901 + 0.1404 + 0.0280 = 2.081 scf/scf natural gas
For N_2: 0.0515 + (79/21)(2.172) = 8.222 scf/scf natural gas

The total cubic feet of combustion products per scf of natural gas burned are the sum of scf CO_2/scf of natural gas, scf H_2O/scf of natural gas, and scf N_2/scf of natural gas:

Total cubic feet of combustion products} = 1.132 + 2.081 + 8.222
= 11.435 scf of products/scf of natural gas

The following values of enthalpies at 60 °F and 1,200 °F are obtained from Table 8.5: for CO_2:

$$\begin{aligned}\Delta H_{CO_2} &= H \text{ at } 1{,}200\ °F - \text{H at } 60\ °F \\ &= 12{,}960 - 243.1 \\ &= 12{,}270\ \text{Btu/lbmol}\end{aligned}$$

Since there are 379 scf per lbmol of any ideal gas and 1.132 scf CO_2/scf natural gas, one obtains

$$\Delta H_{CO_2} = (12,720)(1.132)/(379)$$
$$= 38.0 \text{ Btu/scf of natural gas}$$

For N_2:

$$\Delta H_{N_2} = (8,452 - 194.9)(8.222)/(379)$$
$$= 179.1 \text{ Btu/scf of natural gas}$$

For H_2O (g):

$$\Delta H_{H_2O} = (10,176 - 224.2)(2.081)/(379)$$
$$= 54.6 \text{ Btu/scf of natural gas}$$

Determine the amount of heat required of take the products of combustion from 60 °F to 1,200 °F ($\sum \Delta H$). Since the water present in the combustion product is vapor, then

$$\Delta H_\lambda = \frac{\left(\frac{1,060\ \text{Btu}}{\text{lb}}\right)\left(\frac{18\ \text{lb}}{\text{lbmol}}\right)\left(\frac{2.081\ \text{scfH}_2\text{O}}{\text{scf natural gas}}\right)}{\frac{379\ \text{scf}}{\text{lbmol natural gas}}}$$
$$= 104.8 \text{ Btu/scf of natural gas}$$

Therefore,

$$\sum \Delta H = \Delta H_{CO_2} + \Delta H_{N_2} + \Delta H_{H_2O} + \Delta H_\lambda$$
$$= 38.0 + 179.1 + 54.6 + 104.8$$
$$= 376.5 \text{ Btu/scf of natural gas}$$

The available heat (HA) of natural gas at 1,200 °F in Btu/scf natural gas is

$$\text{HA}_{1200} = \text{HV}_G - \sum \Delta H$$
$$= 1,105 - 376.5$$
$$= 728.5 \text{ Btu/scf of natural gas}$$

The enthalpy change of air going from 200 °F to 1,200 °F is

$$\Delta H_{air} = \text{H at } 1,200\ °\text{F} - \text{H at } 200\ °\text{F}$$
$$= 8,524 - 1,170$$
$$= 7,354 \text{ Btu/lbmol}$$

The heat rate $\dot{Q}$ required to heat 5,000 scfm of air from 200 °F to 1,200 °F in Btu/min is

$$\dot{Q} = (5,000 \text{ scfm})(7,354 \text{ Btu/lbmol})/(379 \text{ scf/lbmol})$$
$$= 97,018 \text{ Btu/min}$$

The fuel requirement is therefore

$$\text{NG} = \dot{Q}/\text{HA}$$
$$= 97,018/728.5$$
$$= 133.2 \text{ scfm}$$

To size the unit, first calculate the volumetric flow rate of the products of combustion (CP) in scfm:

$$q_{\mathrm{CP}} = 11.435 \frac{\text{scf products}}{\text{scf natural gas}} (133.2 \text{ scfm natural gas})$$
$$= 1,523 \text{ scfm}$$

Also calculate the total flue gas flow rate q_{T} in scfm:

$$q_{\mathrm{T}} = 5,000 + 1,523$$
$$= 6,523 \text{ scfm}$$

Determine the total flue gas flow rate at 1,200 °F in acfm using Charles' law:

$$q_{\mathrm{T}} = (6,523)(1,200 + 460)/(60 + 460)$$
$$= 20,823 \text{ acfm}$$
$$= 347 \text{ acfs}$$

The diameter D and length L of the afterburner in feet are

$$D = \left(4_{qT}/v\pi\right)^{1/2}$$
$$= \left[(4)(347)/(20)(\pi)\right]^{1/2}$$
$$= 4.7\ ft$$
$$L = (2.0)(4.7)$$
$$= 9.4\ ft$$

Therefore, the residence time t for the gases in the afterburner in seconds is

$$t = L/v$$
$$= (9.4)/(20)$$
$$= 0.47 \text{ s}$$

Note how small the residence time is. This is a typical residence time for this application, but it is quite fast. This is one of the main reasons that combustion is a particularly complex process to control.

8.5.2 A Permit Review

A regulatory agency engineer must review plans for a permit to construct a direct flame afterburner (Figure 8.5) serving a lithographer. Review is for the purpose of judging whether the proposed system, when operating as it is designed to operate, will meet emission standards. The permit application provides operating and design data. Agency experience has established design criteria that, if met in an operating system, will typically ensure compliance with standards. Operating data from the permit application are provided below.

- Application = lithography
- Effluent exhaust volumetric flow rate = 7,000 scfm (60 °F, 1 atm)
- Exhaust temperature = 300 °F
- Hydrocarbons in effluent air to afterburner (assume hydrocarbons to be toluene) = 30 lb/hr
- Afterburner entry temperature of effluent = 738 °F
- Afterburner heat loss = 10% in excess of calculated heat load
- Afterburner dimensions = 4.2 ft in diameter, 14 ft in length

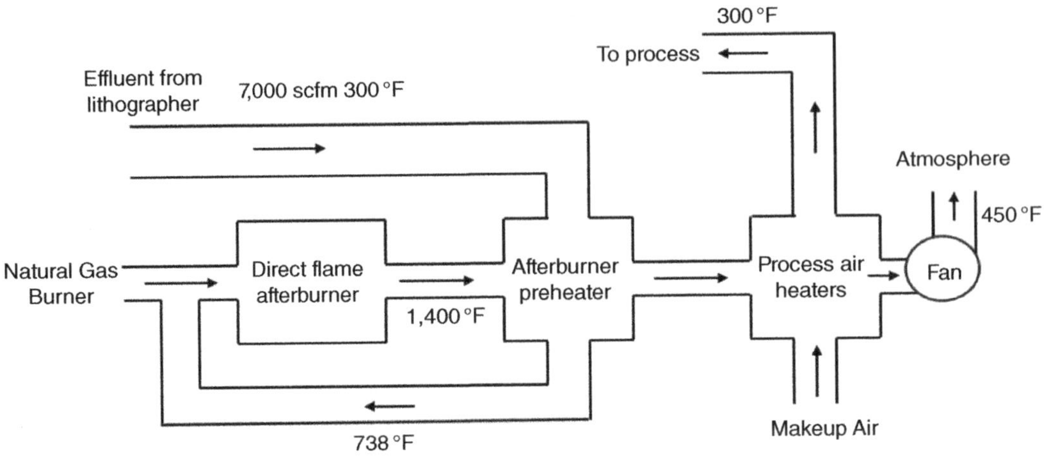

Figure 8.5 Direct flame afterburner system.

Agency Design Criteria

- Afterburner temperature = 1,300 – 1,500 °F
- Residence time = 0.3 – 0.5 sec
- Afterburner velocity = 20 – 40 ft/sec

Standard Data

- Gross heating value of natural gas = 1,059 Btu/scf of natural gas
- Combustion products per cubic foot of natural gas burned = 11.5 scf/scf natural gas
- Available heat of natural gas at 1,400 °F = 600 Btu/scf of natural gas
- Molecular weight of toluene = 92
- Average heat capacity of effluent gases at 738 °F (above 0 °F) = 7.12 Btu/(lbmol·°F)
- Average heat capacity of effluent gases at 1,400 °F (above 0 °F) = 7.38 Btu/(lbmol·°F)
- Volume of air required to combust natural gas = 10.33 scf air/scf natural gas

Solution: The design temperature is already within agency criteria. To determine the fuel requirement for the afterburner, first calculate the total heat load (heating rate) $\dot{Q}$ required to raise 7,000 scfm of the effluent stream from 738 °F to 1,400 °F in Btu/min:

$$\dot{n} = (7{,}000\ \text{scfm}) \bigg/ \left(\frac{379\ \text{scf}}{\text{lbmol}}\right)$$

$$= 18.47\ \text{lbmol/min}$$

$$\dot{Q} = \left[C_{p2}(T_2 - T_b) - C_{p1}(T_1 - T_b)\right]$$

$$= 18.47\left[(7.38)(1{,}400 - 0) - (7.12)(738 - 0)\right]$$

$$= 93{,}780\ \text{Btu/min}$$

Calculate the actual heat load required, accounting for a 10% heat loss, in Btu/min:

$$\begin{aligned}\text{Actual heat load} &= (1.1)\dot{Q}\\ &= (1.1)(93,780)\\ &= 103,200 \text{ Btu/min}\end{aligned}$$

Calculate the flowrate of natural gas required to supply the actual heat required to heat 7,000 scfm of the effluent from 738 to 1,400 °F in scfm:

$$\begin{aligned}\text{NG} &= \dot{Q}/\text{HA}\\ &= (103,200)/(600)\\ &= 172 \text{ scfm}\end{aligned}$$

Determine the total volumetric flow rate through the afterburner q_T by first calculating the volumetric flow rate of the combustion products of the natural gas q_2 in scfm:

$$\begin{aligned}q_1 &= q_{\text{NG}}\left(11.5\,\frac{\text{scf combustion products}}{\text{scf natural gas}}\right) q_{\text{NG}} = \text{NG}\\ &= (172)(11.5)\\ &= 1,978 \text{ scfm}\end{aligned}$$

Also note that the volumetric flow rate of the effluent is 7,000 scfm. The volumetric flow rate of air required to combust the natural gas required q_2 in scfm is

$$\begin{aligned}q_2 &= q_{\text{NG}}\left(10.3\,\frac{\text{scf air}}{\text{scf natural gas}}\right)\\ &= (172)(10.33)\\ &= 1,776 \text{ scfm}\end{aligned}$$

Calculate the total volumetric flow rate through the afterburner q_T in scfm. Since primary air is employed in the combustion of the natural gas, q_2 is not subtracted from q_T. Thus, the q_2 calculation is not required in this solution.

$$\begin{aligned}q_T &= 7,000 + q_T\\ &= 7,000 + 1,978\\ &= 8,978 \text{ scfm}\end{aligned}$$

Applying Charles' law to convert scfm to acfm

$$\begin{aligned}&= \frac{(8{,}978)(1{,}400 + 460)}{60 + 460}\\ &= 32{,}100 \text{ acfm}\end{aligned}$$

To determine whether the afterburner velocity meets the agency criteria, first calculate the cross-sectional area of the afterburner in ft^2:

$$\begin{aligned}S &= \pi D^2/4\\ &= \pi(4.2)^2/4\\ &= 13.85 \text{ ft}^2\end{aligned}$$

The afterburner velocity v in ft/sec is then

$$\begin{aligned} V &= q_T/S \\ &= (32,110)/(13.85) \\ &= 2,318 \text{ ft/min} \\ &= 38.6 \text{ ft/sec} \end{aligned}$$

Thus, the afterburner velocity is within the agency criterion. The residence time t is then

$$\begin{aligned} t &= L/v \\ &= 14/38.6 \\ &= 0.363 \text{ sec} \end{aligned}$$

This also meets the agency criterion.

8.5.3 Paint Baking Oven Application

An installed afterburner is required to incinerate a 3,000 acfm contaminated gas stream discharged from a direct-fired paint baking oven at 350 °F. The following summarizes the data taken from the plans:

Data Sheet

- Exhaust flow rate from oven: 3,000 acfm
- Exhaust gas temperature from oven: 350 °F
- Solvent emission to afterburner: 0.3 lb/min
- Final temperature in afterburner: 1,000 °F
- Gross heating value of natural gas: 1,100 Btu/scf
- Total heat requirement: 26,884 Btu/min
- Natural gas requirement: 35.0 scfm
- Furnace volume: 46.0 ft^3
- Exhaust flow rate from afterburner at 1,000 °F: 6,350 acfm
- Gas velocity through catalytic bed: 8.6 ft/sec
- Number of type A 19 × 24 × 3.75 inch catalyst elements: 4

The following additional information and rules of thumb may be required to review the plans for the catalytic afterburner:

1. Heat will be recovered from the afterburner effluent, but that process will not be considered in this problem.
2. Catalytic afterburner operating temperatures of approximately 950°F have been found sufficient to control emissions from most process ovens.
3. Preheat burners are usually designed to increase the temperature of the contaminated gases to the required catalyst discharge gas temperature without regard to the heating value of the contaminants (especially if considerable concentration variation occurs).
4. A 10% heat loss is usually a reasonable estimate for an afterburner. This may be accounted for by dividing the calculated heat load by 0.9.
5. The properties of the contaminated effluent may essentially be considered identical to those of air.
6. The natural gas is combusted using near stoichiometric (0% excess) external air.
7. The catalyst manufacturer's literature suggests a superficial gas velocity through the face surface of the catalyst element (in this case 19 inches × 24 inches) of 10 ft/sec.

Two key questions are to be considered:

a. Is the operating temperature adequate for efficiency control?
b. Is the fuel requirement adequate to maintain the operating temperature?

Solution:

a. The plans indicate a combustion temperature of 1,000 °F; this is acceptable when compared to a 950 °F rule-of-thumb temperature.

b. To determine whether the fuel requirement is adequate to maintain the operating temperature, first calculate the lbmol/min of gas to be heated from 350 °F to 1,000 °F.

$$\dot{n} = (3,000\ \text{acfm})[(460 + 60)/(460 + 350)] = 1,926\ \text{scfm}$$
$$1,926/379 = 5.08\ \text{lbmol/min}$$

Also determine the heat requirement in Btu/min to raise the gas stream (air) temperature from 350 to 1,000 °F.

H at 350 °F = 2,222 Btu/lbmol
H at 1,000 °F = 6,984 Btu/lbmol

$$\begin{aligned}\dot{Q} &= \dot{n}\Delta H \\ &= (5.08)(6,984 - 2,222) \\ &= 24,190\ \text{Btu/min}\end{aligned}$$

The total heat requirement Q_T in Btu/min is then

$$\begin{aligned}\dot{Q}_T &= \dot{Q}/0.9 \\ &= 24{,}190/0.9 \\ &= 26{,}870\ \text{Btu/min}\end{aligned}$$

The available heat of the natural gas in Btu/scf at 1,000 °F may be calculated from the following equation.

$$\begin{aligned}\text{HA}_{1,000}/\text{HV}_\text{G} &= \left(\text{HA}_{1,100}/\text{HV}_\text{G}\right)_{\text{ref fuel}} \\ \text{HA}_{1,000} &= 1,100\ (745/1,059) \\ &= 774\ \text{Btu/scf natural gas}\end{aligned}$$

The natural gas (NG) requirement in scfm may be calculated from the last two results:

$$\begin{aligned}\text{NG} &= \dot{Q}_T/\text{HA}_{1,000} \\ &= 26{,}888/774 \\ &= 34.7\ \text{scfm}\end{aligned}$$

References

1 S. Turns, *An Introduction to Combustion*, 3rd edition, McGraw Hill, New York City, NY, 2012.
2 A. J. Buonicore and L. Theodore, *Industrial Control Equipment of Gaseous Pollutants*, Vol. 2 CRC Press/Taylor & Francis Group., Boca Raton, FL, 1975.
3 R. Rolke, et al., *Afterburner System Study*. EPA-R2-72-062, RTP, NC, 1972.

4 L. Theodore and R. Allen, *Air Pollution Control Equipment*, A Theodore Tutorial, East Williston, NY, 1993, (Originally Published by the US EPA/APRI, RTP, NC).
5 R. Dupont, K. Ganesan and L. Theodore, *Unit Operation for Environmental Equipment*, Scrivener-Wiley Publications, Beverly, MA, 2017.
6 M. G. Klett and J. B. Galeski, *Flare Systems Study*. EPA-600/2-76-079, TRP, NC, 1976.
7 L. Theodore, *Personal Notes*, East Williston, NY, 1985.
8 L. Theodore, *Ask the Experts: Designing Thermal Afterburners*, CEP, New York City, NY, 2005.
9 L. Theodore, *Chemical Reactor Analysis and Applications for the Practicing Engineer*, John Wiley & Sons, Hoboken, NJ, 2012.
10 L. Theodore, *Chemical Reactor Kinetics*, Theodore Tutorials, East Williston, NY, 1992 (Originally Published by the US EPA/APRI, RTP, NC).
11 L. Theodore, *Heat Transfer Applications for the Practicing Engineer*, John Wiley & Sons, Hoboken, NJ, 2011.
12 A.M. Flynn, T. Akashige, and L. Theodore, *Kern's Process Heat Transfer*, 2nd Edition, Scrivener-Wiley, Beverly, MA, 2019.
13 D. Kern, *Process Heat Transfer*, McGraw-Hill, New York City, NY, 1950.
14 L. Theodore, Heat Transfer, Theodore Tutorials, East Williston, NY, 1993 (Originally Published by the US EPA/APRI, RTP, NC).
15 J. Richardson, *Personal Notes*, 2024.

Suggested Reading

1 M. G. Klett and J. B. Galeski, *Flare Systems Study*, EPA contract No. 68-02-1331, EPA-600/2-76-079, Washington D.C., 1976.
2 R. Rolke, *Afterburner Systems Study*, OPA Contract No. EHS-D-71-3, NTIS, PB212-500, Washington D.C., 1972.
3 J. A. Danielson, *Air Pollution Engineering Manual*, EPA OAP Publication No. AP-40, Washington D.C., 1973.
4 A. C. Stern, *Fundamentals of Air Pollution*, Academic Press, New York City, NY, 1968.
5 L. Theodore and A. J. Buonicore, *Industrial Control Equipment for Gaseous Pollutants*. Vols I and II, CRC Press, Boca Raton, FL, 1975.
6 L. Theodore, *Air Pollution Control Equipment Calculations*, John Wiley & Sons, Hoboken, NJ, 2008.
7 L. Theodore, *Chemical Engineering: The Essential Reference*, McGraw Hill, New York City, NY, 2014.
8 L. Theodore, R. R. Dupont, and K. Ganesan, *Unit Operations in Environmental Engineering*, John Wiley & Sons, Hoboken, NJ, 2017.
9 J. Richardson, R. R. Dupont, P. Farber, K. Ganesan and L, Theodore, *Air Pollution Control Equipment Revisited*, AWMA Conference, Calgary, CA, 2024.
10 J. Richardson and L. Theodore, *Optimizing Air Pollution Equipment Performance: Operation and Maintenance*, John Wiley & Sons, Hoboken, NJ, 2025.

9

Condensers

9.1 Introduction

Condensers have found a wide range of applications in controlling gaseous emissions. Their application in any specific process depends on the amount and type of coolant available, liquid disposal problems, and the volume of the recovered compounds. Condensers are usually used as preliminary devices and for the best control are followed by a more efficient device such as an absorber (Chapter 6), adsorber (Chapter 7), or combustion device (Chapter 8).

Condensation is a process in which the volatile gases are removed from the containment stream and converted into a liquid. Condensation is usually achieved by reducing the temperature of a vapor mixture until the partial pressure of the condensable component equals its vapor pressure. Condensation requires low temperatures to liquefy most pure containment vapors. Condensation is affected by the composition of the containment gas stream. The presence of additional gases, such as air, that do not condense at the same conditions, hinders condensation. Condensation by increasing the pressure is possible, but usually not practical from an economic perspective.

Condensers are normally used in combination with primary control devices. Condensers can be located upstream or (before) an incinerator, adsorber, or absorber. These condensers reduce the volume of vapors that the primary control equipment must handle. Therefore, the size and the cost of the primary control device can be reduced. Similarly, condensers can be used to remove water vapors from a process stream with a high moisture content upstream of a control system. A prime example is the use of condensers in rendering plants to remove moisture from the cooker exhaust gas. When used alone, refrigeration is required to achieve the low temperatures required for condensation. Refrigeration units are used to successfully control gasoline vapors at large gasoline dispensing terminals.

Condensation and subsequent removal of vapors reduce not only organic emissions but also other pollutant vapors and/or gases. Many compounds, because of their relatively high boiling points, readily condense even though they are not highly concentrated. Thus, at a given temperature, if the partial pressure of a compound is increased until it is equal to or greater than its vapor pressure at that temperature, the compound will condense. Alternatively, if the temperature of a gaseous mixture is reduced to the saturation temperature, at which the vapor pressure equals the partial pressure of one of the constituents, condensation will occur.

Condensers have found a wide range of applications in the chemical industry, where their purpose has been to condense concentrated vapors in the primary process, rather than to reduce the concentration of a pollutant.

9.2 Description of Condenser Equipment

Condensation can be accomplished by increasing pressure or decreasing temperature (removing heat). In practice, air pollution control condensers almost always operate through extraction of heat. Condensers differ in the means of removing heat and the type of device used. The two different means of condensing are direct contact (or contact), where the cooling medium with vapors and condensate are intimately mixed and combined, and indirect (or surface), where the cooling medium and vapor/condensate are separated by a surface area of some type.

Contact condensers are simpler, less expensive to install, and require less auxiliary equipment and maintenance. The condensate/coolant from a contact condenser has a volume 10 to 20 times that of a surface condenser. This condensate typically cannot be reused and may pose a waste disposal problem unless the dilution of the pollutant is sufficient to meet regulatory requirements. Some typical contact condensers are shown in Figure 9.1.

Surface condensers form the bulk of the condensers used for air pollution control. Some of the applicable types of surface condensers are shell and tube, double pipe, spiral plate, flat plate, air-cooled, and various extended surface tubular units [1]. This chapter focuses on shell-and-tube condensers, because they are so widely used in industry and have been standardized by the Tubular Exchanger Manufacturers Association (TEMA).

Condensing can be accomplished in either the shell or tubes. The economics, maintenance, and operational ramifications of the allocation of fluids are extremely important, especially if extended surface tubing is being considered. The designer should be given as much latitude as possible in specifying the condenser. A diagram of a generic shell-and-tube heat exchanger is shown in Figure 9.2 [2].

Air-cooled condensers consist of a rectangular bank of high finned tubes, a fan, a plenum for even distribution of air to the rectangular face of the tube bank, a header for vapor inlet and condensate outlet, and a steel support structure. Fins are usually aluminum 0.5 to 0.625 in height that

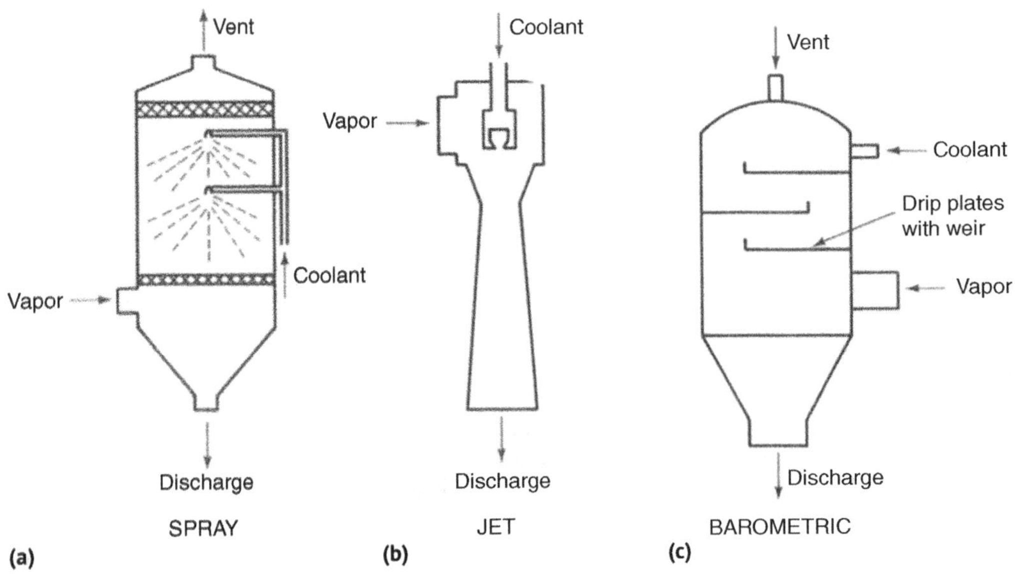

Figure 9.1 Contact condensers: (a) spray; (b) jet; (c) barometric.

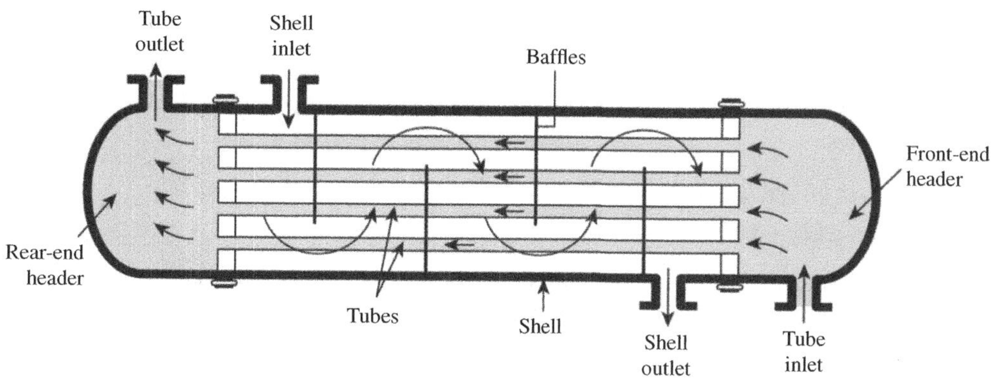

Figure 9.2 Generic diagram of a shell-and-tube heat exchanger.

are applied to a bare tube by tension winding, or soldering, or sold-extruded from the tube itself. Bimetallic tubes can be used to provide process corrosion resistance on the inside and aluminum extruded fins on the outside. Headers for the tube side frequently contain removable gasketed plugs corresponding to each tube end and are used for access to the individual tube-to-tube sheet joint for maintenance.

Nearly all tubular exchangers employ roller expanding of the tube ends into tube holes drilled into tube sheets as a means of providing a leakproof seal. TEMA requires that tube holes be of close tolerance and contain two concentric grooves. These grooves give the tube joint greater strength but do not increase sealing ability and may actually decrease it. Research is currently being done to clarify what is the best procedure to follow for a given set of materials, pressures, temperatures, and loadings. Welding or soldering of tubes to tube sheets may be performed for additional leak tightness or in lieu of roller expanding.

9.3 Process Design Considerations

The main design factors to be considered in condensing operations are the type(s) of compounds in the condenser and their temperatures, volumes, concentrations, vapor pressures and heat capacities. The temperature and amount of coolant available are also important considerations. In the design of a contact condenser, the amount of cooling water to be used is the critical design factor. The amount of heat that must be removed from the noncondensable gases must also be included. The amount of cooling water required is this heat load divided by the allowable difference in water temperature (inlet water temperature minus outlet water temperature). In a contact type of condenser, about 15 pounds of water (1.8 gallons) at 60 °F is required to condense 1 pound of steam at 212 °F and cool the condensate to 140 °F. Dilution and subcooling of the condensate prevents release of volatile compounds. In a typical cylindrical contact-spray-chamber condenser, contact times on the order of 1 second with a cross-sectional velocity of about 400–500 ft/min have been used. Pressure drops on the order of 1 inch of water are typical of these units. The use of a contact condenser can result in a water pollution problem, a factor that sometimes restricts the use of this type of condenser. In the design of a surface condenser, the area of heat exchange is the critical factor.

9.3.1 Contact Condensers

Design of contact condensers involves calculating the quantity of coolant required to condense and subcool the vapor and proper sizing of discharge piping and hotwell. Calculation of coolant flow rate is as follows:

$$\dot{M}_c = \frac{\dot{M}_v H_v + \dot{M}_l C_{pl}(T_i - T_d)}{C_{pc}(T_d - t_i)} \tag{9.1}$$

where $\dot{M}_c, \dot{M}_v, \dot{M}_l$ = flow rate of vapor, liquid condensate, and coolant, respectively, lb/hr
C_{pl}, C_{pc} = heat capacity of condensate and coolant, Btu/lb·°F
T_i, T_d = inlet temperature of vapor and discharge temperature, °F
t_i = inlet temperature of coolant, °F
H_v = latent enthalpy of vaporization of condensate, Btu/lb

The heat is then given by:

$$Q = \dot{M}_c C_{pc}(T_d - t_i) = \dot{M}_v H_v + \dot{M}_l C_{pl}(T_i - T_d) \tag{9.2}$$

Use of a contact condenser requires consideration of coolant availability, liquid waste disposal, or treatment facilities. Contact condensers are relatively efficient scrubbers as well as condensers and have the lowest equipment cost. Contaminants that are not ordinarily condensable or even particulates can be removed from the vapor stream. Reputable manufacturers of the various types of contact condensers should be consulted for sizing and layout recommendations.

9.3.2 Surface Condensers

The heat transfer for a surface condenser is governed by the following relationship [3]:

$$Q = U_o A T_m \text{ or } A = \frac{Q}{U_o T_m} \tag{9.3}$$

where
Q = total heat load, Btu/hr
U_o = overall heat-transfer coefficient, Btu/hr·°F·ft^2
T_m = mean temperature difference, °F
A = surface area, ft^2

The first determination is the heat load. In the simplest case, it is the latent heat to condense plus the subcooling load. If a condensing temperature range exceeding 10 to 20 °F exists, gas cooling or desuperheating and intermediate liquid subcooling must be accounted for. The next step is specifying the coolant flow rate and temperatures to balance the vapor heat load. Temperature crosses where the outlet coolant temperature is higher than the vent and condensate outlet temperature should be avoided. Temperature crosses can be performed in condensers, but they exclude various flow arrangements that could improve efficiency and cost.

After the terminal temperatures of the vapor and coolant have been determined, a mean temperature difference can be established. For isothermal or linear condensing and liquid subcooling, the logarithmic mean temperature difference (LMTD) is applicable. The LMTD is calculated as follows:

$$\text{LMTD} = \frac{t_g - t_l}{\ln\left(\frac{t_g}{t_l}\right)} \tag{9.4}$$

where t_g = greater temperature difference between the hot and cold streams
t_l = lesser temperature difference between the hot and cold streams
ln = natural log

Fluids are usually in counterflow, but in some cases, as with a fluid near its freezing point, parallel flow is used to afford a greater degree of safety from freezing the condenser. Frequently, there is a combination of parallel and counter flow as in a two-pass tube side, single-pass shell tubular condenser. The procedure is to apply a correction to the LMTD. The correction factors have been reduced to a series of graphs available in most heat-transfer texts [2]. Strictly, the correction factor applies only to the subcooling or desuperheating zones, but it is sometimes applied to condensers.

If condensation is nonlinear and has a temperature range exceeding 20 °F, the LMTD is not applicable and a condensing curve or calculations at several points over the temperature range is required to determine a correct T_m. A mixture of miscible vapors or a vapor and noncondensables are examples where the LMTD is not applicable.

The remaining unknown to calculate the required surface is the overall heat-transfer coefficient, U_o. The overall heat-transfer coefficient is the inverse of the sum of the resistances to heat transfer and is given by the following relationship [3]:

$$U_o = \frac{1}{\frac{1}{h_o} + \frac{1}{h_{io}} + r_w + f_o + f_{io}} \tag{9.5}$$

where
h_o = outside or shell-side coefficient, Btu/hr·°F·ft^2
h_{io} = inside or tube-side coefficient corrected to the outside area
r_w = resistance of the surface area or tube wall, hr·°F·ft^2/Btu
f_o = fouling resistance on outside or shell side of surface area
f_{io} = fouling resistance on inside corrected to outside of surface area

Individual tube-side coefficients or resistances are corrected to the outside area. This results in an effective reduction in the individual heat-transfer coefficient. Finned tubing can have an outside-to-inside surface area ratio from 3 to as much as 25. Correcting the inside heat-transfer coefficient to the outside area is performed as follows:

$$h_{io} = \frac{h_i A_i}{A_o} \tag{9.6}$$

where A_i and A_o are the inside and outside surface areas, ft^2

It becomes apparent, therefore, that the use of fin tubes will be most beneficial when there is a major difference in the inside and shell or outside heat-transfer coefficients and the higher coefficient is inside the tubes.

Finned tubes present a resistance to heat transfer not applicable to bare tubes. It results from a temperature gradient in the fin itself. This resistance, due to conduction through the fin, is a function of the fin geometry and material, and is known as fin efficiency.

Determination of heat-transfer coefficients requires an evaluation of the vapor–liquid equilibrium; the simplest case is isothermal or moderate condensing range. The primary resistance to heat transfer is the liquid film, and the Nusselt theory and appropriate equations as originally described by Kern [2] are applicable. The general equation is as follows:

$$h\left(\frac{\mu_f^2}{k_f^3\rho_f^2 g}\right)^{1/3} = 1.5\left(\frac{4G'}{\mu_f}\right)^{-\frac{1}{3}} \tag{9.7}$$

where
μ_f = viscosity of condensate film, lb/ft·hr
k_f = thermal conductivity of film, Btu/hr·ft·°F
ρ_f = density of condensate film, lb/ft^3
g = acceleration of gravity, ft/hr^2
G′ = condensate loading, lb/hr·linear ft

There are various semiempirical modifications to this theoretical value to differentiate between horizontal, vertical, inside tubes, and outside tubes. The literature should be consulted for the actual loading to be use [2–8].

If there is noncondensable gas present or a vapor mixture of miscible and immiscible components with a large condensing range, diffusion of the condensing vapor through another vapor or a gas must be considered. The presence of a noncondensable gas in a vapor stream reduces the film coefficient; the amount of reduction is related to the size of the gas cooling load relative to the total load. The greater the ratio of sensible heat to total, the closer the heat-transfer coefficient approaches that of cooling the gas only. The calculation procedure is essentially that originally established by Colburn and Houge [9]. It's a trial-and-error method performed for at least five points along the condensing curve.

The choice of a coolant will depend upon the particular plant and the efficiency required of the condenser. The most common coolant is the primary plant coolant, usually cooling tower or river water. It has been shown that the vapor outlet temperature is critical to the efficiency of a condenser. In some instances, use of a chilled brine or a boiling refrigerant can achieve collection efficiency that will be sufficient without additional control devices. It is not unusual to specify multiple stage condensing, where condensers, usually two, connected in series use cooling mediums with successively lower temperatures. For example, a condenser using cooling-tower water can be used prior to a unit using chilled water or brine—thereby achieving maximum efficiency while minimizing the use of chilled water.

Careful consideration should be given to the fouling resistances f_o and f_{io} applied to the condenser design. TEMA lists some typical values to be used for various streams. These values are guidelines only and attention should be given to the economics of lower and higher values. The purpose of adding a fouling resistance is to provide consistent performance without excessive cleaning. Theoretically, increased fouling factors will provide greater safety allowance. In practice, however, they can actually decrease the safety allowance. Consider, for example, a typical application where a maximum pressure drop and overall length is specified. The most efficient design will use the length and pressure drop to the limit specified. If the fouling resistance specified is so large that this is not possible, velocities of the respective streams will be lower than desirable and fouling will be greater than in a design using a lower value of fouling resistance. Fouling resistances used indiscriminately can actually be "self-fulfilling." The most effective design feature to minimize fouling is high velocity and efficient use of pressure drop.

The effect of the pressure of noncondensables in condensers is also often a concern. For example, most of the heat transfer in steam evaporators does not occur from pure steam but from vapor evolved in a preceding effect. This vapor usually contains inert gases from air leakage if the preceding effect was under vacuum, from air entrained or dissolved in the feed, or from gases liberated by decomposition reactions. To prevent these inerts from seriously impeding heat transfer, the gases should be channeled past the heating surface and vented from the system while the gas concentration is still quite low. The influence of inert gases on heat transfer is due partially to the effect on ΔT of lowering the partial pressure and hence the condensing temperature of the steam. The primary effect, however, results from the formation of an insulating blanket of gas at the heating surface through which the condensate must diffuse before it can condense. The latter effect can be treated as an added resistance or fouling factor.

9.4 Equipment Design Considerations

In terms of equipment design considerations, the preparation of a condenser or heat exchanger for installation begins upon receipt of the unit from manufacturer. Condensers are shipped domestically using skids for complete units and boxes or crates for bare tube bundles. Units are normally removed from trucks using a crane or forklift. Lifting devices should be attached to lugs provided for that purpose (i.e., for lifting of the complete unit as opposed to individual parts) or used with slings wrapped around the main shell. Shell supports are acceptable lugs for lifting, provided that the complete set of supports are used together; never use nozzles for attachment of lifting cables.

Upon receipt of the unit, the general condition should be noted to determine any damage sustained during transit. Any dents, cracks, or connections out-of-square should be reported to the manufacturer prior to attempting to install the unit. Flanged connections are blanked with plywood, Masonite, or equivalent covers, and threaded connections are blanked with suitable pipe plugs. These closures are to avoid entry of debris into the unit during shipping and handling and should remain in place until actual piping connections are made.

9.4.1 Clearance Provision

Sufficient clearance is required for at least inspection of unit or in-place maintenance. Inspection of condensers requires minimal clearances for the following: access to inspection parts if provided, removal of channel or bonnet covers, and inspection of tube sheets and tube-to-tube sheet joints. If removal of tubes or tube bundles in place is anticipated, provision should be provided in equipment layout. Actual clearance requirements can be determined from the condenser setting plan.

9.4.2 Foundations

Condensers must be supported on structures of sufficient rigidity to avoid imposing excessive strains on nozzles due to settling. Horizontal units with saddle type shell supports are normally supplied with slotted holes in one support to allow for expansion. Foundation bolts in these supports should be loose enough to allow movement.

9.4.3 Leveling

Condensers should be carefully leveled and squared to ensure proper drainage, venting, and alignment with piping. On occasion, condensers are purposely angled to facilitate venting and drainage, and alignment with piping becomes the prime concern.

9.4.4 Piping Considerations

The following guidelines for piping are necessary to avoid excessive strains, mechanical vibration, and access for regular inspection.

1. Sufficient support devices are required to prevent the weight of piping and fittings from being imposed on the condenser.
2. Piping should have sufficient expansion joints or bends to minimize expansion stresses.
3. Forcing alignment of piping should be avoided so that residual strains will not be imposed on nozzles.
4. If external forces and moments are unavoidable, their magnitude should be determined and made known to the manufacturer so that any necessary stress analyses can be performed.
5. Surge drums or sufficient length of piping to condenser should be provided to minimize pulsations and mechanical vibrations.
6. Valves and bypasses should be provided to permit inspection or maintenance in order to isolate the condenser during periods other than complete system shutdown.
7. Plugged drains and vents are typically provided at low and high points of shell-tube sides not otherwise drained or vented by nozzles. These connections are functional during startup, operation, and shutdown, and should be piped up for either continuous or periodic use and never left plugged.
8. Instrument connections are provided either on the condenser nozzles or in the piping close to the condenser. Pressure and temperature indicators should be installed to validate the initial performance of the condenser as well as to demonstrate the need to inspection or maintenance.

9.4.5 Operation

The maximum allowable working pressures and temperatures are indicated on the condenser's nameplate. These values must not be exceeded. Special precautions should be taken if any individual part of the condenser is designed for maximum temperature lower than the condenser as a whole. The most common example is some copper-alloy tubing with a maximum allowable temperature lower than the actual inlet gas temperature. This is done to compensate for the low strength levels of some brasses or other copper alloys at elevated temperatures. An adequate flow of the cooling medium must be maintained at all times.

Condensers are designed for a particular fluid throughput. Generally, a reasonable overload can be tolerated without causing damage. If operated at excessive flow rates, erosion or destructive vibration could result. Erosion could also occur at normally acceptable flow rates if other conditions, such as entrained liquids or particulates in a gas stream or abrasive solids in a liquid stream, are present. Evidence of erosion should be investigated to determine the cause. Vibration can be propagated by other than flow overloads (for example: improper design, fluid maldistribution, or corrosion/erosion of internal flow-directing devices such as baffles). Considerable study and research has been applied in earlier years to develop a reliable vibration analysis procedure to predict or correct damaging vibration. At this point, the developed correlations are considered "state of the art," yet most manufacturers have the capability of applying some type of vibration check when designing a condenser. Vibration can produce severe mechanical damage, and operation should not be continued when an audible vibration disturbance is evident.

9.4.6 Cleaning

Fouling of condensers is the deposition of foreign material on the interior and/or exterior of tubes. Evidence of fouling in operation is increased pressure drop and general decreasing performance. Fouling can be so severe that tubes are completely plugged, resulting in thermal stresses and subsequent mechanical damage of equipment.

The nature of the deposited foul determines the method of cleaning. Soft deposits can be removed by steam, hot water, various chemical solvents, or brushing. Plant experience can determine which method to use. Chemical cleaning should be performed by contractors specialized in the field who will consider the deposit to be removed and the materials on construction. If the cleaning method involves elevated temperature, considerations should be given to thermal stresses induced in the tubes; using steam to clean individual tubes can loosen the tube-to-tube sheet joints.

Mechanical methods of cleaning are useful for soft and hard deposits. There are numerous tools for cleaning tube interiors: brushes, scrapers, and various rotating cutter-type devices. The condenser manufacturer or suppliers of tube tools can be consulted in the selection of the correct tool for the particular deposit. When cutting or scraping deposits, care should be exercised to avoid damaging tubes.

Cleaning of tube exteriors is generally performed using chemicals, steam, or other suitable fluids. Mechanical cleaning is preferred but requires that the tubes be exposed, as in a typical air-cooled condenser, or capable of being exposed, as in a removable bundle shell-and-tube condenser. The layout pattern of the tubes must provide sufficient intersecting empty lanes between the tubes, as in square pitch. Mechanical cleaning of tube bundles, if necessary, requires the utmost care to avoid damaging tubes or fins.

9.4.7 Testing

Proper maintenance requires testing of a condenser to check the integrity of the following: tubes, tube-to-tube sheet joint, welds, and gasketed joints. The normal procedure consists of pressuring of the shell with water or air at the nameplate-specified test pressure and viewing the shell welds and the face of the tube sheet for leaks in the tube sheet joints or tubes. Water should be at ambient temperature to avoid false indications due to condensation. Pneumatic testing requires extra care because of the destructive nature of a rupture or explosion or fire hazards when residual flammable materials are present. Condensers of the straight-tube floating heat construction require a test gland to perform the test. Tube bundles without shells are tested by pressuring the tubes and viewing the length of the tubes and back face of the tube sheets.

Defective tubes can either be replaced or plugged. Replacing tubes requires special tools and equipment; the user should contact the manufacturer or a contractor qualified in repair. Plugging tubes, although a temporary solution, is acceptable provided that the percentage of the total number of tubes per tube pass to be plugged is not excessive. The type of plug to be used is a tapered one-piece or two-piece metal plug suitable for the tube material and inside diameter. Care should be exercised in seating plugs to avoid damaging the tube sheets. If a significant number of tube or tube joint failures are clustered in a given area of the tube layout, their location should be noted and reported to the manufacturer. A concentration of failures is usually caused by something other than corrosion (for example: impingement, erosion, or vibration).

9.4.8 Improving Operation and Performance

Within the constraints of the existing system, improving operation and performance refers to maintaining operation and original or consistent performance. There are several factors previously mentioned which are critical to the design and performance of a condenser: operating pressure, amount of noncondensable gases in the vapor stream, coolant temperature and flow rate, and mechanical soundness. Most condenser manufacturers will provide designs for alternate conditions as a guide to estimating the cost of improving efficiency via alternate coolant flow rates and temperatures as well as alternate configurations (e.g., vertical, horizontal, shell side, or tube side).

9.5 Applications

Three illustrative examples complement the presentation for this chapter concerned with condensers. The reader should note that there are additional illustrative examples available in the literature [2–8].

9.5.1 Heat-transfer Coefficient Value

List approximate numerical values of several gas/vapor heat-transfer coefficients. Solution: The solution is provided in Table 9.1.

9.5.2 Heat Transfer Requirement

Given 10,000 scfm (60 °F, 1 atm) of butanol-contaminated air heated from 300 °F to 1,500 °F, find the heat transfer required to bring about this change of temperature. The enthalpy values of air at 300 °F and 1,500 °F are 1,870 and 10,895 Btu/lbmol, respectively.

Solution: The molar flow rate is

$$\begin{aligned} \dot{n} &= 10{,}000/379 \\ &= 26.4 \text{ lbmol/min} \end{aligned}$$

Table 9.1 List approximate numerical values of several gas/vapor heat transfer coefficients.

Fluid; gas/vapor	h, Btu/(hr·ft^2·°F)
Steam	1,500
Steam, 10% gas	500–700
Steam, 20% gas	300–500
Steam, 40% gas	200–250
Pure light hydrocarbons by molecular weight	250
Mixed light hydrocarbons	150–200
Medium hydrocarbons	80–120
Medium hydrocarbons with steam	100–150

The heat load may now be calculated:

$$\dot{Q} = \dot{n}C_p(T_2 - T_1) = \dot{n}(H_2 - H_1)$$

Substituting, one obtains

$$\begin{aligned} Q &= (26.4)(10{,}895 - 1{,}870) \\ &= 238{,}260 \text{ Btu/min} \\ &= 14.30 \times 10^6 \text{ Btu/hr} \end{aligned}$$

9.5.3 Condenser Area Calculation

A heavy hydrocarbon with a heat capacity, $C_p = 0.55$ Btu/lb·°F, is being cooled in a double pipe heat exchanger from $T_1 = 210\,°F$ to $T_2 = 170\,°F$. The oil flows inside a tube at a rate of 8,000 lb/hr and the tube surface temperature is maintained at 60 °F. The overall heat-transfer coefficient U = 63 Btu / (hr·ft²·°F). Calculate the required heat transfer area, A, in ft².

Solution: First calculate the LMTD.

$$\Delta T_1 = 210 - 60 = 150\ °F$$
$$\Delta T_2 = 170 - 60 = 110\ °F$$

Apply Equation 9.4,

$$\Delta T_{LM} = \frac{\Delta T_1 - \Delta T_2}{\ln\left(\dfrac{\Delta T_1}{\Delta T_2}\right)} = \frac{150 - 110}{\ln\left(\dfrac{150}{110}\right)} = 129\ °F$$

Calculate the heat transferred or the duty. Apply Equation 9.2,

$$Q = \dot{M}_c C_{pc}\,\Delta T = (8{,}000)(0.55)(210 - 170) = 176{,}000 \text{ Btu/hr}$$

Write the describing equation for the area and substitute.

$$Q = UA\Delta T_{LM}$$
$$A = \frac{Q}{U\Delta T_{LM}} = \frac{176{,}000}{(63)(129)} = 21.65 \text{ ft}^2$$

References

1 D. Kern, *Process Heat Transfer*, McGraw-Hill, New York City, NY, 1950.
2 Y. A. Çengel and A. J. Ghajar, *Heat and Mass Transfer: Fundamentals & Applications*, McGraw-Hill, New York City, NY, 2011.
3 L. Theodore, *Heat Transfer Applications for the Practicing Engineer*, John Wiley & Sons, Hoboken, NJ, 2011.
4 L. Theodore, F. Ricci and T. VanVleet. *Thermodynamics for the Practicing Engineer*, John Wiley & Sons, Hoboken, NJ, 2007.
5 L. Theodore and F. Ricci, *Mass Transfer Operations for the Practicing Engineer*. McGraw-Hall, New York City, NY, 2010.

6 L. Theodore and R. Allen, *Air Pollution Control Equipment*, A Theodore Tutorial, Theodore Tutorials, East Williston, NY, 1993, (Originally Published by the APTI/USEOA. RTP, NC). 2009.
7 L. Theodore and J. Barden, *Mass Transfer Operations*, A Theodore Tutorial, Theodore Tutorials, East Williston, NY, 1998, (Originally Published by the USEPA/APTI, RTP, NC).
8 A. M. Flynn, T. Akashige and L. Theodore, *Kern's Process Heat Transfer*, 2nd edition, Scrivener-Wiley, Beverly, MA, 2019.
9 A. P. Colburn and O. A. Hougen, Design of cooler condensers for mixtures of vapors with noncondensing gases, *Ind. Eng. Chem.*, 26(11), New York City, NY, 1178–1182, 1934.

Suggested Reading

1 W. Strauss, *Industrial Gas Cleaning*, Pergamon Press, New York City, NY, 1966.
2 J. A. Danielson, *Air Pollution Engineering Manual*, EPA OPA Publication No. AP-40, Government Printing Office, Washington, D.C., 1973.
3 A. C. Stern, H. C. Wohlers, R. W. Boubel and W. P. Lowry, *Fundamental of Air Pollution*, Academic Press, New York City, NY, 277–280, 1973.
4 A. J. Buonicore and L. Theodore, *Industrial Control Equipment for Gaseous Pollutants*, Volume I, CRC Press, Inc., Boca Raton, FL, 1975.
5 L. Theodore and A.J. Buonicore, *Vapor Control by Condensation: Performance Equations and Design Procedures*, paper#75-23.2, Air Pollution Control Association Meeting, Boston, 1975.
6 W. Connery, N. Kafes, A.J. Bounicore and L. Theodore, *Evaluating Surface Condensers for Air Pollution Control Applications*, Proc. Third An. APCA-AICHE Conf. Energy Environ, 276–82, Pittsburgh, PA, 1975.
7 L. Theodore, F. Ricci and T. VanVleet, *Thermodynamics for the Practicing Engineer*, John Wiley & Sons, Hoboken, NJ, 2007.
8 L. Theodore and F. Ricci, *Mass Transfer Operations for the Practicing Engineer*, McGraw-Hall, New York City, NY, 2010.
9 L. Theodore and R. Allen, *Air Pollution Control Equipment*, A Theodore Tutorial, Theodore Tutorials, East Williston, NY, 1993, (Originally Published by the APTI/USEPA. RTP, NC). 2009.
10 L. Theodore, *Air Pollution Control Equipment Calculations*, John Wiley & Sons, Hoboken, NJ, 2008.
11 L. Theodore, *Chemical Engineering: The Essential Reference*, McGraw-Hill, New York City, NY, 2014.
12 L. Theodore, R. R. Dupon and K. Ganesan, *Unit Operations in Environmental Engineering*, John Wiley & Sons, Hoboken, NJ, 2017.
13 J. Richardson, R. R. Dupont, P. Farber, K. Ganesan and L. Theodore, *Air Pollution Control Equipment Revisited*, AWMA Conference, Calgary, CA, 2024.
14 J. Richardson and L. Theodore, *Optimizing Air Pollution Equipment Performance: Operation and Maintenance*, John Wiley & Sons, Hoboken, NJ, 2025.

Part III

Air Pollution Control Equipment for Particulates

Part III is concerned with particulate air pollution control equipment. Following a chapter concerned with basic principles and fundamentals, air pollution control equipment receive treatment. The devices include:

- Chapter 10: Basic Principles: Particulates
- Chapter 11: Gravity Settlers
- Chapter 12: Cyclones
- Chapter 13: Wet Scrubbers
- Chapter 14: Electrostatic Precipitators
- Chapter 15: Baghouses

Each of the chapters contains a description of the device, process design considerations, equipment design considerations, and three illustrative examples that are concerned with design aspects of the control equations. The chapters concludes with a "Suggested Reading" section.

10

Basic Principles: Particulates

Note: The material presented in this chapter primarily reviews fluid–particle dynamics, particle size, distribution, and size-efficiency calculations. The reader is referred to Chapter 5 (Basic Principles: Gases) for additional material that is applicable to particulate control equipment; topics of importance include conversion of units, physical and chemical properties, and the ideal gas law.

10.1 Introduction

The authors have employed several definitions for particulates: a small, discrete mass of solid or liquid matter; a fine liquid or a solid particle that is found in the air or emissions; any solid or liquid matter that is dispersed in a gas; or insoluble solid matter dispersed in a liquid so as to produce a heterogeneous mixture. Further, particulate matter 10 (PM_{10}) is defined as particulate matter with a diameter less than or equal to 10 micrometers (μm), while particulate matter 2.5 ($PM_{2.5}$) is particulate matter with a diameter less than or equal to 2.5 micrometers (μm).

It should be noted that particle size is the single most important characteristic that affects the behavior of a particle. The range in sizes of particles observed in practice is remarkable. Some of the droplets collected in the demisters of wet scrubbers and the solid particles collected in large-diameter cyclones are as large as raindrops. However, some of the particles created in high-temperature incinerators and metallurgical processes can consist of a few molecules clustered together. These particles cannot be seen by sensitive light microscopes because they are extremely small in size; these sizes approach those of individual gas molecules (which range from 0.2 to 1.0 nm). In fact, particles composed of a few molecules clustered together can exist in a stable form. Some of these industrial processes generate particles in the range of 10–100 nm. However, particles in this size range can grow and agglomerate to yield particles in the 100+ nm range [1, 2].

10.2 Fluid Particle Dynamics

Most industrial techniques used for the separation of particles from gases involve the relative motion of the two phases under the action of various external forces. The collection methods for particulate pollutants are based on the movement of solid particles (or liquid droplets) through a gas. The final objective is their removal in order to comply with applicable standards and regulations and/or their recovery for economic reasons. In order to accomplish this, the particle is subjected to external forces: forces large enough to separate the particle from the gas stream during its residence time in the control unit.

Whenever a difference in velocity exists between a particle and its surrounding fluid, the fluid will exert a resistive force on the particle. Either the fluid (gas) may be at rest with the particle moving through it, or the particle may be at rest with the gas flowing past it. It is generally immaterial which phase (solid or gas) is assumed to be at rest; it is the relative velocity between the two that is important. The resistive force exerted on the particle by the gas is called the *drag*.

In treating fluid flow through pipes, a friction factor term is used in many engineering calculations. An analogous factor, called the drag coefficient, is employed in drag force calculations for flow past particles. Details are available in the literature [3, 4].

If a particle is initially at rest in a stationary gas and is then set in motion by the application of a constant external force or forces, the resulting motion occurs in two stages. The first period involves acceleration, during which time the particle velocity increases from zero to some maximum velocity. The second stage occurs when the particle achieves this maximum velocity and remains constant. During the second stage, the particle is not accelerating, and the rate of change of velocity is therefore zero. The final, constant, and maximum velocity attained is defined as the terminal settling velocity of the particle. Most particles can be shown to reach their terminal settling velocity almost instantaneously [5].

Another important consideration involves the Cunningham correction factor, CCF [6]. At very low values of the Reynolds number, when particles approach sizes comparable to the mean free path of the fluid molecules, the medium can no longer be regarded as continuous. For this condition, particles can fall between the molecules at a faster rate than predicted by the aerodynamic theories that led to the previous standard drag coefficients. To allow for this "slip," Cunningham introduced a multiplying correction fact to Stokes' law (see Chapter 12). This can alter the equations used to describe flow in the laminar regime.

10.3 Particle Sizing

Unfortunately, the term "measurement methods" has come to mean different things to engineers and scientists in the instrumentation field. Traditionally, this term has referred to sampling techniques employed for source characterization purposes. Source characterization includes either the determination of particle size distribution, concentration and flow rate, or chemical composition and other properties, or both. The equipment and procedures employed to measure particle data in a stream (usually air) represents the sampling part of the overall procedure. The analysis of the data is also important, usually providing the aforementioned information on particle size distribution, concentration, and flow rate.

An accurate quantitative analysis of the discharge of particulates from a process must be determined prior to the design and/or selection of recovery/control equipment. If the unit is properly engineered, utilizing the emission data as input to the device, most particulates can be successfully controlled by one or a combination of the methods to be discussed later in this book.

There are many techniques available for measuring the particle size distribution of particulates. The wide size range covered, from nanometers to millimeters, cannot always be analyzed using a single measurement principle. Added to this are the usual constraints of capital costs versus operating costs, speed of operation, degree of skill required, and most important, the end-use requirement.

Physical sizing is traditionally one of the most common methods available for classifying (or sizing) particles. This has been achieved by dry or wet screening with sieves, microscopic analysis, electric-grating techniques, and light-scattering methods.

More realistic determinations of particle behavior in any environment must consider the size, shape, and density. The best technique to accomplish this is aerodynamic sizing. Only by knowing the aerodynamic size of particles is it possible to determine how they will behave in an air stream and the kind(s) of control equipment required to capture them.

Consider, for example, a ping-pong ball and a golf ball. On close examination, they will appear almost equal in size; however, if both were tossed into a moving air stream, they would behave quite differently. Even though the size and shape are similar, the density is quite different, and the behavior of the two objects is far from being similar. Since the density is quite different, the behavior of the objects is far from being similar aerodynamically. This is the primary fallacy in physical sizing.

A common method of specifying large particle sizes is to designate the screen mesh that has an aperture corresponding to the particle diameter. Since various screen scales are in use, confusion may result unless the screen scale involved is specified. The screen mesh generally refers to the number of screen openings per unit of length or area. The aperture for a given mesh will depend on the wire employed. The Tyler and the U.S. Standard Screen Scales (Table 10.1) are the most widely used in the United States. The screens are generally constructed of wire mesh cloth, with the diameters of the wire and the spacing of the wires being specified. These screens form the bottoms of metal pans about 8 inches in diameter and 2 inches high, whose sides are so fashioned that the bottom of one sieve nests snugly on the top of the next.

Table 10.1 Tyler and the U.S. Standard Screen Scales.

Tyler mesh	Aperture, microns	U.S. mesh	Aperture, microns
400	37	400	37
325	43	325	44
370	53	270	53
250	61	230	62
200	74	200	74
170	88	170	88
150	107	140	105
100	147	100	149
65	208	70	210
48	295	50	297
35	417	40	420
28	589	30	590
20	833	20	840
14	1,168	16	1,190
10	1,651	12	1,680
8	2,362	—	—
6	3,327	—	—
4	4,699	—	—
3	6,680	—	—

The clear space between the individual wires of the screen is termed the screen aperture. As indicated above, the term mesh is applied to the number of apertures per linear inch; for example, a 10-mesh screen will have 10 openings per inch, and the aperture will be 0.1 inch minus the diameter of the wire.

Particle size itself is difficult to define in terms that accurately represent the types of particles of interest. This difficulty stems from the fact that particles exist in a wide variety of shapes, not just spheres. In the case of spherical particles, the definition of particle size is easy: it is simply the diameter. For the irregularly shaped particles, there are a variety of ways to define the size. The most convenient of these is to directly relate it to how the particle behaves in a fluid such as air, and the particle size definition that is most useful for evaluating particle motion in a fluid is termed the *aerodynamic diameter*. It takes into account the particle density as shown in Equation 10.1:

$$d_{pa} = d_p \sqrt{\rho_p C_c} \tag{10.1}$$

where

d_{pa} = aerodynamic diameter, μm
d_p = actual diameter, μm
ρ_p = particle density, g/cm^3
C_c = Cunningham correction factor (discussed in more detail later), dimensionless

Note that particles that appear to be different in physical size and shape can have the same aerodynamic diameter. Conversely, some particles that appear to be visually similar can have somewhat different aerodynamic diameters. Thus, the term aerodynamic diameter is useful in describing all particles, including fibers and particle clusters. Although it is not a true size because nonspherical particles require more than one dimension to characterize their size, the aerodynamic diameter provides a simple means of categorizing the sizes of particles with a single dimension and in a way that relates to how particles move in a fluid [7].

10.4 Particle Size Distribution

Particle emissions from both human-made (synthetic) and natural sources do not consist of particles of any one size. Instead, they are composed of particles over a relatively wide size range. It is often necessary to describe this size range. The authors have defined particle size analysis as a determination of the particle size distribution of a sample, and particle size distribution (PSD) as an equation, graph, or table that quantifies, usually by percent, the various sizes of particulates found in a sample.

Particle size distributions are often characterized by a *mean* particle diameter. Although numerous *means* have been defined in the literature, the most common are the arithmetic mean and the geometric mean. The arithmetic mean diameter is simply the sum of the diameters of each of the particles divided by the number of particles measured. The geometric mean diameter is the nth root of the product of the n number of particles in the sample. In addition to the arithmetic and geometric means, a particle size distribution may also be characterized by the "median" diameter. The median diameter is that diameter for which 50% of the particles are larger in size and 50% are smaller in size. Another important characteristic is the measure of central tendency. It is sometimes referred to as the dispersion or variability. The most common term employed is the *standard deviation* [8, 9]. These terms are discussed later in this chapter.

There are various methods for expressing the results of particle size measurements; the most common method is by plotting either (1) a frequency distribution curve or (2) a cumulative distribution curve. These topics are considered in this section.

Ultimately, the decision of how to represent particle size distribution information is left to the user and/or practitioner. Considerations include the following:

1. Choosing an approach
2. Checking on the reasonableness of the choice
3. Ability to draw the proper conclusions from the approach selected

A typical particulate size analysis method of representation employed in the past is provided in Table 10.2. These numbers mean that 40% of the particles by weight are greater than 5 nm in size, 27% are less the 5 nm but greater than 2.5 nm, 20% are less than 2.5 nm but greater than 1.5 nm, and the remainder (13%) are less than 1.5 nm. Another form of representing data is provided in Table 10.3.

One basic way of summarizing data is by the computation of a central value. The most commonly used central value statistic is the aforementioned arithmetic average, or the mean. This statistic is particularly useful when applied to a set of data having a fairly symmetrical distribution. The mean is an important statistic in that it summarizes all the data in the set and because each piece of data is taken into account in its computation. The formula for computing the mean is:

$$\overline{X} = \frac{X_1 + X_2 + X_3 + \cdots + X_n}{n} = \frac{(\sum_{i=1}^{n} X_i)}{n} \tag{10.2}$$

Where

$\overline{X}$ = arithmetic mean

X_i = any individual measurement

n = total number of observations

X_1, X_2, X_3 = measurements 1, 2, and 3, respectively

Table 10.2 Early PSD representation.

—	>5.0	μm	40%
5	>2.5	μm	27%
<2.5	>1.5	μm	20%
<1.5	—	μm	13%

Table 10.3 Size ranges in arithmetic increments.

Size range, μm	Percent in size range, %
0–2	10
2–4	15
4–6	30
6–8	30
8–10	10
>10	5

The most commonly used measure of dispersion, or variability, of sets of data is the standard deviation, s. Its defining formula is given by expression [8, 9]:

$$s = \sqrt{\sum \frac{(X_i - \overline{X})^2}{n - 1}} \tag{10.3}$$

where
s = standard deviation (always positive)
X_1 = value of the ith data point
$\overline{X}$ = mean of the data sample
n = number of observations

(Note: The above term s represents the sample standard deviation. In statistical circles, the symbol σ represents another, but near similar, standard deviation. Unfortunately, σ has been employed in the air pollution literature to represent the sample standard deviation. For this reason, the symbol σ is used in this text to represent the sample standard deviation. Refer to the literature for additional details [8, 9].)

Frequency distribution curves are usually plotted on regular coordinate (linear) paper. The curve describes the amount of material (particles) contained within each size range. When gas-borne particles produced in a given operation are measured, the data generally have a tendency to show a preferential (maximum amount of) particle size. A plot of percent mass versus particle size (d_p) on a linear scale gives a curve with a peak at the preferential size. Such a curve is shown in Figure 10.1. This figure shows a normal probability distribution (see also later discussion) that is symmetric about the preferential size. This curve is rarely encountered for particulates consisting of very fine particles. However, this curve may occasionally be approached for particles such as fumes formed by vapor phase reaction and condensation, or for tar and acid mists.

Particle size data can also be plotted as a cumulative plot. Particle size for each size range is plotted on the ordinate. The cumulative percent by weight (frequency) is plotted on the abscissa. The cumulative percent by weight can be given as cumulative percent less than stated particle size (%LTSS) or cumulative percent greater than stated particle size (%GTSS). The cumulative percent

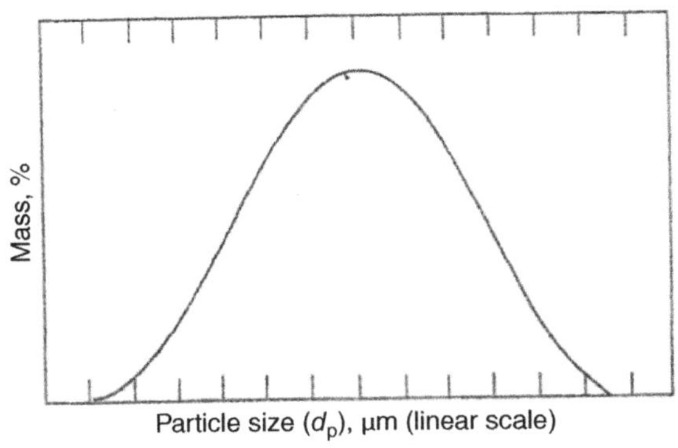

Figure 10.1 Particle size distribution.

by weight can be plotted on either a linear percentage or a probability percentage scale. The particle size range (ordinate) is usually a logarithmic scale.

Frequently, the cumulative distribution is plotted on special coordinate paper called log-probability paper. The particle size of each size range is plotted on a logarithmic ordinate. The percentage by weight larger than d_p is plotted on the probability scale as the abscissa. If the distribution is lognormal (see later discussion), the distribution curve plots out as a straight line. It should be noted that one can just as easily plot percent mass less than d_p (%LTSS) on the abscissa. In fact, %LTSS is preferred because one of the author's initials is carved in the acronym.

When measuring particulate emissions from industrial and/or human-made sources, the graph of the particle size distribution often displays the logarithmic variation of the normal distribution. The normal distribution has a fundamental defect related to its use in particle sizing analysis; implicit in the statement that a random variable is normally distributed is the concept that the values of particle size are at equal distances from the central tendency or preferential size. Suppose that the mean particle size or central tendency of the distribution were 20 μm. It would be equally probable to find either a 15- or 25-μm particle. One might also find a particle the size of 50 μm in the distribution. If the distribution were normal, it would be equally likely to find a particle the size of minus 10 μm (not physically possible). However, if it can be assumed that the logarithm of the particle size is randomly distributed, then this problem is avoided. The ratios of particle size about a central tendency are equally probable, and the ratios are then bounded on the lower end by zero.

The usefulness of the lognormal distribution is more evident when the frequency distribution curve is characteristically skewed. The data can then be plotted as a cumulative plot on log-probability paper. If the distribution follows the lognormal relationship, then the plot will result in a straight line. The linearity of the relationship allows one to describe the distribution statistically with a minimum of two parameters: the geometric mean, d_{gm} and the geometric standard deviation σ_{gm}. The geometric mean value of a lognormal distribution can be read directly from a plot similar to that represented in Figure 10.2. The geometric mean size is the 50% size on the plot.

As discussed previously, the geometric standard deviation is a good measure of the dispersion or spread of a distribution. The geometric standard deviation is the root mean square deviation about the mean value and can be read directly from a plot such as shown in Figure 10.2. For a lognormal distribution (which plots d_p maximum versus percent mass larger than d_p), the geometric standard deviation is given by:

$$\sigma = \sigma_{gm} = \frac{50\% \text{ size}}{84.13\% \text{ size}} \tag{10.4}$$

or

$$\sigma = \sigma_{gm} = \frac{15.87\% \text{ size}}{50\% \text{ size}} \tag{10.5}$$

All one must do is determine the 50% size and the 84.13% size from the plot and divide to determine the geometric standard deviation. Therefore, knowing any two lognormal combination of these parameters (15.87%, 50%, 84.13%, or σ_{gm}) describes the entire range of lognormal particle size distributions.

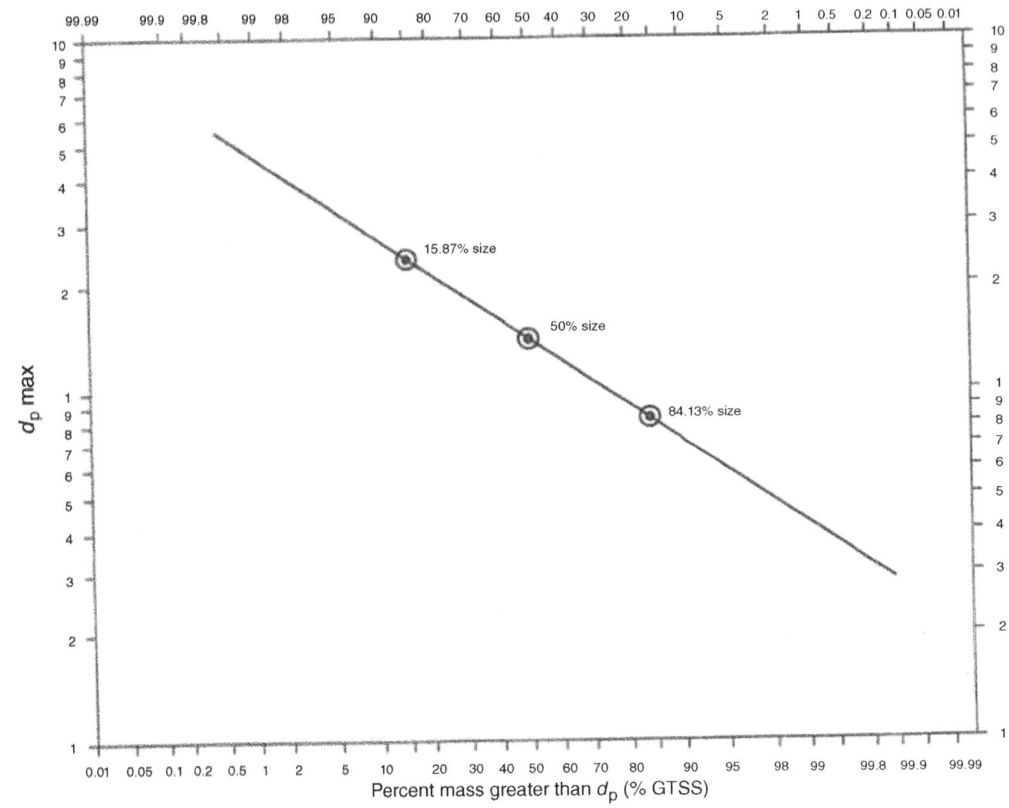

Figure 10.2 Geometric mean and standard deviation from a lognormal distribution plot.

10.5 Particle Collection Efficiency

Efficiency is the other characteristic quantity that warrants further discussion. The efficiency of a particulate control device is usually expressed as the percentage of mass collected by the unit compared with that entering the unit. It may be calculated on a particle number basis:

$$E_N = \left(\frac{\text{particles collected}}{\text{particles entering}}\right) \times 100 \tag{10.6}$$

Or on total mass basis:

$$E = \left(\frac{((\text{inlet loading}) - (\text{outlet loading}))}{\text{inlet loading}}\right) \times 100 \tag{10.7}$$

It is extremely important to distinguish between the two. Larger particles, which possess greater mass and are more easily removed in a control device, will contribute more significantly to the efficiency calculated on a mass or weight basis. Thus, if one considers a volume of aerosol which contains 100 1-μm particles and 100 100-μm particles, and if the efficiency of separation is 90% for 1 μm particles and 99% for 100 μm particles, then on a particle count basis, 90 1-μm and 99 100-μm particles will be removed out of a total of 200. This gives a particle count efficiency of:

$$E_N = \left(\frac{189}{200}\right) \times 100 = 94.5\%$$

On a weight basis, however, if a 1 μm particle has unit mass, a 100 μm particle has 10^6 mass unit. The weight efficiency is then given by

$$E = \left(\frac{90(1) + 99(10^6)}{100(1) + 100(10^6)} \right) \times 100 = 99\%$$

Any expression of the efficiency of a particulate removal device is therefore of little value without a careful description of the size spectrum of particles involved.

It is interesting to note certain observations. The smaller particles, equivalent in mass to a considerably smaller number of large particles, have a much greater impact on visibility, health, and water droplet nucleation than the larger particles. When large tonnages (significant mass discharges) are involved, the high mass efficiencies often reported for particle collection may lead to overly optimistic conclusions about emissions. The small weight percentages of particles that pass through the collector can still represent large numbers of particles escaping to the atmosphere. It would then seem that tonnage collection figures and weight removal efficiencies may not really be that adequate to delineate the entire particulate emission problem.

As noted earlier, the capture efficiency is a strong function of particle size. Combining this size-efficiency material with the particle size distribution information in Section 10.4 allows one to obtain and/or calculate the overall efficiency of a unit. In practice, this is treated analytically by methods other than the somewhat simple example given above. A more detailed presentation is given in the literature [1, 2, 7].

10.6 Particle Collection Mechanisms

The overall collection/removal process for particulates in a fluid essentially consists of four steps [10]:

1. An external force (or forces) must be applied that enables the particle to develop a velocity that will displace and/or direct it to a collection or retrieval section or area.
2. The particle should be retained at this area with strong enough forces so that it is not re-entrained.
3. As collected/recovered particles accumulate, they are subsequently removed.
4. The ultimate disposition of the particles completes the process.

Obviously, the first is the most important step. The particle collection mechanisms discussed below are generally applicable when the fluid is air; however, they may also apply if the fluid is water. In most commercial particulate systems, it is necessary to remove the particles from the gas stream and collect them in layers, dust cakes, or other forms. In the case of electrostatic precipitators (see Chapter 14), the dust is collected on a layer on a vertical collection plate. In the case of fabric filters (see Chapter 15), the particles collect as dust cakes on vertical bags. This is the initial collection step, and most particulate devices use one or a combination of collection mechanisms to accomplish this objective.

The forces listed below are basically the "tools" that can be used for particulate/recovery collection:

1. Gravity settling
2. Centrifugal action
3. Inertial impaction and interception
4. Electrostatic attraction
5. Brownian motion

Note that all of these collection mechanism forces are strongly dependent on particle size. Each mechanism is briefly described below [1, 2, 7].

As indicated above, all particulate devices collect or recover particles by a variety of mechanisms involving an applied force—the simplest of which is gravity. Large particles moving slowly (relatively speaking) in a gas stream can settle out and be collected or recovered. This mechanism is responsible for particle capture in the simplest of all devices—the settling chamber. The distance (d) that the particle travels while settling (traveling) with velocity (v) in time (t) is given by:

$$d = vt; \text{ consistent units} \tag{10.8}$$

Details of gravity forces can be found in Chapter 11.

Centrifugal force is another collection mechanism used for particle capture. The shape or curvature of the collector causes the gas stream to rotate in a spiral motion. Large particles move toward the outside of the wall by virtue of their momentum. The particles lose kinetic energy there and are separated from the gas stream. Particles are then acted on by gravitational forces and are collected. Thus, centrifugal and gravitational forces are both responsible for particle collection in a cyclone. See Chapter 12 for more details.

Inertial impaction occurs when an object (e.g., a fiber or liquid droplet), placed in the path of a particulate-laden gas stream, causes the gas to diverge and flow around it. Larger particles, however, tend to continue in a straight path because of their inertia; they may impinge on the obstacle and be collected (as in Figure 10.3). Since the trajectories of particle centers can be calculated, it is possible to theoretically determine the probability of collision.

Direct interception also depends on inertia and is merely a secondary form of impaction. Note that the trajectory of a particle's center can be calculated; however, even though the center may bypass the target object, a collision might occur since the particle has finite size (Figure 10.4). A collision occurs due to direct interception if the particle's center misses the target object by some dimension less than the particle's radius. Direct interception is, therefore, not a separate principle, but only an extension of inertial impaction.

Inertial impaction is analogous to a small car riding down an interstate highway at 65 mph (mi/hr) and approaching a merge lane where a slow-moving large truck has entered. If the car is not able to get into the passing lane to go around the merged truck, there could be an "impaction" incident. The faster the car is going, the more probable the impaction. The larger the car, the more difficulty it will have going around the truck. Conversely, if the truck enters the interstate road at the same speed as the car, the likelihood of impaction is reduced.

Another primary particle collection mechanism involves *electrostatic forces*. The particles can be naturally charged, or, as in most cases involving electrostatic attraction, be charged by subjecting

Figure 10.3 Inertial impaction.

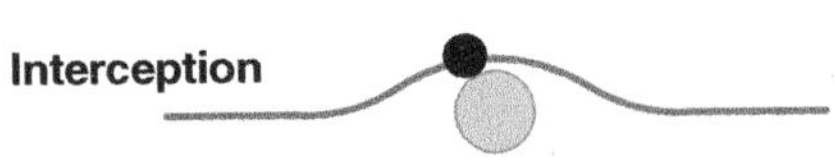

Figure 10.4 Direct interception.

Figure 10.5 Particle capture mechanism for diffusion.

the particle to a strong electric field. The charged particles migrate to an oppositely charged collection surface. This is the collection mechanism responsible for particle capture in both electrostatic precipitators and charged droplet scrubbers. In an electrostatic precipitator, particle collection occurs as a result of electrostatic forces only. The electrostatic force F_E experienced by a charged particle in an electric field is given by

$$F_E = qE_p; \text{ consistent units} \tag{10.9}$$

where
q = particle charge (not to be confused with volume flowrate)
E_p = the collection field intensity (electric field)

Another mechanism that is important for particles is *molecular diffusion*. Many air pollution control technologists consider that diffusion is immaterial for particles with diameters larger than μm which is the upper limit for the Brownian motion range. In this situation, the particles are so small that their individual motion can be affected by collisions on a molecular or atomic level. Collection of these fine particulates is a result of random motion.

Very small particles deflect slightly when they are struck by gas molecules. The deflection is caused by the transfer of kinetic energy from the rapidly moving gas molecule to the small particle. As the particle size, mass, and density increase, the extent of the particle movement decreases. Thus, deflection begins to be effective as a capture mechanism for particles less than approximately 1.0 μm, and it is significant for particles less than 0.5 μm. Because of the bombardment by fluid molecules, particles suspended in a fluid will be subjected to a random and chaotic molecular motion defined as *Brownian motion*. This movement arises in addition to any net motion in a given direction due to the action of other external forces such as gravity. Thus, this effect becomes the predominant collection mechanism for particles in the 0.1–1.0 μm range (see Figure 10.5).

Other factors affecting collection mechanisms include the following:

1. Nonspherical particles
2. Wall effects
3. Multiparticle effects
4. Multidimensional flow

Details are available in the literature [7, 11].

References

1 L. Theodore and R. Allen, *Air Pollution Control Equipment*, A Theodore Tutorial. Theodore Tutorials (originally published by the USEPA/APTI, RTP, NC), East Williston, NY, 1993.

2 L. Theodore, *Air Pollution Control Equipment Calculations*, John Wiley & Sons, Hoboken, NJ, 2008.

3 P. Abulencia and L. Theodore, *Fluid Flow for the Practicing Chemical Engineer*, John Wiley & Sons, Hoboken, NJ, 2009.

4 I. Farag, *Fluid Flow*, A Theodore Tutorial. East Williston, NY: Theodore Tutorials (originally published by USEPA/APTI, RTP, NC, 1995).

5 L. Theodore, *Personal notes*, East Williston, NY, 1966.

6 C. E. Cunningham, *On the velocity of steady fall of spherical particles through fluid medium*, *Proc. Roy. Soc. London Sec. A*. 83, 357, 1910.

7 J. Richardson and L. Theodore, *Optimizing Air Pollution Equipment Performance: Operation and Maintenance*, John Wiley & Sons, Hoboken, NJ, 2025.

8 S. Shaefer and L. Theodore, *Probability and Statistics for Environmental Engineers and Scientists*. CRC Press/Taylor & Francis Group, Boca Raton, FL, 2007.

9 F. Taylor and L. Theodore, *Probability and Statistics*, A Theodore Tutorial, Theodore Tutorials, East Williston, NY, 1992.

10 L. Theodore, *Personal notes*, East Williston, NY, 1970.

11 L. Theodore, *Nanotechnology: Basic Calculations for Engineers and Scientists*, John Wiley and Sons, Hoboken, NJ, 2005.

11

Gravity Settlers

11.1 Introduction

Gravity settlers, or gravity settling chambers, have long been utilized industrially for the removal of solid and liquid waste materials from gaseous streams. Advantages for their use are simple construction, low initial cost and maintenance, low-pressure losses, and simple disposal of waste materials.

The gravity settler was one of the first devices used by man to control particulate emissions. The device basically consists of an expansion chamber in which the particle velocity is reduced to such an extent that the particle may settle out under the action of gravity. One advantage of this device is that the external force leading to separation is provided free by nature. Its use in industry is generally limited to the removal of the larger-sized particles (greater than 40 μm) since the chamber required to remove small particles is extremely large. For example, the area of a chamber required to remove dust from the exhaust gases of a cement kiln burning fuel oil at the rate of 350 gal/hr would be at least of the following order of magnitude: for complete elimination of particles greater than 10 μm: 6,000,000 ft^2; for complete elimination of particles greater than 50 μm: 250,000 ft^2.

Today's demands for cleaner air and stricter emission standards have all but forced this low-efficiency device off the market to be replaced mostly by cyclones (Chapter 12), wet scrubbers (Chapter 13), electrostatic precipitators (Chapter 14), and baghouses (Chapter 15). However, gravity settlers still find use in certain areas, primarily in testing and research or as a precleaner. Settlers can be designed to allow only a certain range of particles to pass through the unit, and these are used in some air sampling tests. They are also useful in studying the size and flow of aerosol particles, knowledge of which can be applied to the re-efficient design of other pollution control devices.

11.2 Description of Gravity Settler Equipment

Gravity settlers are usually constructed in the form of a long, horizontal parallelepiped with suitable inlet and outlet ports. In its simplest form, the settler is an enlargement (large box) in the duct carrying the particle-laden gases; the contaminated gas stream enters at one end, while the cleaned gas exits from the other end (see Figure 11.1). Expansion into the large chamber reduces the velocity of the carrier gas so that particles in the gas settle out by gravity. A typical horizontal-flow gravity settling unit is presented in Figure 11.2. Several thin horizontal collection plates have

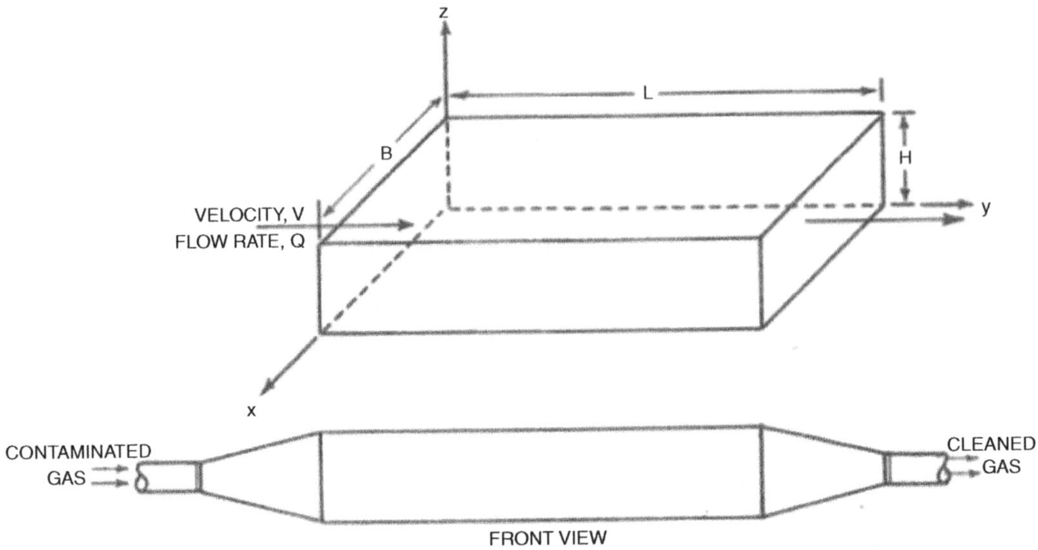

Figure 11.1 Gravity settling box.

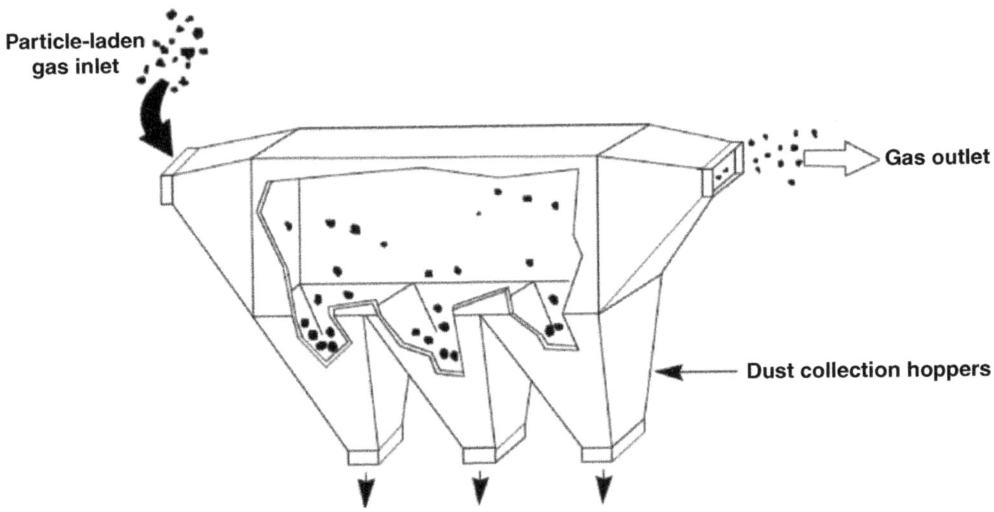

Figure 11.2 Horizontal flow settling chamber.

been introduced to reduce the excessive volume requirements for the collection of small particles (as small as 15 μm). Although the gas velocity increases slightly, the vertical distance for collection decreases more significantly; therefore, the overall collection efficiency increases. The Howard settling chamber is an example of this type of unit, as shown in Figure 11.3. A unit consisting of n shelves will essentially consist of n-1 gravity settlers, each with a height equal to the distance between two adjacent shelves. The vertical distance between shelves may be as little as 1 inch. The gas must be uniformly distributed laterally upon entering the chamber; vertical distribution is also critical. Uniform distribution is usually achieved by the use of gradual transitions, guide vanes, distributor screens, or perforated plates. Some of the disadvantages of the multiple tray settling

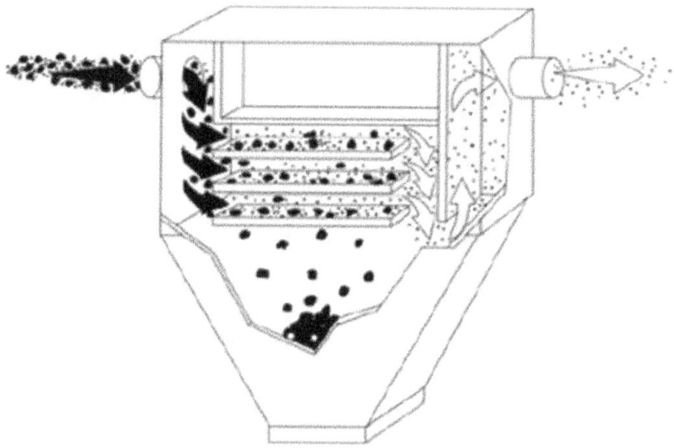

Figure 11.3 Howard settling chamber (multiple tray).

chambers include the cleaning difficulties arising from closely spaced trays, the tendency of the trays to warp during high-temperature operation, and the tendency of the system to be unable to handle dust concentrations exceeding ~1 grain/ft^3.

Since the effective settling rate of the dust decreases with increasing gas turbulence, the velocity of the gas stream in the settling chamber is normally kept as low as possible. For practical purposes, the velocity must not be so high that settled particles are re-entrained or so low that chamber volume becomes excessive. Gas velocities usually range from 1 to 10 ft/sec. In actual practice, the gravitational settling velocities used in design must be based on experience or tests conducted under actual conditions because the terminal settling velocity may be influenced by factors such as agglomeration and electrostatic charge. Pressure drops experienced in settling chambers are quite low, generally less than 0.2 inch of water.

An elutriator is a slight modification of the gravity settler. The unit consists of one or more vertical tubes or towers through which the dust-laden gas passes upward at a given velocity. The large particles that settle at a velocity higher than that of the rising air are collected at the bottom of the tube, while the smaller particles are carried out the top. In order to vary the air velocity, several columns of different diameters used in series to bring about more refined separation (see Figure 11.4).

Other control devices that can be considered under the above broad classification depend, to some extent, on another effect (in addition to gravity) to lead to successful separation process. This other mechanism is an inertial or momentum effect. It arises by changing the direction of the velocity of the gas and imparting a downward motion to the particle. From a calculational point of view, this induced particle motion is superimposed on the motion arising due to gravity. Momentum separators include another effect, in addition to gravity, to enhance particle separation and collection. This is accomplished by changing the direction of velocity of the gas stream so as to impart an additional downward velocity to the particle. Particles as small as 10–20 μm can be collected. The physical arrangement usually involves the insertion of baffles into a gravity settler system. Examples of these devices are shown in Figure 11.5 (a) through (c). They are more compact and require less space than the simple gravity settler; however, pressure losses are slightly higher (ranging from 0.1 to 1.0 inch of water).

When a moving dust-laden gas stream approaches a body, such as a spherical droplet of water, the gas will be deflected around the droplet. The dust particles, by virtue of their greater

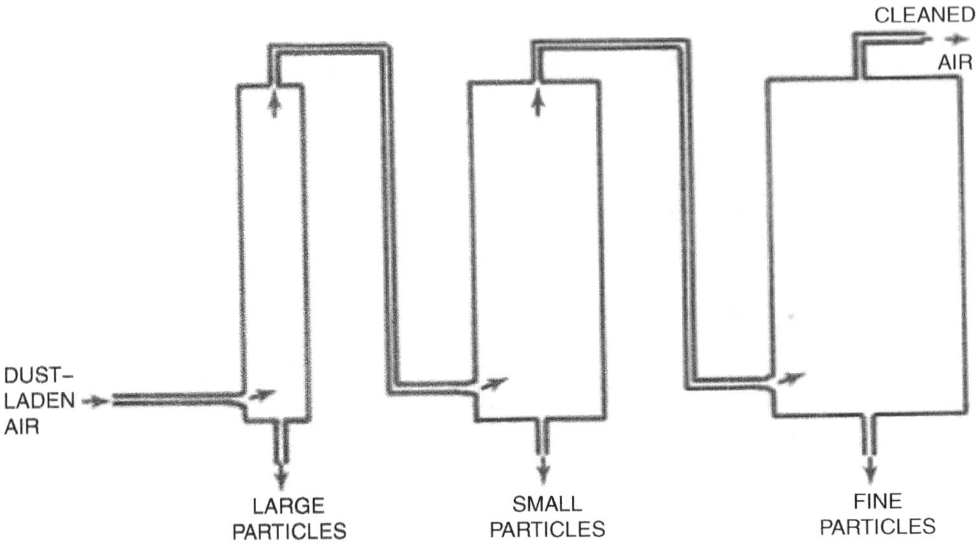

Figure 11.4 Elutriators in series.

inertia, may impact and be collected on the surface of the droplet which is falling under the influence of gravity. The design and operation of the gravity spray tower utilizes this collection phenomenon. The unit consists of a vertical tower or column. The particulate-laden gas is fed to the bottom of the column. A liquid stream (usually water) is introduced at the top of the column after passing through spray nozzles. The water droplets produced fall vertically downward through the column under the influence of gravity. During their fall, the droplets contact and capture the particles in the gas stream. A schematic diagram of this type of unit is shown in Figure 11.6. The system is capable of treating larger volumetric gas flow rates and has few mechanical problems. It operates at low-pressure drops (usually less than 1 inch of water) with air velocities in the neighborhood of 2–5 ft/sec and residence times of approximately 15–30 sec. In addition to the collection of particles (down to the 10–20 μm range), the gravity spray tower can also be utilized to absorb noxious pollutant gases (depending on the liquid reagent used and solubility characteristics) and is often employed as a precooler. It is capable of handling relatively high dust loadings (greater than 5 grains/ft^3). The main disadvantage centers on the liquid effluent treatment and/or waste disposal, as the air pollution problem is replaced by a water pollution problem. Water usage typically ranges between 5 and 20 gal/1,000 ft^3 gas treated (with a typical value around 18 gal/1,000 ft^3). Another disadvantage is the possibility of liquid entrainment in the gas stream leaving the tower; however, a demister, mist eliminator, or entrainment separator installed above the spray nozzles (especially when gas velocities are greater than 6 ft/sec) will often eliminate this liquid entrainment problem. Similar to the gravity settlers, gravity spray towers have rather large space requirements [1].

11.3 Process Design Considerations

The principles and describing equations governing particle collection in a gravity settler are now presented. Although the settler is initially assumed to contain no trays, the analysis can easily be extended to a multiple-tray unit since the capture height between the trays can be treated as

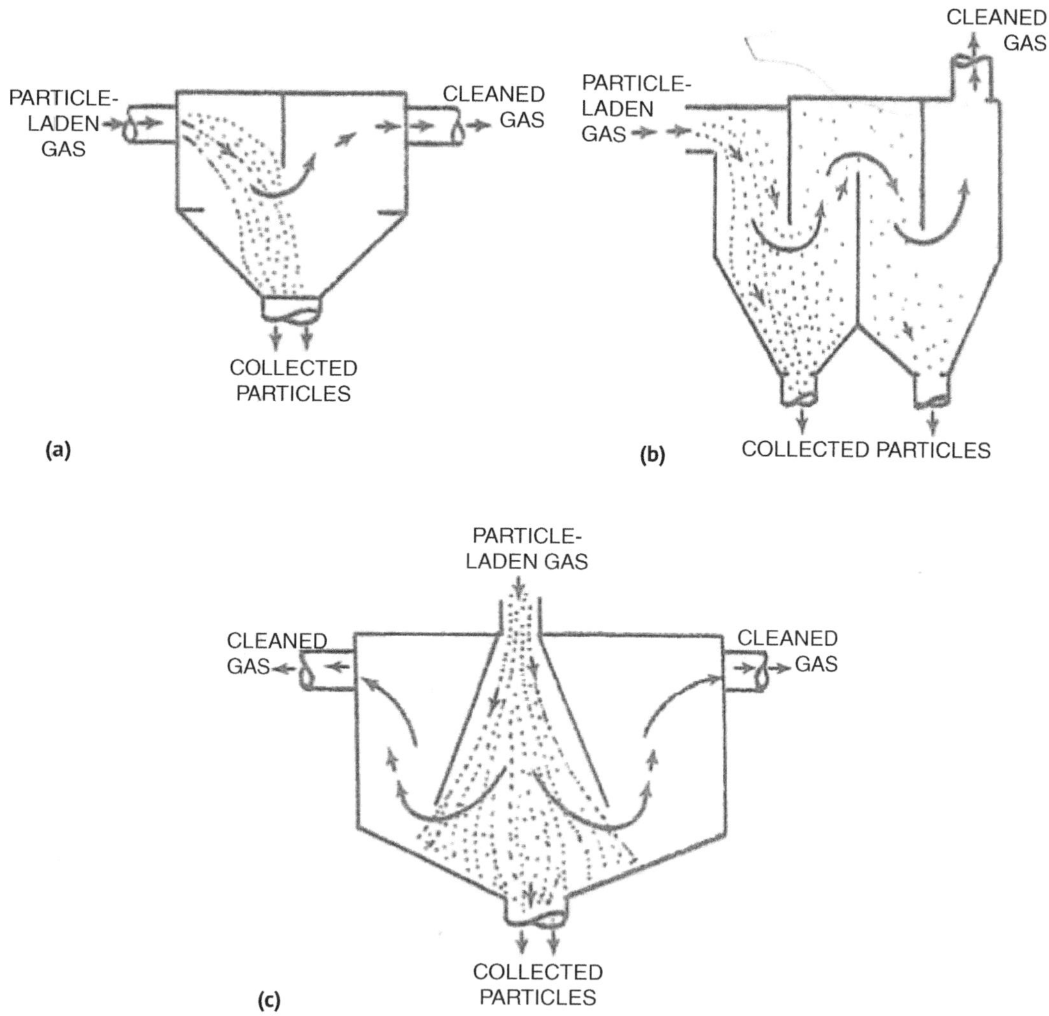

Figure 11.5 Momentum separators.

an individual collector. In addition to operating conditions and physical characteristics, this section will discuss the effect of Reynolds number on particle drag and particle size distribution. Re-entrainment effects are also briefly considered.

The analysis begins by examining the behavior of a single spherical particle in a settler in which the bulk flow air velocity profile is plug, i.e., the gas flow over the whole chamber is uniform (see Figure 11.7). The particle is assumed to be located at the top of the unit—the most difficult inlet (or initial) position for particle capture. At the inlet condition, it is assumed that the vertical velocity component of the particle is at its terminal settling velocity, i.e., the particle requires a negligible period of time (and consequently travels a negligible distance) to achieve a value approaching this constant terminal settling velocity (see Equation 11.8).

For capture to occur, the particle must reach collection surface a′b′c′d′ during its residence time t_r in the unit. For plug flow, t_r is given by:

$$T_r = L/u \tag{11.1}$$

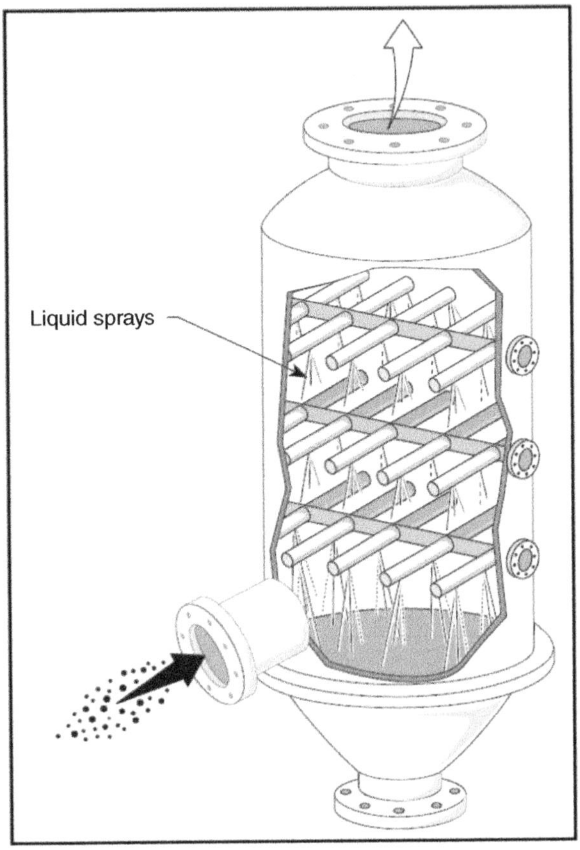

Figure 11.6 Gravity spray tower. *Source*: Environmental Protection Agency/Public domain / https://commons.wikimedia.org/w/index.php?curid=2741152.

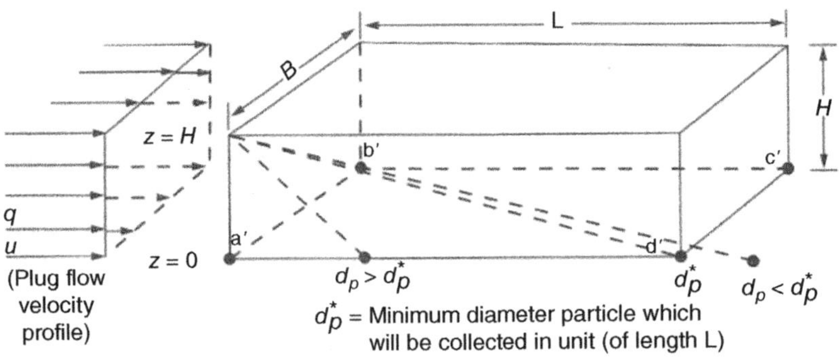

Figure 11.7 Gravity settler nomenclature.

with

$$u = q/BH \tag{11.2}$$

Combining Equations 11.1 and 11.2 gives:

$$T_r = \text{LBH}/q \tag{11.3}$$

where u is the bulk flow velocity, and q is the volumetric flow rate of gas.

The time required for the particle to settle, t_s, a distance, H, while moving at its terminal velocity v_t is

$$t_s = H/v_t \tag{11.4}$$

For capture to occur

$$t_s \leq t_r \tag{11.5}$$

In the limit,

$$t_s = t_r \tag{11.6}$$

Substituting Equations 11.3 and 11.4 into Equation 11.6 gives:

$$H/v_t = \text{LBH}/q \tag{11.7}$$

Or

A terminal settling velocity equation is available for the three Reynolds number ranges [2, 3]. For Stokes' law (without the Cunningham correction factor)

$$v_t = gd_p^2\rho_p/18\mu \tag{11.8}$$

Substituting Equation 11.8 into Equation 11.7 and solving for the particle diameter $\mathrm{d_p}$ gives:

$$d_p = \left(\frac{18\mu q}{g\rho_p BL}\right)^{0.5} \tag{11.9}$$

For the intermediate range

$$v_t = 0.153\left(g\rho_p\right)^{0.71} d_p^{1.14}/\rho^{0.29}\mu^{0.43} \tag{11.10}$$

So that

$$d_p = q^{0.88}\rho^{0.254}\mu^{0.377}/0.193\left(g\rho_p\right)^{0.623}(LB)^{0.88} \tag{11.11}$$

Finally, for Newton's law range, on obtains

$$v_t = 1.74gd_p\rho_p/p \tag{11.12}$$

And

$$d_p = 0.333\left(\frac{\rho}{g\rho_p}\right)\left(\frac{q}{LB}\right)^2 \tag{11.13}$$

The particle diameters calculated above represent *limiting* values since particles with diameters equal to or greater than this value will reach the collection surface and particles with diameters less than this value will escape from the unit. This limiting particle diameter may ideally be thought of as the minimum diameter of a particle that will automatically be captured for the above conditions. This diameter is denoted by d_p^* or d_p (min).

The development described above assumes the bulk gas velocity profile to be plug. However, the velocity profile can be parabolic, as in laminar flow, or, in the general case, arbitrarily distributed [4]. The calculation for the minimum particle diameter is not affected, provided that the volumetric flow rate of gas processed remains the same and the particle concentration remains uniform. This can be observed directly from Equation 11.14. The terminal settling velocity for total collection is the same as a hypothetical volume of fluid passing through LB, the collector area a′b′c′d′. From this analysis, it can be concluded that the performance of a settler depends only on its collection area and is independent of its height. The minimum height, however, is usually established by requiring that the gas velocity through the chamber be low enough to prevent re-entrainment of the collected particles.

Collection efficiencies of 100% were used to derive the equations for d_p^*. The collection efficiency, E, for a monodispersed aerosol (particulates of one size) can be shown to be:

$$E = (v_t BL/q); \tag{11.14}$$

The validity of this equation is observed by again noting that v_t BL represents the hypothetical volume rate of flow of gas passing the collection area, while q is the total volumetric flow rate of gas entering the unit to be treated. An alternate but equivalent form of Equation 11.14 is:

$$E = (v_t/u)(L/H) \tag{11.15}$$

or

$$E = (H^*/H) \tag{11.16}$$

where H^* is the maximum height above the collection area, at inlet conditions, for which particle capture is assured (see Figure 11.8).

If the gas stream entering the unit consists of a distribution of particles of various sizes, then a fractional or grade efficiency curve is frequently specified for the settler. This is simply a curve describing the collection efficiency for particles of various sizes. If a particle of size d_p, will settle a distance H^* in time t, then (H^*/H) represents the fraction of particles of this size that will be collected. If H^* is equal to or greater than H, all the particles of that size or larger will be collected in the settling chamber. A curve of (H^*/H) vs. d_p over the particle size distribution range of the incoming particles entrained in the gas stream is the fractional or size-efficiency curve. Note that H^* represents the distance a particle will settle in residence time t_r.

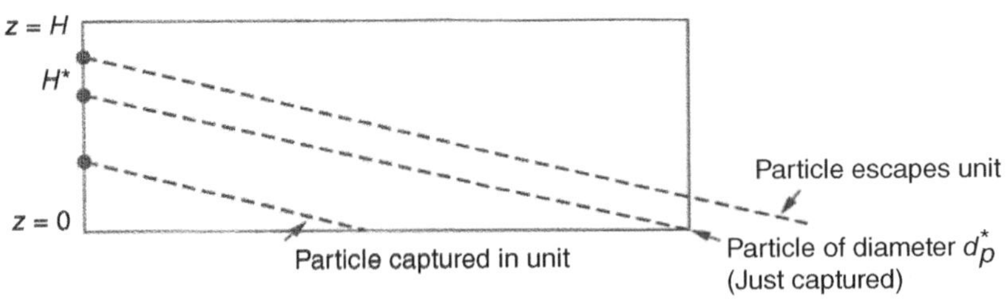

Figure 11.8 Particle collection in a gravity settler.

Combining Equations 11.9 and 11.14 leads to the aforementioned size-efficiency relationship.

$$E = \left[\frac{g\rho_p BLN}{18\mu q}\right] d_p^2 \tag{11.17}$$

The term in brackets in Equation 11.17 is often multiplied by a dimensionless empirical factor to correlate theoretical efficiencies with experimental data. If no information is available, it is suggested that 0.5 be used. Thus, Equation 11.17 can be written:

$$E = 0.5\left[\frac{g\rho_p BLN}{18\mu q}\right] d_p^2 \tag{11.18}$$

The equations above do not apply to large particles.

If the Intermediate law (not Stokes' law) applies, Equation 11.18 becomes (where N has been disregarded):

$$E = 0.153\left(g\rho_p\right)^{0.71} d_p^{1.14}(BL)/\rho^{0.29}\mu^{0.43}p \tag{11.19}$$

Finally, for the Newton's law range, the equation becomes:

$$E = 1.74\left[\frac{d_p g\rho_p}{p}\right]^{0.5}\left(\frac{BL}{q}\right) \tag{11.20}$$

All of the preceding equations were developed with the assumption that the gas flow through the settling chamber is laminar. The equation for determining efficiency when the flow is turbulent can be approximated by:

$$E = 1 - e^{-\left[\frac{Lu}{Hv}\right]} \tag{11.21}$$

The process design variables for a settling chamber consist of length (L), width (B), and height (H). These parameters are usually chosen by the chamber manufacturer in order to remove all particles above a specified size. The chamber's design must provide conditions for sufficient particle residence time to capture the desired particle size range. This can be accomplished by keeping the velocity of the exhaust gas through the chamber as low as possible. If the velocity is too high, dust re-entrainment will occur. However, the design velocity must not be so low as to cause the design of the chamber volume to be exorbitant. Consequently, the units are designed for gas velocities in the range 1–10 ft/s (0.305–3.05 m/s).

The development in this section also provides the theoretical collection efficiency of a settling chamber for a single-sized particle. Since the gas stream entering a unit consists of a distribution of particles of various sizes, a fractional efficiency curve must be used to determine the overall collection efficiency. This is simply a curve or equation describing the collection efficiency for particles of various sizes (see Chapter 10). The overall efficiency can then be calculated using:

$$E = \sum(E_i)(w_i) \tag{11.22}$$

where

E = overall collection efficiency
E_i = fractional efficiency of specific size particle
w_i = mass fraction of specific size particle (in range)

Table 11.1 Pickup velocities of various materials.

Material	Density, g/cm^3	Median particle size, μm	Pickup velocity, ft/s
Aluminum chips	2.72	335	14.2
Asbestos	2.20	261	17.0
Nonferrous foundry dust	3.02	117	18.8
Lead oxide	8.26	14.7	25.0
Limestone	2.78	71	21.0
Starch	1.27	64	5.8
Steel shot	6.85	96	15.2
Wood chips	1.18	1,370	13.0
Wood sawdust		1,400	22.3

The collection of particles from a gas stream involves three distinct phases: deposition on the collecting surface, retention on the surface, and finally, removal of the collected material. Thus far, it has been assumed that once a particle settles to the bottom of the chamber (or onto a tray within the chamber), it is retained there, i.e., re-entrainment effects have been neglected. If the bulk flow gas velocity is sufficiently high, a collected particle may not be retained by the surface but will be picked up and re-entrained into the moving air stream. This can substantially reduce the collection efficiency; thus, the consideration of pickup or suspension velocity for horizontal flow is indeed important. Therefore, in a simple gravity settler, the bulk gas velocity should not exceed this pickup or suspension velocity. Typical pickup velocities are given in Table 11.1. Values for this limiting velocity can also be obtained from semitheoretical considerations. By equating the various forces acting on the particle while neglecting interparticle friction, one can derive pickup velocity equations. If no data for determining the pickup velocity is available, it should be assumed to be 10 ft/s. In this case, the velocity of the gas through the settling chamber (throughput velocity) must be less than 10 ft/s.

11.4 Equipment Design Considerations

The gravity settling chamber is usually of simple design and can be manufactured of almost any material, assuming that it is not affected by temperature or corrosive emissions. Maintenance is minimal, provided that corrosion control is maintained. The hoppers of a settling chamber should be kept empty. With a high inlet dust burden, the chamber's hoppers could fill up quite quickly and should be emptied frequently to eliminate the possibility of re-entrainment and blockage.

Recommended maintenance of a gravity settler requires regular inspection to ensure mechanical soundness of the unit and a level of performance consistent with the original design criteria. A brief general inspection should be performed on a regular basis while the unit is operating. Vibratory disturbance, excessive pressure drop, and decreased efficiency are all signs that thorough inspection and maintenance procedures are required.

Complete inspection requires shutdown of the unit for access to the internals. Scheduling can be determined only from experience and general inspections. Internals and exteriors, where accessible, should be visually inspected for fouling, corrosion, or damage. The nature of any metal deterioration should be investigated to properly determine the anticipated life of the equipment or possible corrective action. Possible causes of deterioration include general corrosion, intergranular corrosion, stress cracking, galvanic corrosion, impingement, or erosion attack.

Within the constraints of the existing system, improving operation and performance refers to maintaining operation at the original or a consistent performance level. There are several factors, previously mentioned, which are related to the design and performance of this unit. Any pressure drop in the settler should be minimized. Flow rates and temperature should be checked regularly to ensure that they are in accordance with the original design criteria. Decreased performance due to fouling will generally be exhibited by a gradual decrease in efficiency and should be corrected as soon as possible. Mechanical malfunctions can also be gradual, but will eventually be evidenced by a decrease in or lack of performance.

More often than not, a gravity settler is a secondary control device since it cannot achieve the collection efficiency required by industry using ordinarily available procedures. One advantage of settlers is that they can reduce the load, improve the performance, and extend the operating life of more expensive devices. Gravity settlers frequently return product or an intermediate to be recycled into the process.

Despite the fact that settling chambers are simple in design and can be manufactured from almost any material, they are infrequently used because of their extremely large space requirements and relatively low efficiency. When settling chambers are used, they are normally followed by a more efficient collection device. They have, on rare occasions, been used following a high-efficiency control device (e.g., an electrostatic precipitation, see Chapter 13) to capture some re-entrained particles.

11.5 Design Applications

Three illustrative examples complement the presentation for the chapter concerned with gravity settlers.

11.5.1 Gravity Settler Design

You have been requested to design a gravity settler to remove all iron particulates from a dust laden gas stream. The following information is given:

$d_p = 35\ \mu\text{m}$; uniform, i.e., no distribution

gas = air at ambient conditions

$$q = 130\ \text{ft}^3/\text{s}$$

u, throughput velocity = 10 ft/s

$$\rho_p = 7.62\ \text{g/cm}^3$$

Solution: First convert d_p and ρ_p to engineering units:

$$d_p = (35\ \mu\text{m})\left(3.281 \times 10^{-6}\frac{\text{ft}}{\mu\text{m}}\right) = 11.48 \times 10^{-5}\ \text{ft}$$

$$\rho_p = \left(7.62\frac{\text{g}}{\text{cm}^3}\right)\left(\frac{1\ \text{lb}}{453.6\ \text{g}}\right)\left(\frac{28{,}316\ \text{cm}^3}{\text{ft}^3}\right) = 475.7\ \text{lb/ft}^3$$

The K value [2, 3, 5] is:

$$\begin{aligned} K &= d_p\left(\frac{g(\rho_p - \rho)\rho}{\mu^2}\right)^{1/3} \\ &= (11.48 \times 10^{-5})\left(\frac{(32.2)(475.7 - 0.0775)(0.0775)}{(1.23 \times 10^{-5})^2}\right)^{1/3} \\ &= 2.28 < 3.3 \end{aligned} \tag{11.23}$$

Stokes' law applies, and the collection area required can be calculated from Equation 11.9:

$$d_p = \left(\frac{18\mu q}{g\rho_p BL}\right)^{0.5} \tag{11.24}$$

Solving for BL, one obtains:

$$\begin{aligned} BL &= \frac{18\mu q}{g\rho_p d_p^2} \\ &= \frac{18(1.23 \times 10^{-5})(130)}{(32.2)(475.5)(11.48 \times 10^{-5})^2} \\ &= 142.5\ \text{ft}^2 \end{aligned}$$

The cross-sectional area for u = 10 ft/s is:

$$\begin{aligned} BH &= q/u = 130/10 \\ &= 13\ \text{ft}^2 \end{aligned}$$

On the basis of the minimum required height for cleaning purposes, H is usually set at 3 ft. Then,

$$\begin{aligned} B &= (13)/H = 13/3 \\ &= 4.33\ \text{ft} \end{aligned}$$

And

$$\begin{aligned} L &= (142.5)/B = 142.5/4.33 \\ &= 32.9\ \text{ft} \end{aligned}$$

The total volume of the settler is therefore

$$\begin{aligned} V &= (BL)H = (142.5)(3) \\ &= 427.5\ \text{ft}^3 \end{aligned}$$

This design is revisited in Application 11.5.3.

11.5.2 Options to Increase the Efficiency of an Existing Settler

The collection efficiency of an installed and operating gravity settler has suddenly decreased from a regulatory required 65 to 40%, and leveled off at 40%. Assuming that there are other gravity settlers available at the plant, provide a series of options—with advantages and disadvantages—that may be implemented to bring the discharge back into compliance.

Solution: This is an open-ended problem, and there are several options available for bringing the operation back into compliance:

1. Inspect the unit to determine the root cause of the problem. Assuming, there are no glaring issues discovered proceed with suggestions below.
2. Reduce the process flow rate.
3. Extend the length of the settler (if possible).
4. Increase the height of the settler (if possible).
5. Increase the width of the settler (if possible).
6. Insert another settler downstream from the present settler.
7. Insert another particle control device downstream from the present unit.
8. Reduce the inlet particulate concentration.
9. Insert shelves in the unit.
10. Insert baffling in the unit.
11. Reduce the process gas flow temperature (if possible).

Additional options are available in the literature [2, 3, 6].

11.5.3 Capture Area

Theodore [7] has claimed that the key process design variable for gravity settlers is the capture area, which is given by:

$$A = BL$$

Explain why, once the capture area is calculated, this base area usually takes the form of a square, i.e., $B = L$.

Solutions: Once the capture area is calculated, the cost of the gravity settler internals may be assumed fixed. Because of material costs, however, the larger the outer casing of the physical system, the higher will be the total cost, all other factors being equal. These material costs generally constitute a significant fraction of the total cost of the settler. If the thickness of the outer casing is the same for alternative physical designs and if labor costs are linearly related to the surface area, then the equipment cost will roughly be a linear function of the outer surface area of the structural shell of the unit. This essentially means that the cost is approximately linearly related to the perimeter P, where:

$$P = 2L + 2B$$

To help minimize the cost (by minimizing the perimeter), one can equate the derivative of the perimeter to zero. Thus, setting:

$$B = A/L$$

gives

$$P = 2L + 2A/L$$

and

$$\frac{dP}{dL} = 2 - \frac{2A}{L^2}$$

Setting this derivative equal to zero leads to

$$A = L^2$$

Since

$$A = BL$$

One can conclude

$$L = B$$

Interestingly, most gravity settlers (as well as electrostatic precipitators) are often designed physically in a form approaching a square box.

References

1 J. Mycock, J. McKenna and L. Theodore. *Handbook for Air Pollution Control Engineering and Technology*, CRC Press/Taylor & Francis Group, Boca Raton, FL, 1995.
2 L. Theodore, *Air Pollution Control Equipment Calculations*, John Wiley & Sons, Hoboken, NJ, 2008.
3 L. Theodore and R. Allen, *Air Pollution Control Equipment*, A Theodore Tutorial. Theodore Tutorials (originally published by the USEPA/APTI, RTP, NC), East Williston, NY, 1993.
4 P. Abulencia and L. Theodore. *Fluid Flow for the Practicing Chemical Engineer*, John Wiley & Sons, Hoboken, NJ, 2009.
5 L. Theodore, personal notes, East Williston, NY, 1971.
6 L. Theodore, personal notes, East Williston, NY, 1968.
7 L. Theodore, personal notes, East Williston, NY, 1973.

Suggested Reading

1 W. Strauss, *Industrial Gas Cleaning*, Pergamon Press, New York, 1966.
2 Stern et al., *Air Pollution*, Academic Press, New York City, NY, 1968.
3 J. Danielson, *Air Pollution Engineering Manual*, EPA OAP Publication No. AP-40. Government Printing Office, Washington, D.C., 1973.
4 Stern et al., *Fundamentals of Air Pollution.*, Academic Press, New York City, NY, 1974.
5 L. Theodore and A. J. Buonicore, *Industrial Air Pollution Control Equipment for Particulates*, Chap. 6, CRC Press, Inc., Boca Raton, FL, 1976.
6 K. Wark and C. Warner, *Air Pollution: It's Origin and Control*, IEP, New York City, NY, 1975.
7 H. Parker, *Air Pollution*, Prentice-Hall, Upper Saddle River, NJ, 1977.

8 R. Conway and R. Ross, *Handbook of Industrial Waste Disposal*, Van Nostrand Reinhold, New York City, NY, 1980.

9 G. Burke, B. Singh and L. Theodore, *Handbook of Environmental Management and Technology*, 2nd edition, John Wiley & Sons, Hoboken, NJ, 2000.

10 L. Theodore, *Air Pollution Control Equipment Calculations*, John Wiley & Sons, Hoboken, NJ, 2008.

11 J. Richardson, R. R. Dupont, P. Farber, K. Ganesan and L. Theodore, *Air Pollution Control Equipment Revisited*, AWMA Conference, Calgary, CA, 2024.

12 J. Richardson and L. Theodore, *Optimizing Air Pollution Equipment Performance: Operation and Maintenance*, John Wiley & Sons, Hoboken, NJ, 2025.

12

Cyclones

12.1 Introduction

Centrifugal separators, commonly referred to as cyclones, are widely used in industry for the removal of solid and liquid particles (or particulates) from gas streams. Typical applications are found in mining and metallurgical operations, the cement and plastics industries, pulp and paper mill operations, chemical and pharmaceutical processes, petroleum production (cat-cracking cyclones), and combustion operations (fly ash collection).

Particulates suspended in a moving gas stream possess inertia and momentum and are acted upon by gravity. Should the gas stream be forced to change direction, these properties can be utilized to promote centrifugal forces to act on the particles. In a conventional unit, the entire mass of the gas stream with the entrained particles enters the unit tangentially and is forced into a constrained vortex in the cylindrical portion of the cyclone.

The aforementioned centrifugal force is widely used when a force greater than that of gravity is desired for the separation of solids and fluids of different densities. A centrifugal force is created by moving a mass in a curved path and is exerted in the direction away from the center of curvature of the path. The centripetal force is the force applied to the moving mass in the direction toward the center of curvature that causes the mass to travel in a curved path. If these forces are equal, the particle continues to rotate in a circular path around the center.

As noted above, these mechanical collectors separate particles from a gas by causing the gas stream to change direction while the particles, because of their inertia, tend to continue in their original direction and become separated from the gas.

The collection efficiency of these collectors depends upon the motion of the gas stream and the physical properties of the suspended particulate. Separation forces are a function of the gas velocity, the rate of change in direction, and of particle size and specific gravity. Their operation depends on separating forces produced by the gas stream motion. The forces developed are primarily a function of the gas stream's speed and rate of curvature, and of particle size and specific gravity. These forces acting on a particle that carry it into a collecting zone can be made as high as 2,500 times gravity [1–3].

There are two main classes of centrifugation equipment: cyclones and centrifuges. Cyclones are generally classified as air recovery/control equipment, and are primarily used for separating gas-solid systems. Centrifuges are primarily employed for liquid–solid separation and are not of concern in this text or air pollution control.

12.2 Description of Cyclone Equipment

In a conventional cyclone unit (see Figure 12.1), the entire mass of the gas stream with the entrained particles tangentially enters the unit and is forced into a constrained vortex in the cylindrical portion of the cyclone.

Upon entering the unit, a particle develops an angular velocity. Due to its greater inertia, it tends to move outward across the gas stream lines in a tangential rather than a rotary direction; thus, it attains a net outward radial velocity. By virtue of its rotation with the carrier gas around the axis of the tube (main vortex) and its higher density with respect to the gas, the entrained particle is forced toward the wall of the unit. Eventually, the particle may reach this outer wall where it is carried by gravity and assisted by the downward movement of the outer vortex and/or secondary eddies, toward the dust outlet at the bottom of the unit. The flow vortex is reversed in the lower (conical) portion of the unit, leaving most of the entrained particles behind. The cleaned gas then passes up through the center of the unit (inner vortex) and out of the collector. The spiral motion of both vortices is in the same direction. The tangential velocity (how fast the gases are swirling) is lowest near the wall and at the center of the cyclone; it reaches a maximum at a point approximately 60–70% of the way from the wall to the center.

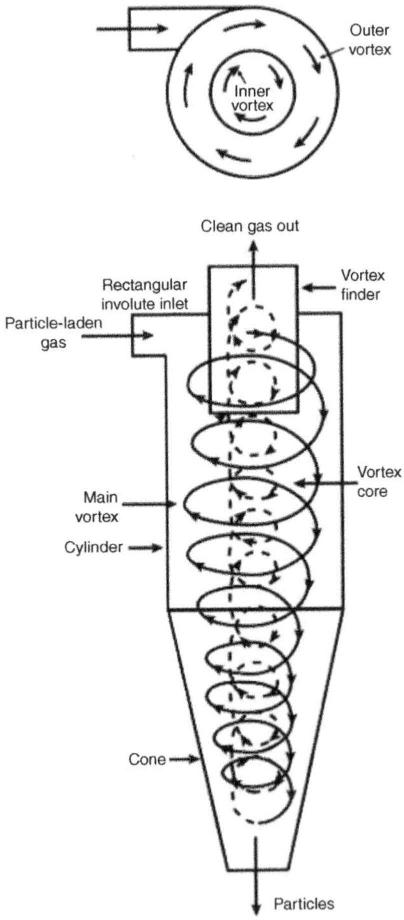

Figure 12.1 Cyclone dust collector diagram.

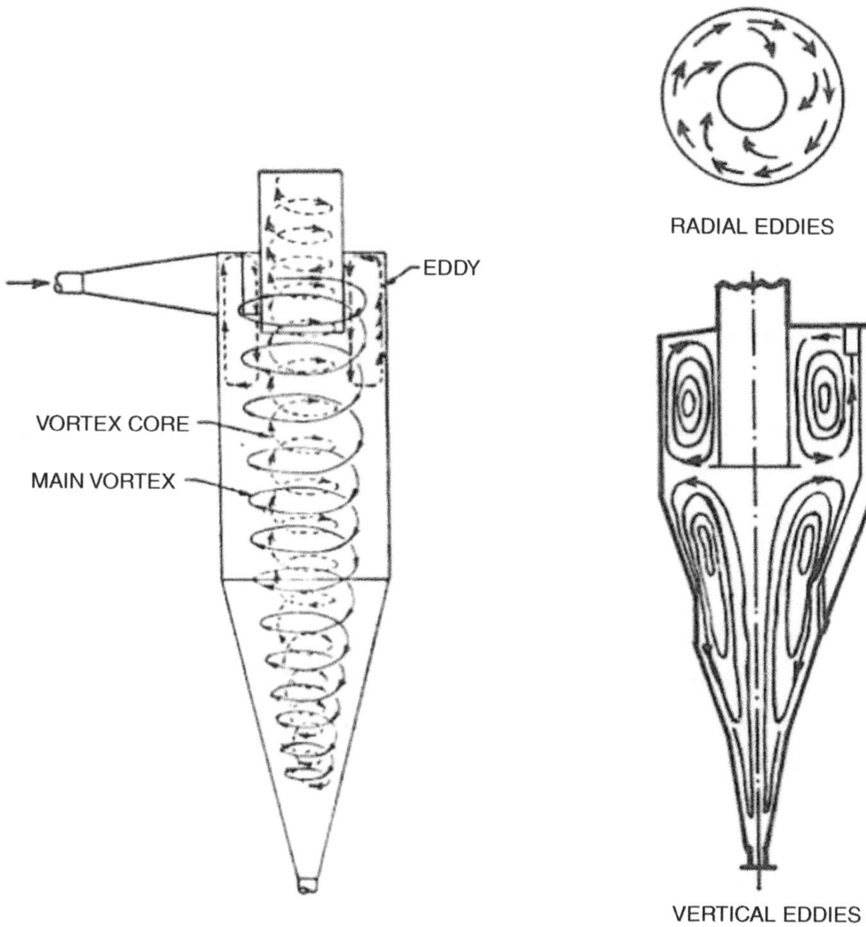

Figure 12.2 Eddy patterns found in typical cyclones.

In addition to the tangential velocity variation, there are also vertical eddies and what is called inward drift (see Figure 12.2). The inward drift is a radial gas flow which moves toward the center of the cyclone, opposing the movement of particles. While vertical eddies can exist in the cone, the most troublesome are those in the annular region near the gas inlet. The eddies, which are caused by the vortices, can carry particles directly from the gas inlet to the gas outlet with a consequent detrimental effect on collection efficiency. Radial eddies formed at the point where the cylinder joins the cone can also be particularly detrimental to collection efficiency.

Eddy currents in the annular region require the gas outlet to extend into the cyclone in order to prevent excessive amounts of dust from passing directly from the inlet to the outlet. Usually, this extension ends just below the bottom of the inlet.

Cyclones may generally be classified into four categories (see Figure 12.3) depending on how the gas stream enters the unit and how the collected particles leave the unit:

a. Tangential inlet and axial dust discharge
   i. Conventional (large diameter)
   ii. High efficiency (small diameter, less than 10 inches)

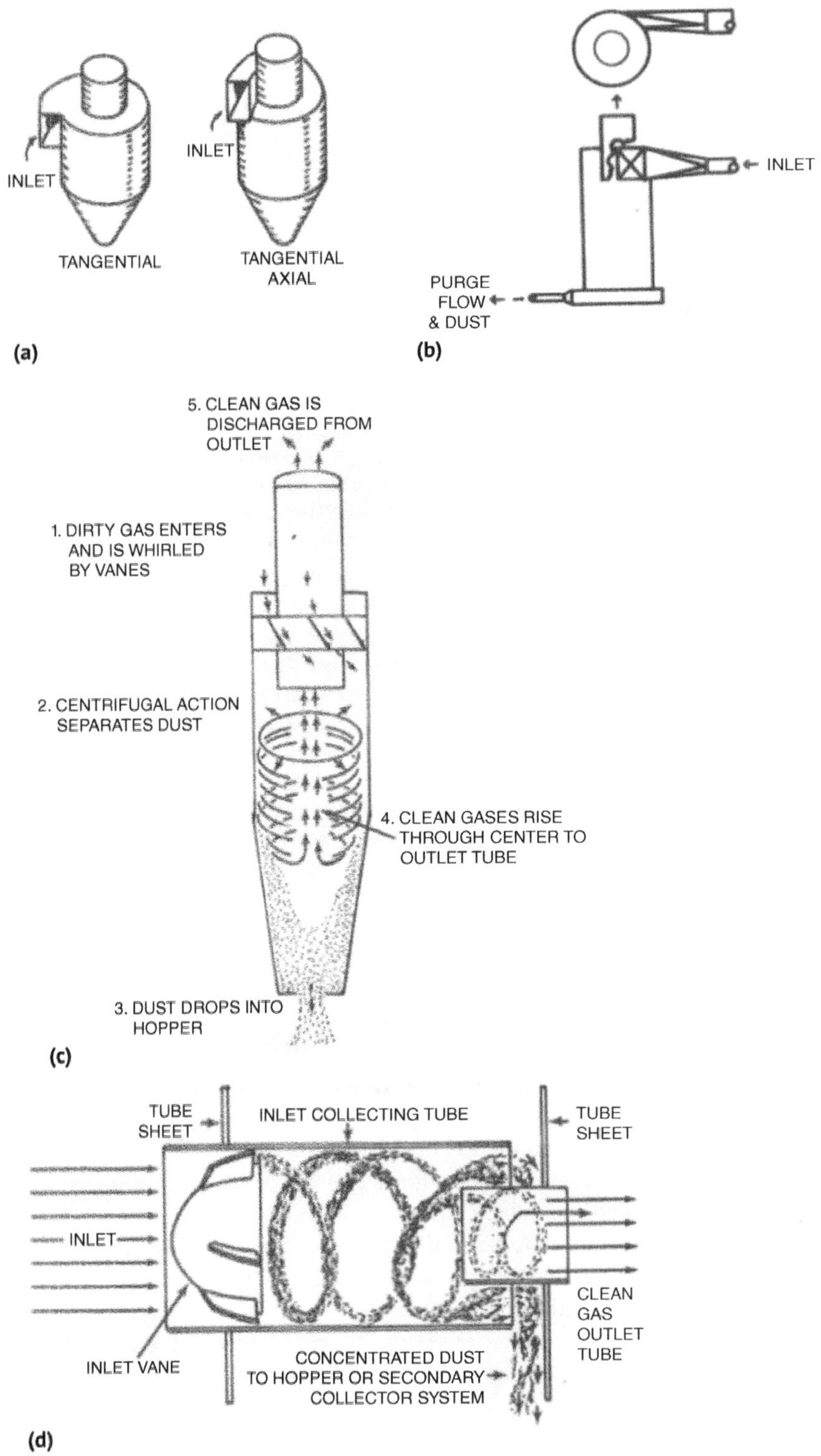

Figure 12.3 General cyclone categories: (a) tangential inlet and axial dust discharge, (b) tangential inlet and peripheral dust discharge, (c) axial inlet and axial dust discharge, and (d) axial inlet and peripheral dust discharge.

b. Tangential inlet and peripheral dust discharge
c. Axial inlet and axial dust discharge
d. Axial inlet and peripheral dust discharge

Types a and c are the most widely used. Large-diameter cyclones, with body diameters 3–5 times the diameter of the inlet duct, are useful where large gas handling capacity and moderate particle collection efficiency are required. The ratio of gas volume to capital investment dollar is greater than that for most cleaning devices.

As can be deduced from the definition of "high-efficiency" cyclones, decreasing the body diameter will increase the efficiency. This is due to increased separation forces caused by the smaller vortex radius. Individual high-efficiency, small-diameter cyclones have a small capacity, and they must be operated in parallel to handle typical gas volumes. They generally have a common gas inlet, dust hopper, and gas outlet and can be arranged in banks of up to several hundred cyclones each. Typical performance of conventional and high-efficiency cyclones is shown in Table 12.1.

The cone portion of the cyclone is not necessary to convert the downward vortex to an upward one, although its presence does reduce the length of cyclone needed to effect this reversal. The primary purpose of the cone is to deliver the collected particles to a central point for easy handling and disposal. If the cone is too small in diameter at the bottom, the vortex core will contact the walls and re-entrain the collected dust. Dust discharge is just as important in reducing this re-entrainment due to the high turbulence and velocities present near the discharge. The static pressure in the vortex core may be slightly negative and tend to draw collected dust up away from the discharge. Mechanical devices (such as rotary valves, double-flop valves, screw conveyers, and dip legs) have been quite successful in minimizing re-entrained dust. To be successful, the mechanism must achieve continuous, complete, and immediate removal of the separated particles and prevent the inflow of gas (infiltration) from the hopper.

As the inlet gas stream enters the annular region at the top of the cyclone, it is squeezed by the existing gas to about half of its inlet width. This causes a significant pressure loss, which can be reduced by the addition of vanes to the annular area. The presence of the vanes, however, reduces the efficiency, apparently due to the prevention of vortex formation in the annulus. Helical and involute inlets (Figure 12.4) are attempts to reduce interference between the incoming gas and the vortex already present in the cyclone.

In addition to cyclones, there are numbers types of inertial separators, including baffle-type separators, chip trap separators, and louvered-type separators. Each is briefly described below.

Table 12.1 Cyclone collection efficiency.

Particle size (μm)	Efficiency (% by weight)	
	Conventional cyclone	High-efficiency cyclone
<5	—	50–80
5–20	50–80	80–90
15–50	80–95	90–99
>50	95–99	95–99

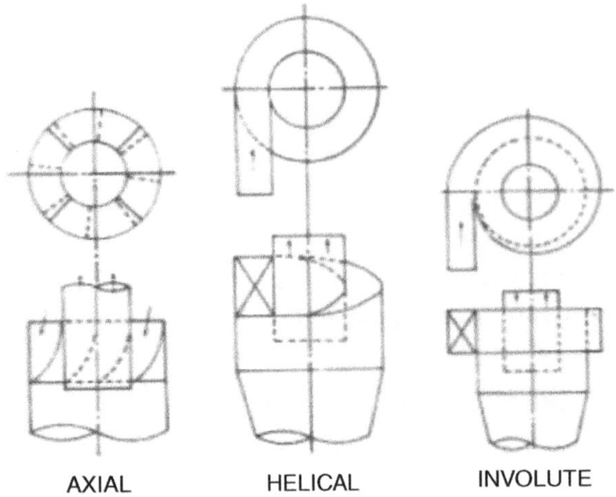

Figure 12.4 Cyclone inlet configuration.

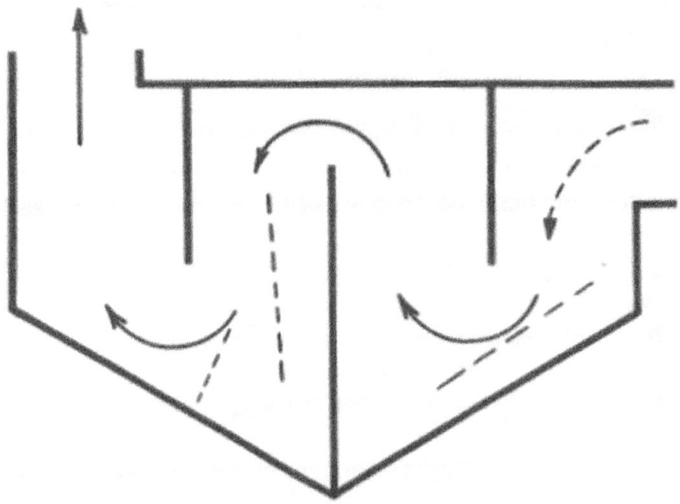

Figure 12.5 Multiple baffle chamber.

The inertial separators depend on the deposition of particulates in a discontinuous fashion, as the gas passes around or through specially shaped baffles or apertures. Some inertial separators operate in the same range of efficiency as a cyclone. The simple baffle-type inertial separator collects only very large particles. The only difference between a baffle-type separator (Figure 12.5) and the settling gravity type is the baffle, which imparts an abrupt change of direction for the dust-laden air. The particles are dropped from the airstream by their inertia [3].

A typical chip trap also relies on an abrupt change of direction in the airstream to drop out the particulate. This change of direction is caused by fixed baffle plates within the unit. The baffle-type collector is used primarily as roughing collectors or as precleaners to more efficient collectors. It

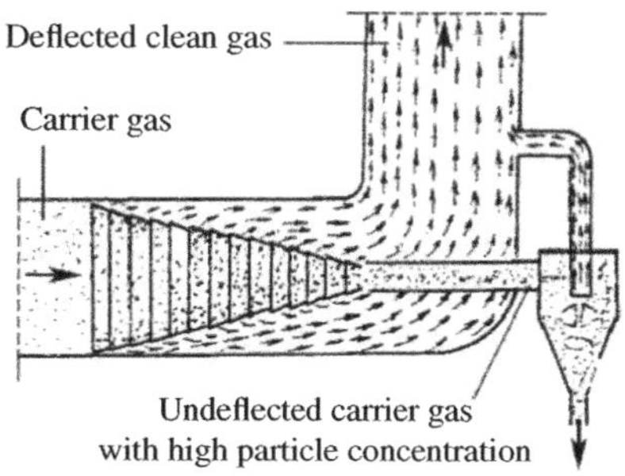

Figure 12.6 Cross-section view of louver type dust collector.

can be used on the same applications as the settling chamber. Since velocities are low, the space requirements are large.

The operating principle of a dust louver-type dust separator can easily be understood by examining a cross-sectional view (Figure 12.6). The air enters through the total air inlet and as the air passes the louver, the direction of flow is reversed. About 90% of the air is drawn off on the clean-air side, with the remaining 10% being drawn off with the dust in the secondary air or dust circuit. The dust is concentrated in a relatively small air volume and can be discharged into the atmosphere, settled into a hopper or into a more efficient secondary collector. The design of louvered separators is governed by angle of the element, shape of the blade, pitch of louvered blade, gauge and type of blade material, ratio of total inlet air to secondary air outlet, and percentage of secondary air volume. The efficiency is a function of louver spacing, with closer spacings yielding higher efficiencies. The louver separator is available in several basic designs [4].

12.3 Process Design Considerations

Although cyclones have been used for many years to collect suspended materials, a satisfactory theory relating the many operating variables has never been universally accepted. The performance of cyclone collectors is based almost entirely on experience and empirical data, probably because of the complexity of the aerodynamics involved in cyclone collector flow. However, the foregoing equations at least identify those factors that can either increase or decrease efficiency. For example, as particle radius increases, efficiency should increase. This is illustrated in Figure 12.7 which shows the basic performance characteristics of a typical multitube mechanical collector. This is known as the collector fractional efficiency curve and is usually established by the supplier of the cyclones. By knowing the designed pressure drop across the cyclone, the particle-size distribution and specific gravity of the dust to be collected, the overall efficiency can be estimated by simple numerical calculation [2].

The most prevalent factor affecting the collection efficiency is the size and nature of the particulate. As the particle size becomes smaller, the separating force decreases, so that for any design and gas volume, there is a limiting particle size below which collection ceases. The nature of the particles also affects the collection efficiency since some types of small particles tend to agglomerate and behave like heavier particles in the separate field. Additional information is presented in Table 12.2.

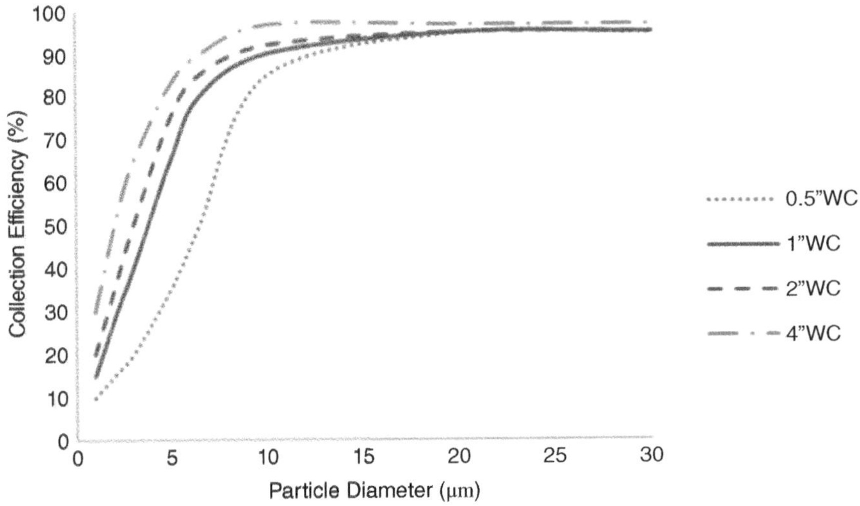

Figure 12.7 Basic performance characteristics of a multitube mechanical collector.

Table 12.2 Effect of process factors on efficiency.

Process factor	Effect on efficiency
Inlet loading	Efficiency decreases as inlet loading decreases
Specific gravity	Efficiency decreases at lower specific gravity
Particle size distribution	Efficiency decreases as percent of dust particle size less than 10 μm increases
Pressure drop	Efficiency decreases as pressure drop decreases

As noted in the previous section, objects moving in circular paths tend to move away from the center of their motion. The object moves outward as if a force is pushing it out. This force is known as centrifugal force. The whirling motion of the gas in a cyclone causes particulate matter in the gas to sense this force and move out to the walls. An expression for this force is as follows:

$$F = \frac{\rho_p d_p^3 v_p^2}{r} \tag{12.1}$$

where
ρ_p = particle density, lb/ft^3 (kg/m^3)
d_p = particle diameter, inches (μm)

The term F is the force that the particulate matter experiences in a cyclone. Equation 12.1 explains several of the cyclone characteristics discussed in the previous section. For example, $\rho_p d_p^3$ is merely proportional to the mass of the particle. The larger the mass, the greater the force. The tendency to move toward the walls is consequently increased and larger particles are more easily collected. The reason why all of the particles do not move to the wall is because of the drag resistance of the air. The buffeting molecules in the gas resist the outward motion and act like an opposing force. Particles move to the wall when the centrifugal force is greater than the opposing drag force.

Also note from Equation 12.1 that as r (the radius of the circular path) decreases, the force increases. This is why smaller cyclones are more efficient for the collection of smaller-sized particles than large cyclones.

These types of considerations, in conjunction with considerations cyclone geometry and vortex formation, have led to the development of numerous design and performance equations. These equations attempt to characterize the behavior of cyclones. Some work well, and some do not. None adequately describe performance under all operating conditions such as at high pressure and high temperatures.

Three important parameters can be used to characterize cyclone performance:

d_{pc} = cut diameter
ΔP = pressure drop
E = overall collection efficiency

Equations involving each of these parameters are provided in this section. The equations should be used with caution, however, since there are strict limitations on their applicability.

12.3.1 Cut Diameter

The *cut diameter* is defined as the size (diameter) of particles collected with 50% efficiency. It is a convenient way of defining efficiency for a control device since it provides information on the effectiveness for a particle size range. A frequently used expression for cut diameter is:

$$d_{pc} = \left(\frac{9\mu B_c}{2\pi N v_i(\rho_p - \rho)}\right)^{0.5} \tag{12.2}$$

where
μ = viscosity, lb/ft·s (Pa·s)
N = effective number of turns (5–10 for the common cyclone)
v_i = inlet gas velocity, ft/s (m/s)
ρ_p = particle density, lb/ft^3 (kg/m^3)
ρ = gas density, lb/ft^3 (kg/m^3)
B_c = inlet width, ft (m)

The cut diameter, d_{pc} or $\left[d_p\right]_{\text{cut}}$, is a characteristic of the control device and should not be confused with the geometric mean particle diameter, d_{gm}, of the size distribution. Figure 12.8 shows a size efficiency curve and points out the cut diameter and the critical diameter, $\left[d_p\right]_{\text{crit}}$, the particle size collected at 100% efficiency. Values of $\left[d_p\right]_{\text{crit}}$ are difficult to obtain from such curves so the cut size is often determined instead.

A number of formulas exist for the calculation of the cut diameter and critical diameter. A value of N, the number of turns, must be known in order to solve Equation 12.2 for $\left[d_p\right]_{\text{cut}}$.

Given the volumetric flow rate, inlet velocity, and dimensions of the cyclone, N can be easily calculated. Values of N can vary from 1 to 10, with typical values in the 4–5 range.

The expression for the cut diameter (Equation 12.2) has been found to agree with some experimental data. However, other experimental work has shown limitations to its application. A high-efficiency cyclone can have a cut diameter of typically 5–10 µm. Equation 12.2 is typical of most of those derived for determining the cut or critical diameter. Note that an increase in the number

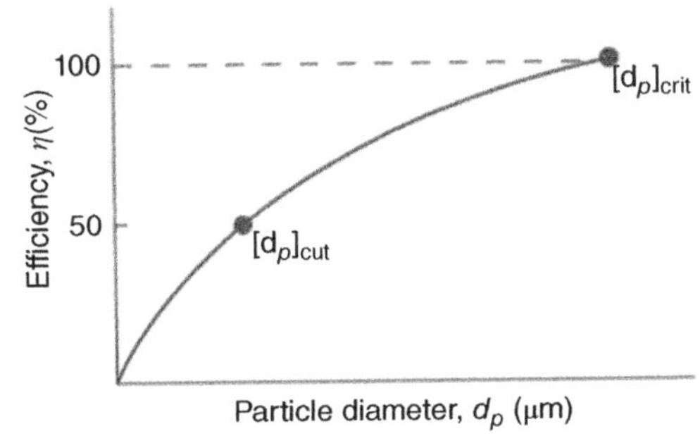

Figure 12.8 Typical size efficiency curve.

of turns, inlet velocity, or the particle density will decrease the cut size as one would expect. A decrease in viscosity will decrease the drag force opposing the centrifugal force and therefore also reduce the cut size (i.e., smaller particles will be collected).

12.3.2 Collection Efficiency

A number of equations have been developed for determining the fractional cyclone efficiency E_i for a given size particle. As noted earlier, fractional efficiency is defined as the fraction of particles of a given size collected in the cyclone, compared to those of that size going into the cyclone.

No efficiency theory or calculation method provides a description for all cyclones. The modification of inlets and outlets, addition of fines, etc., introduce variables that are difficult to treat theoretically. Although theoretical efficiencies can give estimations of cyclone performance, it should be kept in mind that designers of equipment commonly rely on comparative evaluations between similar designs and on experience.

This subsection will describe two methods of calculating cyclone efficiency. The Leith and Licht theory [5] for calculating fractional efficiency will be discussed first. A convenient graphical method for estimating efficiency, developed by Lapple [6] will also be discussed.

The fractional efficiency equation of Leith and Licht has a form similar to many of those developed for particulate control devices. Fractional efficiency is expressed as:

$$E_i = 1 - e^{\left[-2(c\varphi)^{\left(\frac{1}{2N+2}\right)}\right]} \tag{12.3}$$

where

c = cyclone dimension factor

φ = impaction parameter

n = vortex exponent

The exponential form of this equation is quite common since it is employed with other control devices. In this expression, c is a factor that is a function only of the cyclone's dimensions. The symbol φ expresses characteristics of the particles and gas as:

$$\varphi = \frac{\rho_p d_p^2 v_i}{18\mu D_c}(n+1) \tag{12.4}$$

and is known as the inertia or impaction parameter. Note that ρ_p times v_i essentially expresses the particle's initial momentum. The value of n is dependent on the cyclone diameter and temperature of the gas stream. The calculations involved in this method are straightforward, although tedious. The original reference should be consulted if it is to be applied.

An older, and more popular, method of calculating cyclone fractional efficiency and overall efficiency was developed by Lapple. Lapple first computed the ratios $d_p / \left[d_p\right]_{cut}$, the particle diameter versus the cut diameter ratio as determined from Equation 12.2 or Figure 12.1. He found that cyclone efficiency correlates in a general way with this ratio. For a typical cyclone, efficiency will increase as the ratio increases as shown in Figure 12.9.

Lapple's supposed development of this method receives treatment in the "Applications" section of this chapter.

As a universal curve for common cyclones, the preceding correlation has been found to agree reasonably well with experimental data. To calculate fractional efficiencies, the procedure presented below should be completed.

Lapple Calculation Procedure

d_p range	wt fraction in range	$d_p / \left[d_p\right]_{cut}$	E_i for each d_p from experiment or Lapple's method, %	wt fraction × E_i

The sum of these products in the rightmost section of the box will give the overall efficiency. More details are available in the literature [7].

12.3.3 Pressure Drop

The pressure drop across a cyclone is an important parameter to the purchaser of such equipment. Increased pressure drop means greater costs for power to move an exhaust gas through the control device. With cyclones, an increase in pressure drop usually means that there will be an improvement in collection efficiency (one exception to this is the use of pressure recovery devices attached to the exit tube; these reduce the pressure drop but do not adversely affect collection efficiency).

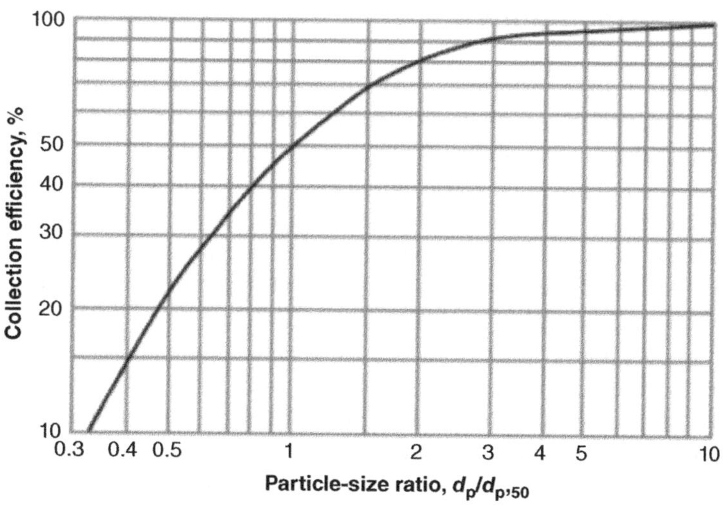

Figure 12.9 Cyclone collection efficiency versus particle size ratio.

For these reasons, there have been many attempts to predict pressure drops from design variables. The idea is that having such an equation, one could work back and optimize the design of new cyclones.

An expression occasionally used is:

$$\Delta P = \frac{0.0027q^2}{k_c D_c^2 B_c H_c \left(\frac{L_c}{D_c}\right)^{\frac{1}{3}} \left(\frac{Z_c}{D_c}\right)^{\frac{1}{3}}}; \text{ consistent units} \tag{12.5}$$

where q = volumetric flow rate.

(In this case, k_c is a dimensionless factor descriptive of cyclone inlet vanes. It is equal to 0.5 for cyclones without vanes, 1.0 for vanes that do not expand the entering gas or touch the outlet wall, and 2.0 for vanes that expand and touch the outlet wall.) This equation, when compared to experimental data, was found to have a poor correlation coefficient.

The most popular of the empirical pressure drop equations has the form:

$$\Delta P = K_c \rho v_i^2; \text{ consistent units} \tag{12.6}$$

where K_c is a proportionality factor.

If ΔP is measured in inches of water, K_c can vary from 0.013 to 0.024, with 0.024 the norm. Velocities for cyclones range from 20 to 70 ft/s (6–21 m/s), although common velocities range from 50 to 60 ft/s (15–18 m/s). At velocities greater than 80 ft/s (24 m/s), turbulence increases in the cyclone, and efficiency will actually decrease. Also, at high loads of particulate matter and high velocities, scouring of the cyclones.by the particles will rapidly increase. To minimize erosion in such cases, a cyclone should be designed for lower inlet velocities.

Pressure drops for single cyclones vary depending on both size and design. Common ranges are:

- Low-efficiency cyclones 1–4 in H_2O (2.5–10 cm H_2O)
- Medium-efficiency cyclones 4–6 in H_2O (10–15 cm H_2O)
- High-efficiency cyclones 6–10 in H_2O (15–25 cm H_2O)

12.4 Equipment Design Considerations

There are many operating variables in cyclone performance. These include characteristics of both the gas and the particles. Gas operating variables include temperature, pressure, and composition. Dust characteristics include size, size distribution, shape, density, and concentration.

As the temperature of a gas increases, its density decreases while its viscosity increases. Since the gas density is negligible compared to the particle density, this has no direct effect on efficiency. A higher temperature will increase the inlet velocity, increasing the particle velocity toward the wall. However, the increasing viscosity decreases the particle velocity toward the wall. The net effect of this is that within the normal operating range of 40–700 °F, the collection efficiency is essentially constant. As the temperature rises above 1,000 °F, the viscosity effect dominates, causing a decrease in efficiency. The gas composition can also affect gas viscosity and density.

The influence of gravity on the dust separation in a cyclone is slight, so that the efficiency is almost independent of the position of the cyclone. Separations remain satisfactory whether the cyclone is vertical, horizontal, or even upside-down with the dust carried off upward from the cyclone. A good cyclone separates the dust satisfactorily in any position. However, coarse particles of dust may keep rotating in the conical part of the cyclone, never reaching the outlet. This difficulty can arise no matter how the cyclone is positioned, but less so if the cyclone is in its normal vertical position.

To operate efficiently, cyclone dust collectors should be airtight. Gaskets must be used to close the gaps between flanges. For axial entry collectors, header sheets must be sealed without any breaks in the welds. All doors, ports, and poke holes should also be sealed to prevent re-entrainment. Since there are always some fine dust particles suspended in the air, even slight leaks can cause re-entrainment and escape of dust particles from the cyclone.

Structurally, the design of a cyclone must have an axial gas outlet, a dust discharge, and an inertial forces for gas inlet that will produce the gas rotation necessary to create the dust separation. As discussed in Section 12.2, these three elements can be combined in a number of ways.

Collection efficiency in a cyclone is primarily determined by the pressure drop and/or the inlet velocity. The pressure drop can be increased or decreased by varying the diameter of the cyclone body or by varying the volumetric flow rate per tube. These features must be designed into the system. If the cyclone is operated at a lower volumetric flow rate, dampers should be used so that the gas velocity will be increased.

Since cyclones have no moving parts, fine tuning them is very difficult. Spiral vanes are used in axial entry cyclones, allowing some control of volumetric flow rate by moving a vane in and out of a constricted opening in the collector element. When fully in, maximum rotation is induced, resulting in greater centrifugal action. When fully out, much of the gas bypasses the vanes.

Dampers at the inlet are used to control turndown of gas flow. This can be accomplished if the collector is sectionalized and allows for the designed pressure drop to be maintained. The excess tubes can be capped off if the turndown is permanent.

Efficiency can also be improved by arranging cyclones in series or parallel; however, these gains can be offset by increased maintenance problems. Multicyclone arrangements plug more easily and, when common hoppers are used, uneven flow distribution can lead to re-entrainment problems.

A summary of changes in performance characteristics produced changes in cyclone design and exhaust gas properties is given in Table 12.3; quantitative equations are provided in Table 12.4.

Table 12.3 Changes in performance characteristics.

Cyclone and process design change	Pressure drop	Efficiency	Cost
Increase cyclone size (D_c)	Decreases	Decreases	Increases
Lengthen cylinder (L_c)	Decreases slightly	Decreases	Increases
Lengthen cone (Z_c)	Decreases slightly	Increases	Increases
Increase exit tube diameter (D_e)	Decreases	Decreases	Increases
Increase inlet area, maintaining velocity	Increases	Decreases	Decreases
Increase velocity	Increases	Increases	Increase operating cost
Increase temperature, maintain velocity	Decreases	Decreases	No change
Increased dust concentration	Decreases for large changes	Increases	No change
Increasing particle size and/or density	No change	Increases	No change

Table 12.4 Effect of operation variables on cyclone operation and performance.

Variable	Effect of efficiency	Effect of pressure drop
Flow rate	$\frac{100-E_1}{100-E_2}=\left(\frac{q_2}{q_1}\right)^{0.5}$	$\frac{\Delta P_1}{\Delta P_2}=\frac{q_1^2\rho_1}{T_1}\left[\frac{T_2}{q_2^2\rho_2}\right]$
Gas density	$\frac{100-E_1}{100-E_2}=\left(\frac{\rho_p-\rho_2}{\rho_p-\rho_1}\right)^{0.5}$	$\frac{\Delta P_1}{\Delta P_2}=\frac{q_1^2\rho_1}{T_1}\left[\frac{T_2}{q_2^2\rho_2}\right]$
Particle density	$\frac{100-E_1}{100-E_2}=\left(\frac{\rho_p-\rho_2}{\rho_p-\rho_1}\right)^{0.5}$	Negligible
Gas viscosity	$\frac{100-E_1}{100-E_2}=\left(\frac{\mu_2}{\mu_1}\right)^{0.5}$	Negligible
Dust loading	$\frac{100-E_1}{100-E_2}=\left(\frac{c_{i2}}{c_{i1}}\right)^{0.5}$	$\frac{\Delta P_d}{\Delta P_c}=\frac{1}{0.013c_i^{0.5}+1}$

Note: where 1 = condition 1; 2 = condition 2; q = gas volumetric flow rate, ft^3/s; E = efficiency; ΔP = pressure drop, in H_2O (subscript d indicates the gas stream under loading conditions and subscript c indicates no dust loading or clean conditions); ρ = gas density, lb/ft^3; T = absolute temperature, °R; μ = gas viscosity, lb/ft·s; c_i = inlet dust concentration, $grains/ft^3$.

As noted earlier, the key operating variables in cyclone performance are the gas temperature, pressure, and composition, and dust characteristics, including size and distribution, shape, density, and concentration.

Increasing the gas temperature decreases its density and increases its viscosity. The direct effect on efficiency by changes in gas density can be neglected since the gas density is minute compared to the dust density. If the collector pressure is held constant, the tube capacity will increase because of the lower gas density. Higher temperature increases the inlet velocity, and this increases the particle velocity toward the wall. However, because of the temperature rise, the increasing gas viscosity tends to decrease the particle velocity toward the wall. The net effect of these factors caused by the temperature rise is essentially negligible in the normal operating range of 40 to 700 °F. Thus, at constant pressure drop, efficiency is essentially constant with temperature. As the temperature extends above 1,000 °F, the viscosity effects become more predominant and there is gradual decline in efficiency. Gas composition can also affect both gas viscosity and density.

12.5 Process Design Conditions

Because of their simplicity of construction and lack of moving parts, cyclones can be fabricated with a variety of materials, including ceramic, or lined with various materials, including ceramics. They are applicable over a broad range of temperatures from below ambient to above 2,000 °F. Cyclones can also be used at high gas pressures. Generally, the only criterion for good performance is the requirement for proportionate increase in power input to maintain separation velocities. The collector is insensitive to inlet concentration, and in fact, the efficiency can increase with increasing particle concentration because of particle interactions. Cyclones find most use when operating under the following situations: where dust is coarse, where dust loading is above 3 g/ft^3. where high efficiency is not a critical factor, and where stack opacity is not a criterion.

Cyclones are the most economical of the four generic classes of particulate collectors. They are also the most versatile. As unsophisticated as this type of collection device is, it does the job, and that is often the only requirement for success.

12.6 Design Applications

Three illustrative examples below complement the presentation in this chapter concerned with cyclones.

12.6.1 Cyclone Selection

You have been assigned the job of selecting and specifying a cyclone unit to be used to reduce an inlet fly ash loading (with the particle size distribution given in Table 12.5) from 3.1 gr/ft^3 to an outlet value of 0.06 gr/ft^3. The flow rate from the coal-fired boiler is 100,000 acfm. Fractional efficiency data provided by a vendor are presented in Figure 12.10 for three different types of cyclones (multiclones).

Which types of cyclones are required to meet the above specifications given above? The optimum operating pressure drop is 3.0 in H_2O at this condition, and the average inlet velocity may be assumed to be 60 ft/s.

As described earlier, multiple-cyclone collectors (multiclones) are high efficiency devices that consist of a number of small diameter cyclones operating in parallel with a common gas inlet and outlet. The flow pattern differs from a conventional cyclone in that instead of bringing the gas in at the side to initiate the swirling action, the gas is brought in at the top of the collecting tube, and swirling action is then imparted by a stationary vane positioned in the path of the incoming gas. The diameters of the collecting tubes usually range from 6 to 24 inches with pressure drops in the 2–6 inch range. Properly designed units can be constructed and operated with a collection

Table 12.5 Particle size distribution data.

Particle diameter range, μm	Weight fraction, w_i
5–35	0.05
35–50	0.05
50–70	0.10
70–110	0.20
110–150	0.20
150–200	0.20
200–400	0.10
400–700	0.10

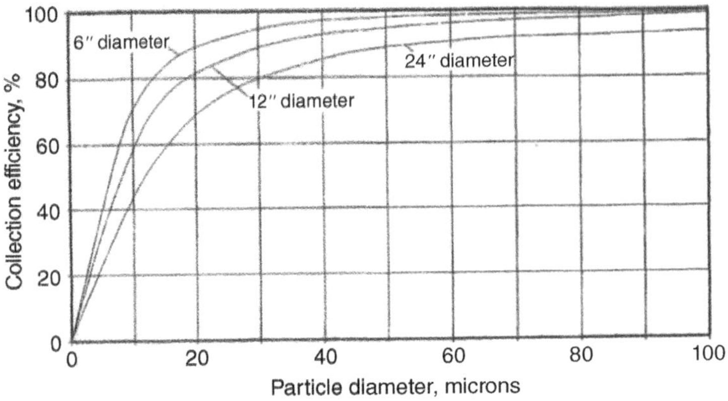

Figure 12.10 Fractional efficiency data.

Table 12.6 Average particle size.

Particle diameter range, μm	Average particle diameter, μm
5–35	20
35–50	42.5
50–70	60
70–110	90
110–150	130
150–200	175
200–400	300
400–700	550

efficiency as high as 90% for particulates in the 5–10 μm range. The most serious problems encountered with these systems involve plugging and flow equalization.

Solution: Calculate the required collection efficiency E_R:

$$\begin{aligned} E_R &= \left[(3.1 - 0.06)/3.1\right](100) \\ &= 98\% \\ &= 0.98 \end{aligned}$$

Calculate the average particle size associated with each size range (see Table 12.6). Table 12.7 provides the overall efficiency E_6 for the 6 inch tubes. Since $E_6 > E_R$, the 6 inch tubes will do the job. Table 12.8 is generated for the 12 inch tubes. Since the overall efficiency $E_{12} < E_R$, the 12 inch tubes will not do the job. Thus, it will be necessary to use the 6 inch tubes for a conservative design.

The number of cyclones required to meet the specifications stated in the problem may also be calculated. Since the gas flow to a multiclone is axial (usually from the top), the cross-sectional area available for inlet flow is given by the annular area between the outlet tube and cyclone body. The outlet tube diameter is usually one-half the body diameter. Since the outlet tube diameter is

Table 12.7 6 inch tube efficiency.

Average particle diameter, μm	Weight fraction, w_i	Efficiency for 6 inch tubes, %	$E_i w_i$ for 6 inch tubes, %
20	0.05	89	4.45
42.5	0.05	97	4.85
60	0.10	98.5	9.85
90	0.20	99	19.8
130	0.20	100	20
175	0.20	100	20
300	0.10	100	10
550	0.10	100	10
			$E_6 = 98.95$

Table 12.8 12 inch tube efficiency.

Average particle diameter, μm	Weight fraction, w_i	Efficiency for 12 inch tubes, %	$E_i w_i$ for 12 inch tubes, %
20	0.05	82	4.1
42.5	0.05	93.5	4.67
60	0.10	96	9.6
90	0.20	98	19.6
130	0.20	100	20
175	0.20	100	20
300	0.10	100	10
550	0.10	100	10
			$E_{12} = 97.97$

one-half the body diameter, the inlet cross-sectional area (for axial flow) for each 6 inch (0.5 ft) tube is

$$A = 0.785(0.5^2 - 0.25^2)$$
$$= 0.147 \text{ ft}^2$$

The velocity in each tube is 60 ft/s. The number of tubes n is therefore given by

$$(60)(60)(0.147)(n) = 100{,}000$$

Solving for n, one obtains

$$n = 190 \text{ tubes}$$

required in this multiple-cyclone unit. A 15 × 15, 14 × 14, or 12 × 16 design is recommended. Note, these all include excess tubes to account for margin of error. It's always advised to have a few more than not enough.

12.6.2 Breakeven Efficiency

Gas (50,000 acfm) containing a dust at a loading of 2.0 gr/ft^3 is emitted from a copper plant in Butte, Montana. A cyclone is employed for particle capture, and the dust captured from the cyclone is worth \$0.01/lb of dust. For the sake of simplified calculation, assume that the efficiency of collection E is related to the system pressure drop ΔP, by the formula:

$$E = \frac{\Delta P}{\Delta P + 7.5}$$

where ΔP s in units of lb$_f$/ft^2. If the fan is 55% efficient (overall) and electric power costs \$0.18/kW·hr, at what collection efficiency is the cost of power equal to the value of the recovered material? What is the pressure drop in inches of H_2O at this condition?

Solution:

Take as a basis 1.0 minute.

Cost of recovered particulate ER, is

$$\begin{aligned} \text{ER} &= \left(50,000\,\text{ft}^3/\text{min}\right)\left(2.0\,\text{gr/ft}^3\right)(1\,\text{lb}/7,000\,\text{gr})(\$0.01/\text{lb})\text{E} \\ &= 0.143\,\text{E}, \$/\text{min} \end{aligned}$$

Cost of power to recover E_p is given by:

$$\begin{aligned} E_p &= \left(50,000\ \text{ft}^3/\text{min}\right)\left(\Delta\text{P lb}_\text{f}/\ \text{ft}^2\right)(1\ \text{min}\cdot\ \text{kW}/44,200\ \text{ft}\cdot\text{lb}_\text{f})(1/0.55) \\ &\quad \times\ (\$\ 0.18/\text{W}\cdot\text{hr})(1\ \text{hr}/60\ \text{min}) \\ &= 0.006\Delta\text{P},\ \$/\text{min} \end{aligned}$$

Since

$$E = \Delta P/(\Delta P + 7.5)$$

equates E_R with E_P and replacing E with the equation above. This yields

$$(0.143)\left[\Delta P/(\Delta P + 7.5)\right] = 0.006\ \Delta\text{P}$$

Solving by trial and error (or quadratically), one obtains

$$\begin{aligned} \Delta P &= 16.3\ \text{lb}_\text{f}/\text{ft}^2 \\ &= 3.1\ \text{in}\ \text{H}_2\text{O} \end{aligned}$$

The breakeven efficiency is therefore

$$\begin{aligned} \text{E} &= 16.3/(16.3 + 7.5) \\ &= 0.685 \\ &= 68.5\% \end{aligned}$$

12.6.3 THEO

Company THEO designs and predicts the performance of a special product line of cyclones using Lapple's "cut" diameter method. Company officials have decided to computerize the preceding calculations. This will require converting Lapple's figure (see Figure 12.9) to equation form.

Dr. Otto, one of the foremost authorities in the world on air pollution control equipment, has been hired to generate an equation describing fractional collection efficiency E as a function of either d_p or (d_{pc}/d_p). Tied down by numerous other consulting activities, Dr. Otto has decided to subcontract this job to you. He reasons that since you have taken an air pollution control equipment course, you now qualify as an authority in the field. Present a cyclone efficiency equation to THEO at your earliest convenience. Justify the form of your final equation.

Solution: This is obviously an open-ended problem: a problem for which there are several (if not an infinite number of) answers. One of the crudest approaches to estimate the overall efficiency of a unit in terms of d_p and d_{pc} $(d_{p,\text{cut}})$ is [8]:

$$E = \frac{d_p}{d_p + d_{pc}} \tag{12.7}$$

If the numerator and denominator are divided by d_p in this equation, one obtains:

$$E = \frac{1}{1 + \frac{d_{pc}}{d_p}}$$

Setting

$$Z = \frac{d_{pc}}{d_p}$$

Equation 12.7 becomes:

$$E = \frac{1}{1 + Z}$$

Assume a model of the form:

$$E = \frac{1}{1 + aZ^b}$$

So that

$$E + EaZ^b = 1$$

and

$$\frac{1 - E}{E} = aZ^b$$

Taking logarithms

$$\log\left(\frac{1 - E}{E}\right) = \log a + b \log Z$$

This is an equation of the form:

$$Y = MX + B$$

which is a linear equation!

Regressing the “data” from Figure 12.5 gives $a = 1.0$ and $b = 2.0$. For $a = 1.0$ and $b = 2.0$, a regression coefficient of 0.99946 was obtained. Therefore, one may exactly represent Lapple’s supposed “experimental field” data by [9].

$$E = \frac{1}{1 + \left(\frac{d_{pc}}{d_p}\right)^2} \tag{12.8}$$

The reader is left with the exercise of obtaining or generating another solution.

References

1 L. Theodore and R. Allen, *Air Pollution Control Equipment*, A Theodore Tutorial, Theodore Tutorials, East Williston, NY, 1993 (originally published by the APTI/USEOA. RTP, NC).
2 L. Theodore, *Air Pollution Control Equipment Calculations*, John Wiley & Sons, Hoboken, NJ, 2008.
3 A. J. Buonicore and L. Theodore, *Study of Industrial Air Quality Control Methods for Particulates*, Proceedings and Am. Environ. Eng. and Sci. Conference, Louisville, KY, 1972.
4 IGCI Gravity, *Louver and Dynamic Mechanical Collector*. Publication M-3, February, 1970.
5 D. Leith and W. Licht, Cyclone performance and design, *Atmos. Environ.*, 7, 527–549, 1973.
6 C. Lapple, Cyclone efficiency, *Chem. Eng.*, 58, 144, 1951.
7 L. Theodore, *Engineering Calculations: Cyclone Recovery/Collection Systems*, Chemical Engineering Progress, 18–20, New York City, NY, September 2005.
8 L. Theodore and R. Allen, personal notes, East Williston, NY, 1979.
9 L. Theodore and V. DePaola, Predicting cyclone efficiency, *J. Air Poll. Control Assoc.*, 10, 1132–1133, 1980.

Suggested Reading

1 B. van der Hegge Zigneu, *Proceedings of the Royal Academy*, Amsterdam, Netherlands, 1929.
2 A. ter Linden, *Investigations into Cyclone Dust Collectors*, Proceedings of the Institute Mechanical Collectors, Amsterdam, Holland, 1949.
3 A. Walker, *Cyclone Theory and Applications*, Lecture for Technical Staff Personnel, Research-Cottrell, Somerville, NJ, 1957.
4 IGCI Gravity, *Louver and Dynamic Mechanical Collector*. Publication M-3, February, 1970.
5 K. Wark and C. Warner, *Air Pollution; Its Origin and Control*, IEP, New York City, NY, 1975.
6 H. Parker, *Air Pollution*, Prentice-Hall, Upper Saddle River, NJ, 1977.
7 L. Theodore, *Air Pollution Control Equipment Calculations*, John Wiley & Sons, Hoboken, NJ, 2008.
8 L. Theodore, *Chemical Engineering: The Essential Reference*, McGraw Hill, New York City, NY, 2014.
9 R. Dupont, K. Ganesan and L. Theodore, *Unit Operation for Environmental Equipment*, Scrivener-Wiley Publications, Beverly, MA, 2017.
10 J. Richardson, R. R. Dupont, P. Farber, K. Ganesan and L. Theodore, *Air Pollution Control Equipment Revisited*, AWMA Conference, Calgary, CA, 2024.

13

Wet Scrubbers

13.1 Introduction

Wet scrubbers have found widespread use in cleaning contaminated gas streams because of their ability to effectively remove both particulate and gaseous pollutants. Specifically, wet scrubbing describes the technique of bringing a contaminated gas stream into intimate contact with a liquid. The particulate collection mechanisms involved in the wet scrubbing operation may include some or all of the following [1–4].

- Inertial impaction
- Direct interception
- Diffusion (Brownian movement)
- Electrostatic forces
- Gravitational forces
- Condensation
- Thermal gradient

Each of these mechanisms is briefly described below.

Inertial impaction occurs when an object (the droplet), placed in the path of a particulate-laden gas steam, causes the gas to diverge and flow around it. Larger particles, however, tend to continue in a straight path because of their inertia; they may impinge on the obstacle and be collected (as in Figure 13.1a). Direct interception also depends on inertia and is merely a secondary form of impaction. A collision occurs due to direct interception if the dust particle's center misses the target object by some dimension less than the particle's radius (see Figure 13.1b). Direct interception is, therefore, not a separate principle, but only an extension of inertial impaction. Diffusion is another extension of the impaction principle. Very small particles (submicron) suspended in a gas stream have an individual oscillatory motion known as Brownian movement (see Figure 13.1c). In this case, the particle and target collide as a result of relative motion within the limited space. As in all diffusional processes, the rate of diffusion is favored by large areas for diffusion, thereby necessitating small liquid droplets with high surface-to-volume ratios for high collection efficiencies. While collision or impaction may be the result of either inertia or Brownian movement, the results are the same. Electrostatic forces result when particles and liquid droplets become electrically charged (see Figure 13.1d). In addition, when only the particle or obstacle is charged, a charge may be induced on the uncharged component, resulting in a polarization force that can also effect particle removal. The electrostatic charge may be acquired, for example, by liquid droplets during their formation. An electrical charge may also be induced by flame ionization, friction, or the presence of

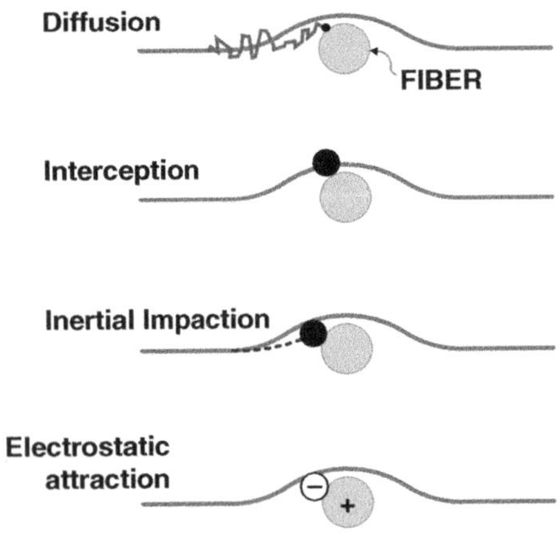

Figure 13.1 Filtration collection mechanisms: (a) diffusion, (b) interception, (c) inertial impaction, (d) electrostatic attraction.

charged matter. The effect of the electrostatic mechanism may be significant when the charge on the particle or obstacle is high and when gas velocity is low. Gravitation forces can also cause a particle, as it passes an obstacle, to fall from the streamline and settle on the surface of the obstacle. Such forces come into play, however, only when dealing with larger particles (usually greater than 40 μm). Condensation effects may also come into play. Condensation occurs if the gas or air is rapidly cooled below its dew-point. When moisture is condensed out of the gas stream, fogging occurs, and the dust particles can serve as condensation nuclei. The dust particles can become larger as a result of the condensed liquid, and the probability of removal by impaction is increased. Particle collection may also be affected by thermal gradients. Such forces can drive particles from hotter to colder regions. The motion can be caused by unequal gas molecular collision energy on the surfaces of the hot and cold sides of the particle; it is directly proportional to the temperature gradient.

13.2 Description of Wet Scrubber Equipment

Wet scrubbers are constructed with such a multiplicity of designs that no single type can be considered representative of the category as a whole. Some units simply consist of an existing dry-type collector modified by the introduction of a liquid phase to assist in particulate removal and to prevent particulate re-entrainment; other units are specifically designed to operate as wet collectors.

It is difficult to generalize about relationships among operating parameters such as pressure drop and liquid flow rate. It is also difficult to classify wet scrubbers according to particulate collection mechanisms since the usual case finds many of the collection mechanisms working simultaneously in each type of scrubber. However, in a very general sense, wet scrubbers may be loosely categorized by pressure drop (or energy consumption). Low-energy scrubbers are those with typical pressure drops less than 5″ WC; medium-energy scrubbers are those with typical pressure drops from 5 to 15″ WC; and high-energy scrubbers are those with typical pressure drops greater than 15″ WC. Spray chambers and spray towers, for example, provided the lowest pressure drop and, correspondingly, the lowest collection efficiencies; they would be classified as low-energy scrubbers. The medium-pressure-drop group could include centrifugal fan wet scrubbers, atomizing impingement collectors, and certain packed-bed scrubbers. The most familiar of the high-pressure-drop (high-energy) group is the venturi-type collector. Many scrubbers can conceivably be listed in more than one of the categories listed above, and each type of collector with its eventual application, determines whether it will be classified as a low-, medium-, or high-energy scrubber. Only conventional scrubbers are considered here.

Another important design consideration for wet scrubbers is concerned with suppressing the steam plume that is generated from the atomizing droplets. Water scrubber systems removing pollutants from high-temperature processes (i.e., combustion) can generate a supersaturated water vapor which becomes a visible white plume as it leaves the stack. Although not actually an air pollution problem, such a plume may be objectionable for aesthetic reasons. Regardless, there are several ways to avoid or eliminate the steam plume. The most obvious way is to specify control equipment which does not use water to interact with high-temperature gas streams. Should this not be possible or practical, a number of suppression methods are available involving:

1. Mixing the supersaturated gas stream with heated and relatively dry air.
2. Condensing the moisture by direct contact with water, and then mixing the condensed gas stream with heated ambient air.
3. Condensing the moisture by direct contact with water, and then reheating the scrubber exhaust gases well above the dew point of the gas prior to discharge from the stack.

Eleven classes of wet scrubbers are briefly described below. Additional description is available in the literature [1–4].

13.2.1 Spray Tower

The most common low-energy scrubbers are gravity spray towers in which liquid droplets are made to fall through rising exhaust gases and are drained at the bottom of the chamber (see Figure 13.2). The droplets are usually formed by liquid atomized in spray nozzles. The spray is directed into a chamber shaped to conduct the gas through the finely dispersed liquid. The relative velocity between the droplets and the gas in a vertical tower is eventually the terminal settling velocity of the droplets. However, to avoid spray droplet re-entrainment, the terminal settling velocity of the droplets must be greater than the velocity of the rising gas stream. In practice, the vertical gas velocity typically ranges from 2 to 5 ft/s. For higher velocities, a mist eliminator must be used in the top of the tower.

Spray towers are suited for both particle collection and mass transfer (gas absorption), as are all wet scrubbers. Operating characteristics include low-pressure drop (typically less than 1 to 2 inches of water, exclusive of mist eliminator and gas distribution plate), ability to handle spray liquids having a high solids content, and liquid requirements ranging from 3 to 20+ gal/1,000 ft^3 of gas treated. Spray towers are capable of handling large gas volumes and are often used as precoolers to reduce gas stream temperatures. Gas rates from 800 to 2,500 lb/hr ft^2 are typical. Gas retention times within the tower typically range from 20 to 30 sec.

The chief disadvantage of spray towers is their relatively low scrubbing efficiency for dust particles in the 0- to 5-μm range. Particles larger than 10 μm can typically be removed with efficiencies up to about 70%. The efficiency may be improved by the addition of high-pressure sprays, ranging from 100 to 400 psig. Depending on the spray pressure utilized, efficiencies can be increased by as much as 20%. The use of baffles can also increase collection efficiency. An important advantage of spray towers is their capability to treat relatively high dust concentrations without fear of any plugging. Although the units have a relatively large space requirement, they are inexpensive and are used in practice primarily for the collection of coarse dusts (greater than approximately 25 μm) and as a precooler.

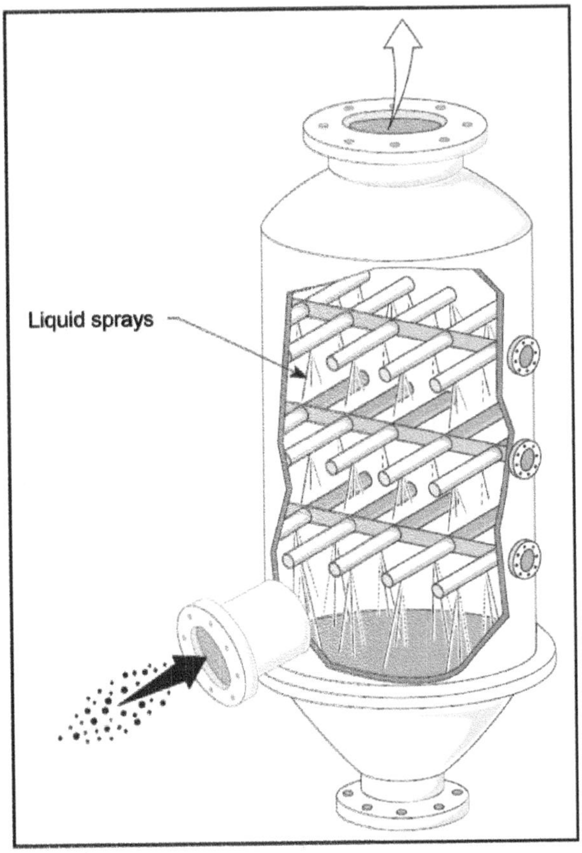

Figure 13.2 Gravity spray tower.

13.2.2 Venturi Scrubbers

To achieve high collection efficiency for particles by impaction, a small droplet diameter and high relative velocity between the particle and droplet are required. This is accomplished in a venturi scrubber by introducing the scrubbing liquid at an angle to a high-velocity gas flow in the venturi throat (vena contracta). Very small water droplets are formed, and high relative velocities are maintained until the droplets are accelerated to their terminal velocity. Gas velocities through the venturi throat typically range from 12,000 to 24,000 ft/min. The velocity of the gases alone causes the atomization of the liquid. The energy expended in the scrubber (except for the small amount used in the sprays and mist eliminators) is accounted for by the gas stream pressure drop through the scrubber. Another important factor to the effectiveness of the venturi scrubber is the conditioning of the particulates by condensation. If the gas in the reduced-pressure region in the throat is fully saturated, or supersaturated (preferably), some condensation will occur on the particulates in the throat due to the Joule–Thompson effect. Because of the cooling effect of the scrubbing liquid, condensation will be more pronounced if the gas is hot.

The size of the droplets generated in scrubber units affects both the collection efficiency and pressure drop through the unit, i.e., small droplet sizes requiring high-pressure atomization give greater collection efficiencies but produce larger pressure drops through the venturi than do large droplet sizes. Various correlations are available in the literature to estimate the mean liquid drop diameter from different types of atomizers under different operating conditions [1–4]. These correlations are applicable to fluids within a certain range of operating conditions and properties such as the volume ratio of liquid to gas, the relative velocity of gas to liquid, the type of nozzle, the surface tension of the liquid, and so on. It is important to select a correlation which takes these factors into consideration when using some of these correlations to estimate droplet diameter.

The venturi itself is only a gas conditioner and must be followed by a separating section for the elimination of entrained droplets (see Figure 13.3). Water is injected into the venturi in quantities ranging from 6 to 15 gal/1,000 ft^3 of gas. Very high collection efficiencies are achievable with operating pressure drops ranging from 6 to 70 inches of water. Pressure drops of 25- to 30-inch H_2O are not at all uncommon. For example, a 10-inch H_2O pressure-drop venturi can typically remove particles as small as a few micrometers with virtually 100% efficiency, while a 60-inch H_2O pressure-drop venturi is often required to remove particles as small as 0.3 to 0.4 μm. Since collection efficiency is directly related to pressure drop, variable-throat venturi scrubbers have been introduced to maintain constant pressure drop with varying gas flows. In these systems, the scrubbing efficiency and pressure drop may be adjusted by changing the position of the disk located in the venturi.

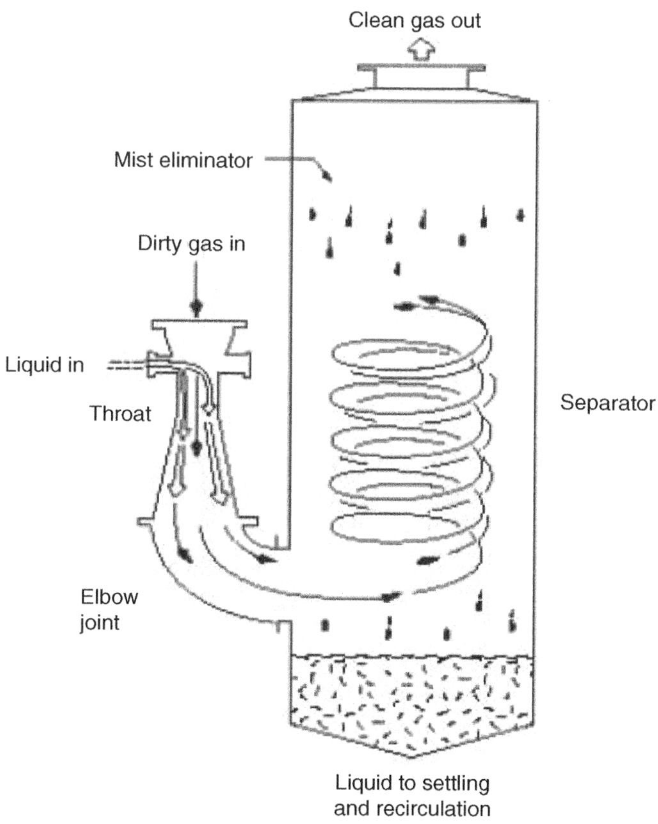

Figure 13.3 Venturi scrubber with cyclone separator.

13.2.3 Centrifugal Collectors

As noted, particulate collection efficiency is improved by increasing the relative velocity between the droplets and the gas stream. The particulates are assumed to travel at the same velocity as the gas stream. An increase in this relative velocity may be achieved by utilizing the centrifugal force of a spinning gas. The spin may come from introducing gases to the scrubber tangentially or by directing the gas stream against stationary swirl vanes. Although normally designed for specific installations, cyclonic spray scrubbers can accommodate gas entrance velocities up to 100+ ft/s. In centrifugal systems, the principal benefit is derived from the wetted walls hindering re-entrainment of collected particulates. Water rates of 2 to 5 gal/1,000 ft^3 of pressure drops from 2 to 6 inches of water are typical. The liquid spray may also be directed outward from sprays set in a central pipe manifold. Sprays in this fashion enhance collection efficiency by increasing particle impact on the spray droplets. Sprays directed in from the walls, are more easily serviced. However, since they can be made accessible from the outside of the scrubber, an unsprayed section is usually provided to give liquid droplets containing collected particles sufficient time to reach the walls of the collector and prevent them from exiting the unit in the cleaned gas stream. Rotating motion given to the gas stream by fixed vanes and impellers, with the scrubbing liquid introduced centrally either as spray or a liquid stream.

13.2.4 Dynamic Collectors

Dynamic collectors serve primarily as air movers (centrifugal blowers). Spray nozzles at the inlet offer an opportunity for impaction. Liquid droplets from the nozzles and uncaptured dust must pass through the many specifically shaped blades of the rotating fan wheel. Centrifugal forces and impingement on the blades are utilized for further collection and water separation. Problems have arisen, however, due to instability attributed to sludge buildup on the fan wheel. Slurry is removed and discharged separately from the cleaned air. The design pressure drop is about 6 inches of water with a maximum pressure drop of 9 inches. Water requirements typically range from 0.5 to 1.5 gal/1,000 ft^3 of gas, and power requirements range from 1 to 2 hp/1,000 acfm of gas treated. The collection efficiency is similar to other scrubbers with a 6-inch pressure drop. The chief advantages are compactness, moderate power requirements, and low water consumption.

13.2.5 Packed-bed Scrubbers

Packed-bed scrubbers find application in the removal of pollutant gases and vapors. However, such systems have also been considered for the removal of particulate matter. There are four general ways in which these beds may be operated to affect particulate removal. Cross-flow packed scrubbers will efficiently remove the relatively large particulates; they are also used for the removal of liquid particulates. The cocurrent flow packed scrubber is somewhat more efficient than the cross-flow and is effective in removing both solid and liquid particulates. For the collection of particulates as small as 3 to 5 μm, packed-countercurrent scrubbers have been found to be most efficient; these scrubbers operate generally at much higher liquid rates than either the cross-flow or the cocurrent flow types. Bed depth in packed countercurrent scrubbers is typically 2 to 6 ft; gas velocities typically range between 3 and 6 ft/s. Cross-flow beds are usually less than 2 ft in depth.

The difficulty of removal of particulate material increases rapidly with a decrease in particulate size. A 1 μm particle is only 10 times larger in diameter than a 0.1 μm particle, but it will have 100 times more surface area and 1,000 times more weight. As particle size decreases, the surface area/mass ratio increases so rapidly that eventually the surface properties can dominate over the mass properties. When this happens, higher velocities and more acute changes in direction are required to permit the particle to be separated from the gas stream by impingement. Cross-flow, cocurrent flow, and countercurrent flow packed scrubbers are not capable of achieving a high-enough gas velocity to remove effectively particulates smaller than 3 to 5 μm in diameter. To remove particles in this small size range, much more energy must be used to attain the velocities needed, and to avoid flooding, the liquid should be introduced in the same direction as the gas (i.e., the bed is operated in a cocurrent flow mode). Liquid requirements for cross-flow operation typically range from 1 to 4 gal/1,000 ft^3 of gas, with pressure drop from 0.1 to 0.25″ WC per foot of bed depth. Liquid requirements for countercurrent-packed scrubbers usually range from 10 to 20 gal/1,000 ft^3 of gas, with pressure drops typically 0.2 to 1.5″ WC per foot of bed depth. However, some adaptations, which make use of highly efficient turbulent layer above the bed, require considerably less scrubbing liquid, typically 2 to 2.5 gal/1,000 ft^3 of gas at a pressure drop ranging from 4 to 6″ WC. In cocurrent flow, a liquid requirement of 7 to 15 gal/1,000 ft^3 of gas is fairly common. Coarsely packed beds are used for removing coarse dusts and mists 10 μm and larger, with velocities through the bed on the order of 400 ft/min. Finely packed beds may be used for removing contaminants in the smaller size range, but because of pressure drop considerations the velocity through the bed must be kept relatively low, preferably below 50 ft/min. Finely packed beds have a greater tendency to plug, and their applications are generally limited to gas streams with relatively low grain loadings.

13.2.6 Moving-bed Scrubbers

Moving-bed (fluid-bed) scrubbers incorporate a zone of movable packing where gas and liquid can mix intimately. The system uses packing consisting of low-density polyethylene or polypropylene spheres about 1½ inches in diameter; these are kept in continuous motion between the upper and lower retaining grids. Such action keeps the spheres continually clean and considerably reduces any tendency for the bed to plug. Pressure drops typically range from 3 to 5″ WC (per stage), and collection efficiencies can be in excess of 99% for particles down to 2 μm. Particle collection may be enhanced by using several moving-bed stages in series.

13.2.7 Atomizing-inertial Scrubbers

The atomizing-inertial category of wet scrubbers probably represents the largest grouping of commercial dust collectors. Available designs vary more than in any other category. For the sake of convenience, this category has been further divided into those that utilize the kinetic energy of the gas stream to fragment or atomize the liquid on which the particle impacts and those that utilize a perforated plate with an impingement baffle over each perforation.

13.2.8 Orifice-type Wet Scrubbers

In orifice-type wet scrubbers (sometimes referred to as self-induced spray scrubbers), the gas stream comes into contact with a pool of liquid at the entrance to a constriction (submerged orifice). Liquid is entrained and carried into the restriction where greater liquid-particulate interaction occurs, resulting in a high frequency of particulate impaction on the droplets. Upon leaving the restriction, most of the water droplets (those large enough) are separated by gravity since the gas velocity is reduced from what is was in the restriction. Smaller droplets are subsequently removed by centrifugal force and impingement on baffles located in the upper part of the unit.

13.2.9 Impingement Plate Scrubbers

Impingement plate scrubbers utilize perforated plates with an impingement baffle over each perforation. The intention here is to expand the impingement area of the liquid through use of the gas stream's kinetic energy. Gas flowing up is divided into thousands of jets by the orifices. Gas velocities of 15 to 20 ft/s through the orifices are common. Each jet aspirates liquid from the blanket and creates a wetted surface on the baffle, located at the point of maximum jet velocity. The directed impingement on the wetted target dynamically precipitates particles and entraps them in the scrubbing liquid. On impingement, each jet forms minute gas bubbles which rise through and create turbulence in the liquid blanket. This provides extremely close gas-liquid contact for maximum cleaning. Continuous violent agitation of the blanket by the bubbles prevents settling of entrapped particles and flushes them away in the scrubbing liquid.

13.2.10 Atomizing-mechanical Scrubbers

In mechanically induced scrubbers, high-velocity sprays are generated at right angles to the direction of gas flow by a partially submerged rotor. The dirty gas stream passes through the area of the collector that contains mechanically produced droplets. Scrubbing is achieved by impaction because of both high radial-droplet velocity and vertical gas velocity. Liquid atomization occurs at both the rotor and the outlet wall.

Power and liquid requirements range from 3 to 10 hp and from 4 to 5 gal/1,000 ft^3 of gas, respectively, for the high-velocity design, depending on particle size and the desired collection efficiency. The chief advantages of these scrubbers are relatively low liquid and small space requirements, high scrubbing efficiency, and high dust-load capacity. The rotor, however, is susceptible to erosion from large particles and abrasive dusts. In addition, high-energy scrubbing applications usually require a mist eliminator.

13.3 Process Design Considerations

A number of performance and design equations have been developed from basic particle movement principles (theory) to explain the action of wet scrubbing systems. Many of these start from firm scientific concepts but give only qualitative results when predicting collection efficiencies or pressure drops. The interaction of particulate matter having a given particle size distribution with water droplets having another size distribution is not easy to express in quantitative terms. As a result of this complexity, experimentally determined parameters are usually required in order to perform engineering calculations.

One of the more popular and widely used collection efficiency equations is that originally suggested by Johnstone [5].

$$E = 1 - e^{-kR(\psi)^{0.5}} \tag{13.1}$$

where

E = efficiency (fractional)

ψ = inertial impaction parameter (dimensionless)

R = liquid-to-gas-ratio (gal/ 1,000 acf or gpm/1,000 acfm)

k = correlation coefficient, the value of which depends on the system geometry and operating conditions (typically 0.1–0.2 acf/gal)

The term ψ is given by:

$$\psi = \frac{C d_p^2 \rho_p v_t}{9 \mu_G d_0} \tag{13.2}$$

where

d_p = particle diameter (ft)
ρ_p = particle density (lb/ft^3)
v_t = throat velocity (ft/s)
μ_G = gas viscosity (lb/ft·s)
d_0 = mean droplet diameter (ft)
C = Cunningham correction factor

Values for the Cunningham correction factor (see Chapter 10 for more details) are provided in Table 13.1. This correction is usually neglected in scrubber calculations, but the effect becomes more pronounced as the particle size decreases, particularly below 1 μm.

The design of wet scrubbers usually focuses on those parameters affecting collection efficiency and pressure drop. In most cases, the scrubber must be designed to guarantee a specified collection efficiency, which, in turn, is strongly dependent upon pressure drop (among other parameters). The system pressure drop also dictates the power requirements and the size of auxiliary equipment such as the fans [5, 6].

The pressure drop for gas flowing through a venturi scrubber can be estimated from knowledge of liquid acceleration and frictional effects along the wall of the equipment. Frictional losses depend largely on the scrubber geometry and are usually determined experimentally. The effect of liquid acceleration is, however, predictable. Equation 13.3 for estimating pressure drop through venturi scrubbers given as a function of throat gas velocity and liquid-to-gas-ratio assumes that all the energy is used to accelerate the liquid droplets to the throat velocity.

$$\Delta P = 5 \times 10^{-5} v_t^2 R \tag{13.3}$$

where

ΔP = pressure drop (in H_2O)

Table 13.1 Cunningham correction factors for air at atmospheric pressure.

Particle diameter	Temperature		
	70 °F	**212 °F**	**500 °F**
0.1	2.88	3.61	5.14
0.25	1.682	1.952	2.525
0.5	1.325	1.446	1.711
1.0	1.160	1.217	1.338
2.5	1.064	1.087	1.133
5.0	1.032	1.043	1.067
10.0	1.016	1.022	1.033

Another somewhat simpler equation that applies over a fairly wide range of Rs is available in the literature [7–9].

$$\Delta P^* = 0.8 + 1.2R \tag{13.4}$$

where ΔP^* is a dimensionless pressure drop equal to the pressure drop divided by a velocity head.

Studies by Hesketh [10] show that the pressure drop predictions obtained from throat velocity measurements may be subject to error at low velocities if Equation 13.3 is applied for all ranges of velocities. Hesketh's equation for venturi scrubbers that have liquid injected before the throat is given by:

$$\Delta P = \frac{v_t^2 \rho A_t^{0.133} R^{0.78}}{1{,}270} \tag{13.5}$$

where

ρ = gas density, lb/ft^3

A_t = throat cross-sectional area, ft^2

Contact power theory [11] is an empirical approach relating particulate collection efficiency and pressure drop in wet scrubber systems. The concept is an out-growth of the observation that particulate collection efficiency in spray-type scrubbers was determined mainly by pressure drop for the gas plus any power expended in atomizing the liquid. Contact power theory assumes that the particulate collection efficiency of a scrubber is solely a function of the total pressure loss in the unit. The total power loss (p_t) is assumed to be composed of two parts: the power loss of the gas passing through the scrubber (p_G) and the power loss of the spray liquid during atomization (p_L). These two terms are given here in equation form:

$$p_G = 0.157\Delta P \tag{13.6}$$

where

p_G = contacting power based on gas stream energy input, HP/1,000 acfm

ΔP = pressure drop across the scrubber, in H_2O

and

$$p_L = 5.83 \times 10^{-4} P_L R \tag{13.7}$$

where

p_L = contact power based on liquid energy input (HP/1,000 acfm)

P_L = liquid inlet pressure (psia)

R = liquid-to-gas ratio (gal/1,000 acf)

The total power is then

$$p_t = p_G + p_L \tag{13.8}$$

To correlate contacting power with scrubber collection efficiency, the latter is best expressed as the number of transfer units. The number of transfer units is defined by analogy to mass transfer and given by:

$$N_t = \ln\left(\frac{1}{1-E}\right) \tag{13.9}$$

The relationship between the number of transfer units and collection efficiency is by no means unique. The number of transfer units for a given value of contacting power (HP/1,000 acfm) or vice versa, varies over nearly an order of magnitude. For example, at 2.5 transfer units ($E = 0.918$), the contacting power ranges from approximately 0.8 to 10.0 hp/1,000 acfm, depending on the scrubber and the particulate.

For a given scrubber and particulate properties, there is usually a relationship between the number of transfer units and the contacting power. Semrau plotted the number of transfer units (N_t) for a series of scrubbers and particulates against total power consumption; a linear relation on a log–log plot was obtained. This relationship is independent of the type of scrubber and can be expressed by [8]:

$$N_t = \alpha p_t^{\beta} \tag{13.10}$$

where α, β = parameters for the type of particulates being collected (see Table 13.2).

Available data indicate that venturis often operate with pressure drops in the inches of H_2O range. Liquid-to-gas ratios for venturi scrubbers are usually in the range of 5–20 gal/1,000 ft^3 of gas. At many facilities, liquid-to-gas ratios ranging from 7 to 45 gal/1,000 ft^3 of gas have been reported. In many cases, a minimum ratio of 7.5 gal/1,000 ft^3 is needed to ensure that adequate liquid is supplied to provide good gas "sweeping." Gas velocities for venturi scrubbers are in the 100–400 ft/s range. The low end of this range, 100–150 ft/s, is typical of power plant applications, while the upper end of the range has been applied to lime kilns and blast furnaces.

Table 13.2 Constant for use with Equation 13.10.

Aerosol	Scrubber type	α	β
Raw gas (lime dust and soda fume)	Venturi and cyclonic spray	1.47	1.05
Prewashed gas (soda fume)	Venturi, pipeline, and cyclonic spray	0.915	1.05
Talc dust	Venturi	2.97	0.362
Black liquor recovery furnace fume	Orifice and pipeline	2.70	0.362
Cold scrubbing water humid gases	Venturi and cyclonic spray	1.75	0.62
Hot fume solution for scrubbing (humid gases)	Venturi, pipeline, and cyclonic spray	0.74	0.861
Hot black liquor for scrubbing (dry gases)	Venturi evaporator	0.522	0.861
Phosphoric acid mist	Venturi	1.33	0.647
Foundry cupola dust	Venturi	1.35	0.621
Open-hearth steel furnace fume	Venturi	1.26	0.569
Talc dust	Venturi	1.16	0.655
Copper sulfate	Solivore (A) with mechanical spray generator	0.90	1.14
	Solivore (B) with hydraulic nozzles	0.562	1.06
Ferrosilicon furnace fume	Venturi and cyclonic spray	0.87	0.459
Odorous mist	Venturi	0.363	1.41

13.4 Equipment Design Considerations

Although wet scrubbers are relatively simple devices, they do require proper care to ensure long service life and trouble-free operation. Properly designed and installed units can do an excellent job in helping plants comply with air pollution control regulations. However, with most processing equipment, periodic reviews and inspections must be made to ensure that scrubbers are operating most efficiently. Improper performance of the equipment can usually be traced to any of the following:

1. Inadequate scrubber system design and/or operation.
2. Process conditions have changed so that the scrubber system cannot operate efficiently.
3. Scrubber system mechanical condition has deteriorated and performance is below design specifications.

Assuming that the proper unit was selected for the given application, most scrubber problems usually involve spray nozzle plugging, liquid circulation restrictions, and/or entrainment of droplets from the unit. Addressing these problems will improve operation and increase performance. Such problems may include the following:

Wet/Dry Zone Buildup. The scrubber design may improperly allow dry, dust-laden gas to contact the juncture of the scrubbing liquid and the vessel, causing dust buildup. Proper design prevents this contact by extending ductwork sections sufficiently into the scrubber and thoroughly wetting all scrubber surfaces through reliable means (usually gravity flush and sometimes sprays).

Nozzle Plugging. Nozzles plug through improper selection, too small orifices through which too dense scrubbing liquid must pass, improper header design, drawing off a sump that also settles (and concentrates) solids, erratic pump operation, chemical scaling, sealing, and mechanical failure.

Flow Imbalance. The headers external to the scrubber must send the required flow to the proper location at the proper rate. Many problems can be solved through the simple adjustment of flow using existing valves or dampers.

Buildup (Scaling). Scaling is the plating out of deposits on a scrubber surface and is most often related to the chemical composition, solubility, temperature, and pH of the scrubbing liquid. Proper control begins with the scrubber design and process control.

Localized Corrosion. Corrosion is a major factor in reducing the operating life of a scrubber, whether properly designed or not. Wells or pockets of liquid should be avoided and points of particulate buildup should be adequately flushed.

Instrumentation-Fitting Blockage. One problem that can often cause major trouble is instrumentation blockage or pluggage. Often, a standard fitting is not adequate in a scrubber. Specially designed fittings and connections are often recommended.

Sump Swirling. Sump swirling problems are most often associated with cyclonic devices. The swirling of the scrubbing liquid can cause severe wear and drainage problems unless arrested by antiswirl plates in the scrubber or rapid continuous draining.

Entrainment. Entrainment occurs when the droplet separator is not functioning properly. Nearly all scrubbers produce entrainment; only the properly designed systems reduce it to acceptable levels prior to discharge.

Re-entrainment. Re-entrainment occurs beyond the droplet removal section through improper draining or erratic flow patterns. It can also occur in stacks with very high velocities or where fittings protrude into a high-velocity area.

Liquid–Gas Maldistribution. The gas and liquid must be properly distributed for the given application. Each affects the other and is aggravated by the influence of baffles (designed or accidental), buildup, mechanical failure, wear, scaling in headers, or improper design.

Thermal Shock. Thermal shock may occur. When hot gases meet cold scrubber liquor, proper design permits gradual cooling rather than abrupt changes. Thermal shock is a relatively simple problem to prevent (i.e., through the use of multiple cooling zones).

Loss of Seal. The juncture of the scrubber liquid circuit with its surroundings is often a liquid seal. This seal may be at the top of a quencher or from an overflow connection. These lines must have seals able to prevent gas movement to or from the ambient surroundings. Loss of seal can cause entrainment or plugging, and instrumentation malfunction.

Wear. Wear can be tolerated unless it is localized. Unfortunately, a scrubber's functioning parts are also the wear parts. Expect to replace high wear points at frequent intervals. Excessive wear is often a result of excessively high solids concentration.

Vibration. This is most common in pumps and the fans associated with wet scrubber systems. Vibration problems are controlled by monitoring and scheduled preventive maintenance.

As in any air pollution control system, a scrubber system is subjected to a harsh environment. The abrasiveness of some types of particulate causes equipment such as fan wheels and valves to wear away. Stack gases may also contain a variety of components that, when wetted, produce compounds that corrode equipment. This is particularly true of any gas stream from a boiler that fires sulfur-bearing fuel and of gases from many chemical processes. All equipment components must be designed and maintained with these considerations in mind.

Regarding recent developments, several specialty devices are receiving more attention. Although on the market for many years, their unique collection mechanism has recently attracted more interest. Five such units include the *dynawave* unit, the *multi-microventuri* (MMV), the *electrodynamic venturi* (EDV), the *condensing scrubber*, and the *collision scrubber*. These are briefly described below.

The dynawave unit operates in a manner that creates a "froth" zone through which contaminated gas must pass. There is extreme turbulence in the zone and efficient collection of submicrometer particles occurs. The MMV scrubber consists of a "stacked" bank of staggered tubular elements arranged in a pattern similar to a bank of heat exchanger tubes. The thin spacing causes successive microventuri flows at the "pinch" sections leading to high efficiencies. The EDV concept consists of a sequence of fundamental functions, namely saturation, condensation, ionization, and filtration. In condensing scrubbers, water vapor in the flue gas stream is condensed onto particulates causing the particle to grow to a size that enables easier collection by conventional means. At the same time, the temperature and volume of the contaminated gases is reduced providing a lower capital and operating cost for the downstream collection systems and induced draft fan. Finally, in the collision scrubber two gas streams are made to collide head-on at the discharge of opposite venturi throats containing water droplets. The collision action shreds the water droplets into finer ones that can more effectively collect submicron particles. The process also produces a larger liquid surface area for gas absorption.

Some of the advantages and disadvantages of scrubber equipment are provided below:

Advantages

1. No secondary dust sources.
2. Small space requirements.
3. Ability to collect gases as well as particulates (especially "sticky" materials).
4. Ability to handle high-temperature, high-humidity gas streams.
5. Capital cost is low.
6. For some processes, the gas stream is already at high pressures (hence high-pressure drops to achieve high particulate removal efficiencies are not a listed to the scrubber system).

Disadvantages

1. May create water disposal problems.
2. Product is collected wet (energy is required to for additional separation).
3. Corrosion problems are more severe than with dry systems.
4. Steam plume opacity may be objectionable.
5. Pressure drop and horsepower requirements may be high.
6. Solids buildup at the wet–dry interface may be a problem.

After the scrubber system is installed, it is important that all of the major items of equipment, connecting pipe, and auxiliaries be inspected, cleaned, tested, and calibrated (if necessary) before startup.

General preoperation practices include checks to ensure that piping is free of debris, oil levels are correct, fans and pumps rotate in the proper direction, and alignments appear proper. Operating instructions should be reviewed with all appropriate plant personnel. A typical checkout list is provided by Richardson and Theodore [11].

Wet scrubbers are probably the most widely used and most often abused of all air pollution control devices. The proper operation and maintenance of these systems, in many cases, can enhance particulate removal control efficiency and reduce operating costs. An effective preventive maintenance program will also maximize the time between scheduled shutdown periods, significantly minimize unscheduled breakdowns, extend equipment life, and reduce maintenance and repair costs. Proper system design and adherence to a routine maintenance program will go a long way toward eliminating many of the problems that have plagued the operation of particulate scrubber systems.

13.5 Applications

Three illustrative examples complement the information presented in this chapter concerned with wet scrubbers.

13.5.1 Throat Area Calculation

A consulting firm has been requested to calculate the throat area of a venturi scrubber to operate at a specified collection efficiency. Pertinent data are given below.

- Volumetric flow rate of process gas stream = 11,040 acfm (at 68 °F)
- Density of dust = 187 lb/ft^3
- Liquid-to-gas ratio =2 gal/1,000 ft^3
- Average particle size = 3.2 μm (1.05×10^{-5} ft)

- Water droplet size = 48 μm (1.575×10^{-4} ft)
- Johnstone scrubber coefficient k = 0.14
- Required collection efficiency = 98%
- Viscosity of gas = 1.23×10^{-5} lb/(ft·s)
- Cunningham correction factor = 1.0

Solution: Calculate the inertial impaction parameter ψ from Johnstone's equation:

$$E = 1 - e^{-kR(\psi)^{0.5}} \tag{13.1}$$

Substituting:

$$0.98 = 1 - e^{-(0.14)(2)(\psi)^{0.5}}$$

Solving for ψ, one obtains:

$$\psi = 195.2$$

From the calculated value of ψ above, back calculate the gas velocity v_t at the venturi throat:

$$\psi = \frac{Cd_p^2 \rho_p v_t}{9\mu_G d_0}$$

$$v_t = \frac{9\psi d_0 \mu_G}{\rho_P d_P^2} = \frac{(9)(195.2)(1.575 \times 10^{-4})(1.23 \times 10^{-5})}{(187)(1.05 \times 10^{-5})^2} \tag{13.2}$$

$$= 165.1 \text{ ft/s}$$

Calculate the throat area S using the gas velocity at the venturi throat v:

$$S = q/v = (11{,}040)/[(60)(165.1)]$$

$$= 1.114 \text{ ft}^2$$

Note that approximately 1.0 ft^3 of throat area is generally required to treat 10,000–20,000 acfm [12].

13.5.2 Collection Efficiency Calculation

A venturi scrubber is employed to reduce the discharge of fly ash to the atmosphere. The unit is presently treating 215,000 acfm of gas, with a concentration of 4.25 gr/ft^3, and operating at a pressure drop of 32-inch H_2O. Experimental studies have yielded the following particle size collection efficiency data presented in Table 13.3.

Table 13.3 Size efficiency data for problem 13.5.2.

Particle diameter, μm	Weight fraction, w_i	Collection efficiency, %
5	0.00	30
10	0.00	42
20	0.02	86
30	0.05	93
50	0.08	97
75	0.10	98.7
100	0.75	99.9+

Table 13.4 Data calculations for problem 13.5.2.

Particle diameter, μm	Weight fraction, w_i	Collection efficiency, E_i%	w_iE_i%
5	0.00	30	0.00
10	0.00	42	0.00
20	0.02	86	1.72
30	0.05	93	4.65
50	0.08	97	7.76
75	0.10	98.7	9.87
100	0.75	99.9+	75.00
			$E_T = 99.00$

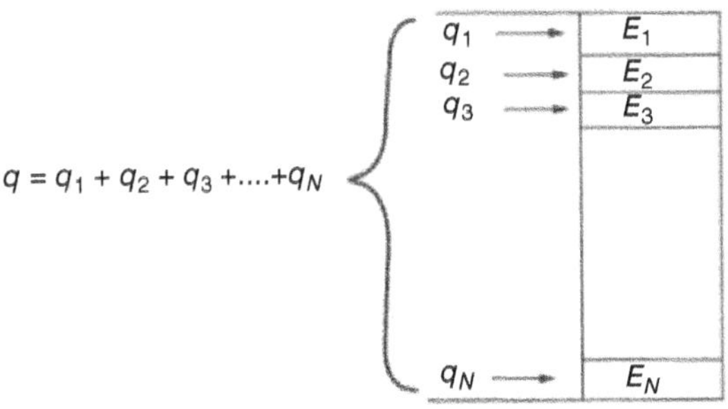

Figure 13.4 N scrubbers in parallel.

Estimate the overall collection efficiency of the unit.

Solution: The overall efficiency of the unit can be calculated from:

$$E_T = \sum w_i E_i$$

The solution is presented in Table 13.4.

13.5.3 Overall Efficiency in Multiple Scrubbers in Parallel

Calculate the overall efficiency of N scrubbers in parallel, assuming that the volumetric flow rates in each scrubber are q_1, q_2, ..., q_N and the corresponding efficiencies are E_1, E_2, ..., E_N, respectively. Assume that the gas is sufficiently well mixed that the particle concentration (particles/volume) is the same at the inlet of each scrubber. Express the result in terms of the q values and the corresponding E values (see Figure 13.4).

Solution: If c_1 is entering and c_{10} is leaving,

$$E_i = 1 - \frac{c_{i0}}{c_i}$$

Table 13.5 Data for three scrubbers in parallel.

Scrubber	*q*, acfm	*c*, gr/ft^3	*E*
1	2,500	2.0	0.996
2	5,000	4.0	0.985
3	2,500	2.0	0.996

or

$$c_{i0} = c_i(1 - E_i)$$

Thus,

$$\begin{aligned} E &= 1 - \frac{c_{10}\,q_1 + c_{20}\,q_2 + \ldots + c_{N0}\,q_N}{c_1\,q_1 + c_2\,q_2 + \ldots + c_N\,q_N} \\ &= 1 - \frac{c_1(1 - E_1)q_1 + c_2(1 - E_2)q_2 + \ldots + c_N(1 - E_N)q_N}{c_1\,q_1 + c_2\,q_2 + \ldots + c_N\,q_N} \\ &= \frac{\sum q_i c_i E_i}{\sum q_i c_i} \end{aligned}$$

If $c_1 = c_2 = \cdots = c_i$, the equation above reduces to

$$\begin{aligned} E &= \frac{\sum q_i E_i}{\sum q_i} \\ &= \frac{\sum q_i E_i}{q} \end{aligned}$$

Also calculate the overall efficiency of three venturi scrubbers operating in parallel treating 10,000 acfm of gas. Data are provided in Table 13.5.

Employ a modified form of the equation developed above

$$E = 1 - \frac{\sum q_i c_i E_i}{\sum q_i c_i}$$

For the flow rates,

$$\begin{aligned} \text{Inlet} &= (2)(2{,}500)(2.0) + (5{,}000)(4.0) = 30{,}000 \text{ gr/min} \\ \text{Outlet} &= (2)(2{,}500)(2.0)(0.004)(5{,}000)(4.0)(0.015) = 340 \text{ gr/min} \end{aligned}$$

Thus,

$$\begin{aligned} E_0 &= \frac{30{,}000 - 340}{30{,}000} \\ &= 0.9887 = 98.87\% \end{aligned}$$

References

1 L. Theodore and A. J. Buonicore, *Control of Particulate Emissions*, USEPA/APTI Manual for Course 413, RTP, NC, 1979.
2 L. Theodore, *Air Pollution Control Equipment Calculations*, John Wiley & Sons, Hoboken, NJ, 2008.
3 L. Theodore, *Chemical Engineering: The Essential Reference*, McGraw Hill, New York City, NY, 2012.
4 L. Theodore, R. R. Dupont and K. Ganesan, *Unit Operations in Environmental Engineering*, John Wiley & Sons, Hoboken, NJ, 2017.
5 H. F. Johnstone and M. H. Roberts, *Ind. Eng. Chem.*, 41, 2417, 1949.
6 H. F. Johnstone, R. B. Field and M. C. Tassler, *Ind. Eng. Chem.* 46, 1601, 1954.
7 L. Theodore and A. J. Buonicore, *Industrial Air Pollution Control Equipment for Particulates*, CRC Press, West Palm Beach, FL, 1976.
8 W. Strauss. *Industrial Gas Cleaning*. Pergamon Press, New York City, NY, 1966.
9 L. Theodore, *Scrubber Design*, Prac. Environ. Eng. Sci. Conf., 4, 365, Washington, D.C., 1974.
10 H. Hesketh, Scrubber design, *J. Air. Pollut. Control. Assoc.*, 24, 938–942, 1974.
11 K. Semrau, Contact power theory, *JAPCA*, 10, 200–202, 1960.
12 L. Theodore, *personal notes*, East Williston, NY, 1993.

Suggested Readings

1 H. F. Johnstone and M. H. Roberts, *Ind. Eng. Chem.*, 41, 2417, 1949.
2 H. F. Johnstone, R. B. Field and M. C. Tassler, *Ind. Eng. Chem.* 46, 1601, 1954.
3 W. Strauss, *Industrial Gas Cleaning*, Pergamon Press, New York City, NY, 1966.
4 Calvert et al., *Wet Scrubber Systems Study*, EPA Report No. PB-213 017, Washington D.C., 1972.
5 J. A. Danielson, *Air Pollution Engineering Manual*, EPA OAP Publication No. AP-40, Washington D.C., 1973.
6 L. Theodore and A. J. Buonicore, *Industrial Air Pollution Control Equipment for Particulates*, CRC Press, West Palm Beach, FL, 1976.
7 A. C. Stern, *Fundamentals of Air Pollution*, Academic Press, New York City, NY, 1974.
8 K. Wark and C. Warner, *Air Pollution: It's Origin and Control*, IEP, New York City, NY, 1975.
9 L. Theodore, *Air Pollution Control Equipment Calculations*, John Wiley & Sons, Hoboken, NJ, 2008.
10 L. Theodore, *Chemical Engineering: The Essential Reference*, McGraw Hill, New York City, NY, 2012.
11 L. Theodore, R. R. Dupont and K. Ganesan, *Unit Operations in Environmental Engineering*, John Wiley & Sons, Hoboken, NJ, 2017.
12 J. Richardson, R. R. Dupont, P. Farber, K. Ganesan and L. Theodore, *Air Pollution Control Equipment Revisited*, AWMA Conference, Calgary, CA, 2024.
13 J. Richardson and L. Theodore, *Optimizing Air Pollution Equipment Performance: Operation and Maintenance*, John Wiley & Sons, Hoboken, NJ, 2025.

14

Electrostatic Precipitators (ESPs)

14.1 Introduction

Electrostatic precipitation is one of the most popular methods in use for removing fine solids and liquids from gas streams. The first successful commercial-scale precipitators were installed by Frederick G. Cottrell on two California chemical plants in 1907. One of these units treated 5,000 cfm of gas and recovered precious metals and sulfuric acid. In 1910, Cottrell installed a successful precipitator on a 250,000-cfm smelter. The technology grew rapidly, and successful applications were soon found on a variety of industrial processes and in the power generation industry. Precipitators have been used to treat over 5,000,000 cfm of gas and have attained removal efficiencies approaching 99.99%.

Compared to other methods of gas cleaning, electrostatic precipitators are as elegant as they are efficient. Instead of performing work on the entire gas stream in the cleaning process, the forces in a precipitator are applied directly to the suspended particles themselves. The result is high collection efficiency at a moderate cost in power. This chapter is by far the longest due to a combination of complexity, unique nature, and widespread use of the ESP systems.

There are three fundamental steps in the electrostatic precipitation process:

1. Charging the particles suspended in the gas stream
2. Collecting the charged particles
3. Removal of the collected particles into an external receptacle

Charging is accomplished by applying a high dc voltage to an electrode system. The presence of a grounded electrode near the charged electrode gives rise to the formation of corona and a unidirectional electric field. Particles traveling between the two electrodes acquire charge from the corona discharge and are driven toward the collecting electrode by the electric field. All commercial precipitators use negative polarity.

Particles arriving at the collecting electrode do not have to be captured immediately for successful operation and may actually be recollected several times before they are removed. Current flows in the collection process are large enough to describe the operation as electrical in nature, the term "electrostatic" being one of convention rather than description. Precipitators may be single stage (or Cottrell type), or two stage, in which the charging and collecting fields are formed independently. Also, the gas stream may flow through tubular collecting electrodes or between parallel plates. The parallel-plate single-stage precipitator is by far the most common.

Particles remain on the collecting electrode, or plate, until they are dislodged by a mechanical blow, called rapping, and fall by gravity into an external vessel, usually some form of hopper. Rapping is not as simple as it may sound and indeed must be as effective as the charging and collecting steps for efficient precipitation.

14.2 Description of Electrostatic Precipitation Equipment

The heart of the electrostatic precipitation process is the discharge electrode system. It must produce a strong, uniform corona while maintaining the correct distance and alignment with respect to the collecting electrodes to prevent imbalances in the electric field and to avoid unnecessary arcing discharges. Discharge electrode sizes and shapes vary mainly by manufacturer, but variations among different applications or in different sections of the same precipitator are feasible. Round, straight wires about 0.1 inch in diameter are the most common discharge electrode in use in the earlier ESP designs. They were hung individually and freely with a suspension weight at their bottom end, or were held in a structural framework which is rigidly attached to the precipitator structure. Modern discharge electrodes utilize a rigid frame design and may be mast-type, or formed elements, where rigidity, mechanical strength, and corona properties are all embodied into the same member.

The physical arrangement of precipitators differs depending on the type of application. Wire and cylinder electrodes are used in some applications; however, for reasons of space economy, most commercial precipitators use plates as collection electrodes. The majority of precipitators are constructed so that the charging and collection steps take place within the same region. Precipitators of this type are termed single stage. For some applications, charging takes place in one section which is then followed by a section consisting of alternately charged plates. The collecting electric field is established independently of the corona field, and such precipitators are termed two-stage.

Electrostatic precipitation occurs in the space between the discharge electrode and the collection surface. A high-voltage, pulsating direct current is applied to an electrode system consisting of a small-diameter discharge electrode that is usually negatively charged, and a collecting plate electrode that is grounded. This produces a unidirectional, nonuniform electric field whose magnitude is highest near the discharge electrode.

Resistivity is related to the ability of a particle to take on a charge. In most industrial applications, the resistivity of the particle is such that the charge on the particle is only partially discharged on contact with the grounded collection electrode. A portion of the charge is returned and contributes to the intermolecular cohesive and adhesive forces that hold the particles to the collection surfaces. The dust layer builds up on the collection plate to a thickness of 0.5 in (1.27 cm). If the dust layer becomes too thick, it is possible for the accumulated layer to act as an insulator, reducing the flow of the electric field lines.

The negative gas ion movement has two main charging effects on dust particles in the interelectrode region. These effects are called field charging and diffusion charging. Each type of charging is used to some extent in particle charging, but one dominates depending on the particle size. Field charging dominates for particles with a diameter >1.0 μm, while diffusion charging dominates for particles with a diameter <0.3 μm. A combination of these mechanisms occurs for particles in the range between 0.3 and 1.0 μm in diameter. It is also possible to charge particles by electron charging. In this case, free electrons that do not combine with gas ions are moving at an extremely fast rate. These electrons hit the particle and impart a charge. However, this effect is responsible for very little particle charging.

In field charging, as particles enter the electric field, they cause local dislocation of the field. Negative gas ions traveling along the electric field lines collide with the suspended particles and impart a charge on them. The ions continue to bombard the particle until the charge on the particle is sufficient to divert the electric lines away from the charged particle. This prevents new ions from colliding with the dust particle. When a particle no longer receives an ion charge, it is said to be saturated. Saturated charged particles then migrate to the collection electrode and are collected.

Diffusion charging is associated with the random Brownian [1] motion of the negative gas ions. The random motion is related to the thermal velocity of the gas ions; the higher the temperature, the more movement occurs. Negative ions collide with the particles because of the random thermal motion of the ions and impart a charge on the particle. The charged particles migrate to the collection electrode. The mechanism is important for charging particles in the submicron range. As noted above, in the intermediate-size range from 0.3 to 1.0 μm in diameter, both field and diffusion charging are important.

In practice, the applied voltage is increased until it produces a discharge (corona), which can be seen as a luminous blue glow around the discharge electrode. The corona is a discharge phenomenon in which gaseous molecules are ionized by electron collisions in the region of a high electric field. The intense electric field close to the discharge electrode accelerates the free electrons that are present in the gas. These electrons acquire sufficient velocity to ionize gas molecules upon collision, producing a positive ion and an additional free electron (see Figure 14.1).

The additional free electrons create more positive ions and free electrons as they collide with additional gas molecules. This process is called avalanche multiplication and occurs in the corona glow region (see Figure 14.2). Avalanche multiplication will continue until the local electric field strength decreases to the point where there is insufficient energy to perpetuate ionization. The sluggish positive ions migrate back to the negative discharge electrode and form new free electrons on impaction with the discharge wire or gas space around the wire. The electrons produced during the avalanche multiplication process follow the electric field toward the grounded collection electrode.

With avalanche multiplication, the electrons leave the corona region and enter the interelectrode region. The magnitude of the electric field is diminished and the free electrons' velocity decrease.

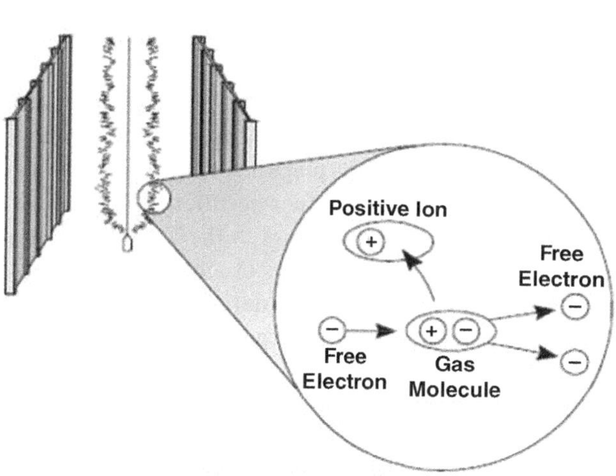

Figure 14.1 Generation of corona.

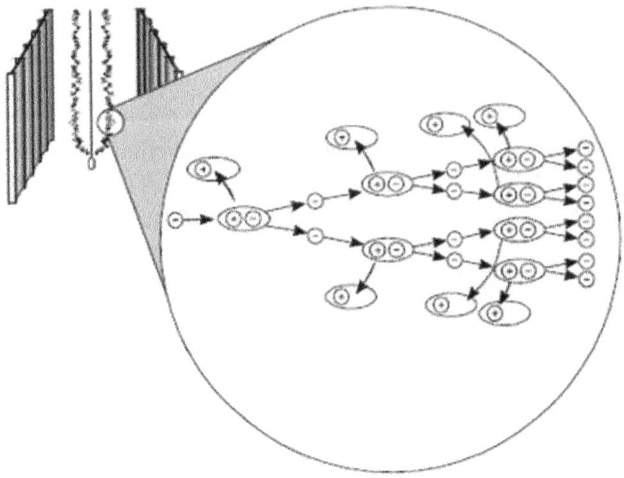

Figure 14.2 Avalanche multiplication.

When electrons impact gas molecules in the interelectrode region, they are captured, and negative gas ions are created. These negative ions serve as the principal mechanism for charging the dust.

Negative gas ions migrate toward the grounded collection electrode. A space charge, which is a stable concentration of negative gas ions, forms in the interelectrode region. Increases in the applied voltage will increase the field strength and ion formation until sparkover occurs. Sparkover refers to internal sparking between the discharge and collection electrodes. It is a sudden rush of localized electric current through the gas layer between the two electrodes. Sparking causes an immediate short-term collapse of the electric field. In general, it is very desirable to operate at voltages high enough to cause some sparking but not at a frequency such that the electric field constantly collapses. The average sparkover rate for optimum precipitator operation is between 50 and 100 sparks per minute. At this spark rate, the gains in efficiency associated with increased voltage compensate for decreased gas ionization due to the collapse of the electric field. For optimum efficiency, the electric field strength should be as high as possible. This is accomplished by applying a high voltage to the discharge electrode with the consequent high corona current flow from the discharge electrode to the collection electrode.

High-voltage rectifiers providing pulsating DC waveforms are in use almost exclusively because of the higher voltage and current attainable under sparking conditions when compared to pure direct current. With few exceptions, the discharge electrode system is subdivided into discrete sections, each being energized by a separate transformer–rectifier (TR) set. This "sectionalization" is important in matching corona currents and voltages to the TR set, and to promote reliability and stability under arcing conditions. A TR set consists of a high-voltage transformer and bridge rectifier, with typical secondary winding root mean square (rms) ratings between 53–66 kV and 250–2,000 mA. Most TR sets can be connected to the precipitator discharge electrode system in either full-wave or double half-wave.

An important adjunct to the discharge electrode system is the automatic regulation of the high-voltage input to the precipitator, because only in rare cases does nature permit ideal operation. The earliest precipitators had no means of voltage regulation, but state of the art advancements such as the silicon-controlled rectifier, solid-state construction, and digital control circuits have enabled precipitators to perform efficiently under the most adverse conditions.

In dry, parallel-plate precipitators, the collecting electrodes (plates) are suspended from the top of the precipitator, parallel to and in proper alignment with the discharge electrodes. These plates and their attachment must be strong enough to dislodge particulate when rapped, and durable enough to withstand millions of rapping blows without fatigue failure. For these reasons, collecting plates are typically made of light-gauge metal and rigidly fastened to the precipitator structure only at their top ends.

Most designs incorporate baffles to provide quiescent zones where the probability of particle collection is enhanced. These baffles are integrated into vertical stiffeners, which are required because collecting plate heights of up to 50 ft are occasionally in use.

Hoppers are best thought of as temporary holding bins to store collected particulate until permanent disposal can be scheduled. They take a variety of shapes, and are also sectionalized to facilitate handling large quantities of dust. It is a mistake to think the precipitation process stops at the hoppers. Removing collected material from the hopper is just as important as getting the material to fall into the hopper in the first place.

The discharge electrodes, collecting plates, and hoppers are all contained and supported by the casing, or shell. This structure must provide a gas-tight envelope in which the process takes place and must also hold the two electrode systems in proper alignment, sometimes under cyclical load conditions where the gas temperature (and hence the inner shell temperature) may vary over several hundred degrees Fahrenheit. In nearly all economical designs, essentially all the important auxiliary equipment are attached somewhere directly to the casing. Insulators that support the discharge electrodes system are made almost exclusively from porcelain of fused alumina, and are contained in individual or grouped insulator compartments or all the high-tension support insulators may be housed in a top housing or penthouse. In many cases, the manufacturer requires these insulators to be heated and ventilated at certain times during operation of the precipitator.

The widest variation in design among manufacturers comes from rapping. In the simplest sense, rappers are either impulse (single blow) or vibrating (multiple blow) types. One impulse type consists of swing hammers striking the lower portions of the collecting electrodes or an intermediate portion of a rigid frame discharge electrode system. The swing hammers are actuated by a cam shaft driven by an electric motor, and swing from the peak of the cam action by gravity to strike the electrode systems in a horizontal direction. Another type of impulse rapper is the drop hammer, which may be actuated by an electromagnetic solenoid located on top of the precipitators, or a motor-driven mechanical linkage. The rapping blow is gravity actuated or may be spring-assisted, striking the tops of the systems in a vertical direction, or nearly any combination among these practices. Vibrating types are used almost exclusively on the discharge electrode systems, especially on industrial processes where weighted wire geometries are most common. Vibrators may be electric or pneumatic, the latter being extremely powerful and potentially destructive. Some precipitator designs utilize what is called "gentle rapping" where a gentle rapping of the plates at regular intervals causes the collected dust to move down the plate over a period of time before being discharged into the collecting hopper.

Other key components in a precipitator include the high-voltage bus and guard that deliver the TR output to the discharge electrodes; gas distribution diffusers at the inlet and outlet face; access systems, and a host of other systems, subsystems, and components necessary to support the operation by providing electrical power distribution, control, insulation, and so on. Most of the above features are shown in Figure 14.3.

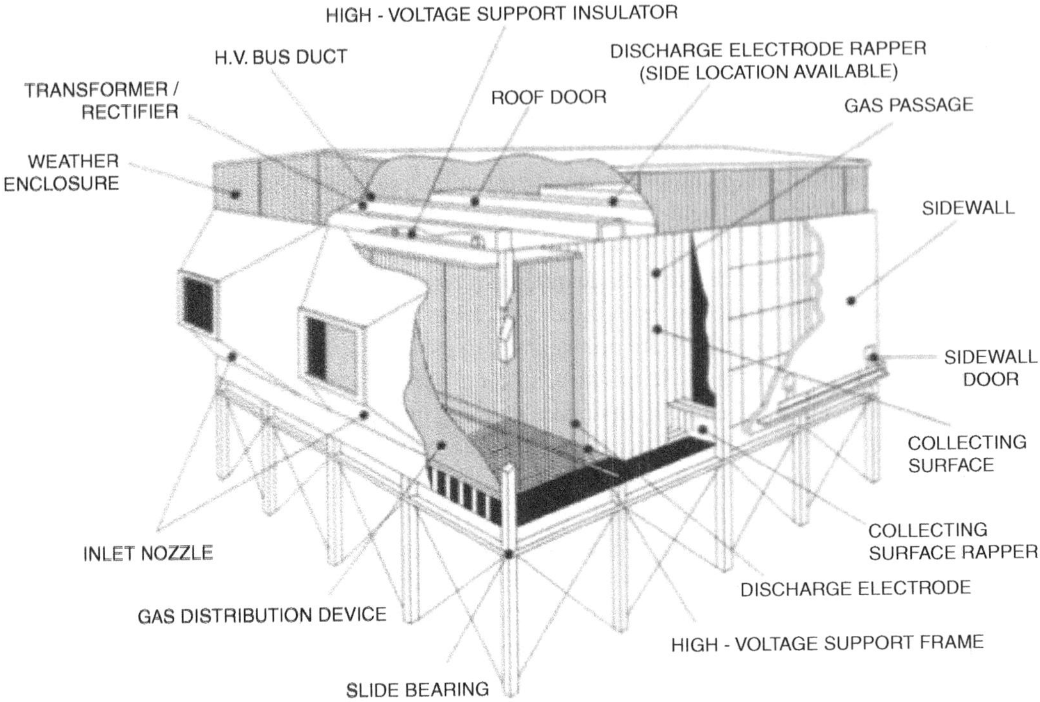

Figure 14.3 Electrostatic precipitator diagram.

As noted above, there are basically two types of electrostatic precipitators (ESPs): high-voltage single-stage and low-voltage double-stage. The high-voltage single-stage precipitator is the more popular type and has been used successfully to collect both solid and liquid particulate matter in industrial facilities such as smelters, steel furnaces, cement kilns, municipal incinerators, and utility boilers. Low-voltage double-stage precipitators are limited almost exclusively to the collection of liquid aerosols discharged from sources such as meat smokehouses, pipe-coating machines, asphalt paper saturators, and high-speed grinding machines.

Low-voltage double-stage precipitators were originally designed for air purification in conjunction with air conditioning systems (they are also referred to as electronic air filters). Double-stage ESPs have been used primarily for the control of finely divided liquid particles. Controlling solid or sticky materials is usually difficult, and the collector becomes ineffective for dust loadings greater than 0.4 grains per standard cubic foot (7.35×10^{-4} gr/m^3). Therefore, double-stage precipitators have limited use for particulate emission control. The low-voltage two-stage precipitator differs from the high-voltage single-stage precipitator in terms of both design and amount of applied voltage. As noted earlier, the double-stage ESP has separate particle charging and collection stages. The ionizing stage consists of a series of small (0.007 inch diameter) positively charged wires equally spaced 1 to 2 inches from parallel ground tubes that charge the particles suspended in the airflow through the ionizer. The direct-current potential applied to the wires is approximately 12–13 kV. The second stage consists of parallel metal plates less than 1 inch (2.5 cm) apart. The liquid particles receive a positive charge in the ionizer stage and are collected at the negative plates in the second stage. Collected liquids drain by gravity to a pan located below the plates. The two major types of high-voltage single-stage ESP configurations are tubular and plate, and particles are both charged and collected in a single stage.

Tubular precipitators consist of cylindrical collection electrodes with discharge electrodes located in the center of the cylinders. Dirty gas flows into the cylinder where precipitation occurs. The negatively charged particulates migrate to and are collected on grounded collecting tubes. The collected dust or liquid is removed by washing the tubes with water sprays located directly above the tubes. These precipitators are generally referred to as water-walled ESPs. Tubular precipitators are generally used for collecting mists or fogs. Tube diameters typically vary from 0.5 to 1 ft (0.15–0.31 m), with lengths usually ranging from 6 to 15 ft (1.85–4.6 m).

Summarizing, plate electrostatic precipitators are used more often than tubular ESPs in industrial applications. High voltage is used to subject the particles in the gas stream to an intense electric field. Dirty gas flows into a chamber consisting of a series of discharge electrode (wire) space along the center line of adjacent plates. Charged particles migrate to and are collected at oppositely charged collection plates. Collected particles are usually removed by rapping (dry precipitator) or by a liquid film (wet precipitator). Particles fall by force of gravity into hoppers where they are stored prior to removal and final disposal.

14.3 Process Design Considerations

There are many ways to describe the design of any industrial machine, and precipitators are no exception. However, in almost all cases, precipitator design can be effectively divided into the four following aspects, each equally important over the life of the plant:

1. Performance
2. Reliability
3. Operability
4. Maintainability

Performance design begins with sizing, or selection of the correct amount of collecting surface area, its arrangement, how it is energized, and how it is rapped. A fundamental knowledge of sizing is important to the operator for understanding the basic variables that control precipitator performance. From empirical observations by Anderson [2–3] in 1919 and theoretical derivation by Deutsch [4–5] in 1922, the first expression for precipitator performance (Deutsch–Anderson equation) took the form

$$E = 1 - e^{-w\left(\frac{A}{q}\right)} \tag{14.1}$$

where

E = collection efficiency
A = total collecting surface area
q = gas flow rate
w = migration velocity

Since E and q are usually known, the amount of collecting surface A can be calculated from a judicious selection of the term w. For decades, designers selected w by analogy, that is, from experience with similar types of dusts and the performance that resulted. For each of these cases, w was presumed constant. Sizing by analogy and some inherent inaccuracies in the Deutsch–Anderson equation related to particle size and uniformity led to some disappointing results, especially for very high efficiency requirements. It is virtually impossible to find a truly analogous situation, since industrial dusts are composed of many different-size particles, and w is in fact not constant but dependent upon many process variables.

An attempt to account for the sensitivity of w on process variables, especially particle-size distribution, appeared in 1957 and was later revised by Allander, Matts, and Ohnfeldt [6], who derived the expression

$$E = 1 - e^{-\left(\frac{Aw_k}{q}\right)^n} \tag{14.2}$$

The second exponent (in this, the so-called modified Deutsch equation) provides a more accurate prediction of performance at high-efficiency levels but becomes too pessimistic in certain situations. A more useful relationship, developed by Feldman in 1975 [7], accounted for such variables as particle-size distribution at the precipitator inlet, changes in size distribution as particles are collected in the precipitator, and operating voltage levels.

The aforementioned migration velocity is sensitive to the voltage and/or the electric field. Therefore, the precipitator must be designed using the maximum electric field for maximum collection efficiency. The migration velocity is also dependent on particle size; generally, larger particles are collected more easily than smaller ones. The particle migration velocity can be estimated by the following equation:

$$w = \frac{q_p E_p}{6\pi\mu r} \tag{14.3}$$

where

q_p = particle charge (charges)
E_p = strength of field in which particles are collected, V/m
μ = gas viscosity, Pa·s
r = radius of the particle, μm

However, most ESPs are designed using a particle migration velocity based on field experience rather than theory. Typical particle migration velocity rates such as those listed in Table 14.1

Table 14.1 Typical precipitation rate parameters for various applications.

Application	Particle migration velocity	
	(ft/s)	(cm/s)
Utility fly ash	0.13–0.67	4.0–20.4
Pulverized coal fly ash	0.33–0.44	10.1–13.4
Pulp and paper mills	0.21–0.31	6.4–9.5
Sulfuric acid mist	0.19–0.25	5.8–7.62
Cement (wet process)	0.33–0.37	10.1–11.3
Cement (dry process)	0.19–0.23	6.4–7.0
Gypsum	0.52–0.64	15.8–19.5
Smelter	0.06	1.8
Open-hearth furnace	0.16–0.19	4.9–5.8
Blast furnace	0.20–0.46	6.1–14.0
Hot phosphorus	0.09	2.7

Application	Particle migration velocity	
	(ft/s)	(cm/s)
Flash roaster	0.25	7.6
Multiple hearth roaster	0.26	7.9
Catalyst dust	0.25	7.6
Cupola	0.10–0.12	3.0–3.7

have been published by various ESP vendors. These values can be used to estimate the collection efficiency of the ESP.

Probably the best way to gain insight into the process of electrostatic precipitation is to study the aforementioned Duetsch–Anderson equation. This equation is used to determine the collection efficiency of the precipitator under ideal conditions. The simplest form of the design equation has been used extensively for many years for theoretical collection efficiency calculations. Unfortunately, while the equation is scientifically valid, there are a number of operating parameters that can cause the results to be in error by a factor of >2.

The Deutsch–Anderson equation neglects three significant process variables:

1. It completely ignores the fact that dust re-entrainment may occur during the rapping process.
2. It assumes that the particle size and, consequently, the migration velocity are uniform for all particles in the gas stream.
3. It assumes that the gas flow rate is uniform everywhere across the precipitator and that particle sneakage through the hopper section does not occur.

Therefore, this equation should be used only for making preliminary estimates of precipitation collection efficiency.

When the desired collection efficiency and gas flow rate are specified, the required collecting area can be determined from the above Deutsch–Anderson equation once an appropriate precipitation rate parameter has been chosen. Later, better correlations with field data on high-efficiency ESPs have been obtained by raising the exponential term in Equation 14.1 to a power m using existing values of w:

$$E = 1 - e^{-\left(\frac{wA}{q}\right)^m} \tag{14.4}$$

This provides a more accurate prediction of performance at high efficiency levels, but can become too pessimistic in certain situations. Typical values of m range between 0.4 and 0.7, with 0.5 as the norm. Equation 14.4 is referred to as the Matts–Ohnfeldt equation.

These and many other models have been proposed to predict particulate collection in an electrostatic precipitator, and while special advantages can be argued for each depending on which critical variables they account for, the fact remains that no model exists that accounts for all the variables that describe migration velocity in all situations. This is attributable mainly to the large number of variables that exist and because many of them are interrelated. In the final analysis, the goal is still to determine the correct amount of collecting surface area, and that usually depends on the proper selection of w.

As discussed above, large particles, over 1.0 μm in diameter, are charged mainly by direct collision with ions and free electrons moving toward the collecting plate along electric field lines. This field charging mechanism is very efficient for large particles, but the velocity of large charged particles toward the collecting plate is impeded by viscous drag forces. Small particles, less than 0.5 μm, become charged mainly by random thermal motion of ions. This diffusion charging mechanism is less efficient, but charged small particles can proceed toward the collecting plate with relative ease since the drag forces opposing their motion are small. The ability of a precipitator to collect the large and small particle sizes is essentially equal, but more difficult for the intermediate-sized particles. A precipitator working on a dust consisting mainly of particles in the intermediate-size range will be less efficient than an equally sized precipitator working on either predominantly large or small particles.

Many parameters must be taken into consideration in the design and specification of electrostatic precipitators. The focus of the remainder of this section will be on typical design parameters such as specific collection area and aspect ratio. Gas flow distribution, resistivity, and electrical sectionalization are treated in the next section.

The specific collection area (SCA) is defined as the ratio of collection surface area to the gas flow rate into the collector. The importance of this term is that it represents the A/q relationship in the Deutsch–Anderson equation:

$$\text{SCA} = \frac{\text{total collecting surface (ft}^2\text{)}}{\text{gas flow rate (1,000 acfm)}}$$

Or in metric units:

$$\text{SCA} = \frac{\text{m}^2}{1{,}000\,\frac{\text{m}^3}{\text{hr}}} \tag{14.5}$$

An increase in the SCA of a precipitator design will in most cases increase the collection efficiency of the precipitator. Most conservative designs call for an SCA of 350–400 ft^2 per 1,000 acfm (20–25 m^2 per 1,000 m^3/hr) to achieve >99.5% particle removal. The general range of SCA is between 200 and 800 ft^2 per 1,000 acfm (11–45 m^2 per 1,000 m^3/hr) depending on precipitator design conditions and desired collection efficiency.

The aspect ratio (AR), the ratio of the total length to height of the collector surface, can be calculated by:

$$\text{AR} = \frac{\text{effective length}}{\text{effective height}} \tag{14.6}$$

Having a precipitator chamber many times larger in length than in height would be ideal. However, space limitations and cost could be prohibitive. The aspect ratio for ESPs can range from 0.5 to 2.0. For >99.5% collection efficiency, the precipitator design should have an aspect-ratio of greater than 1.0.

No critically reviewed design procedure exists for ESPs. However, one suggested "general" design procedure provided by Theodore [8] is presented below:

1. Determine or obtain a complete description of the process, including the volumetric flow rate, inlet loading, particle size distribution, maximum allowable discharge, and process conditions.
2. Calculate or set the overall collection efficiency.
3. Select a migration velocity (based on experience).
4. Calculate the ESP size (capture area).
5. Select the field height (experience).
6. Select the plate spacing (experience).

7. Select a gas throughput velocity (experience).
8. Calculate the number of gas passages in parallel.
9. Select (decide) on bus sections, fields, energizing sets, specific current, capacity of energizing set for each bus section, etc.
10. Design and select hoppers, rappers, etc.
11. Perform a capital cost analysis, including materials, erection, and startup costs.
12. Perform an operating cost analysis, including power, maintenance, inspection, capital and replacement, interest on capital, dust disposal, etc.
13. Conduct a perturbation study to optimize economics.

14.4 Equipment Design Considerations

Outwardly simple in design, the electrostatic precipitator takes part in a process that is complex. Many things can influence its behavior, but most of these influences can be overcome if the precipitator is well designed and properly operated. Operation is, in fact, equally important to equipment design. Three equipment factors involved in the general area of ESP operation, maintenance, and performance—namely, hoppers, collecting plates, rappers, doors, alignment, gas flow distribution, resistivity, and electrical sectionalization—are briefly treated below. Additional details are available in the literature [9–18].

Hoppers. Inadequate ash removal systems can permit buildups in hoppers to the point where they overflow. Of all the causes for misalignment and subsequent discharge electrode arc erosion, overfull hoppers is by far the biggest offender. Weighted wire systems are least sensitive to misalignment by this mechanism, while rigid frame designs appear to be most sensitive. In either case, grounded bus sections and diminished precipitator capacity determine whether there is any electrode damage or not.

Major causes of ash removal problems include low gas temperatures, promoting condensation of tacky condensable matter or free water vapor, insufficient insulation, inadequate hopper heat, internal obstructions, air in-leakage, inoperative vibrators, shallow valley angles, faulty level detectors, and evacuation system faults. Evacuation system capacity problems are more prevalent on industrial precipitators employing mechanical conveyors than on dusting equipment using pneumatic conveyors.

The most direct attack against formation of condensables uses external heat and better insulation. In the worst cases, it may be necessary to apply heat to almost all of the hopper surface instead of the nominal 15% of surface found on many installations. If hang-up at the very bottom is the common fault, heating and insulating the hopper throat and transition pieces can be helpful. Additional heat should be applied in vertical zones and thermostatically controlled. Location of thermostats is important, because if housed in a normally high temperature zone, they may never activate.

Collecting plates. Misalignment and fatigue failures are the most common problems associated with collecting plates. Fatigue failures generally result from the use of rapping forces that the plate attachments were not designed to survive. Manufacturers should be consulted for repairs if fatigue failures of this nature occur.

Rappers. Problems associated with internal swing-hammer rappers include bearing failure due to wear and abrasion by dust in the gas stream, misalignment of the hammers and anvils, and wear or breakage of the anvils themselves. Swing hammer failures on collecting plates can be detected only by changes in precipitator operation, principally energization. On discharge electrode frames, failure of the shaft insulator will stop shaft rotation, which can be detected fairly easily. Failures of electromagnetic drop hammers are associated primarily with the packing gland or boot seal. Gas leaking by the gland can soften the rod tube and corrode mounting bolts. Better packing and seal materials and the use of clamps on the seals can eliminate the problem. For any rapping equipment, frequent inspection is good maintenance practice. On vertical drop hammers, there is a popular notion that binding between the rod and guide tube is detrimental to delivery of the rapping blow. Actually, measurements have shown that the attenuation caused by this binding is less than 5%, and the condition, if it is noted, can probably be left unmodified. These studies show that a dust layer not much thicker than this page, allowed to accumulate between the hammer and the anvil, can attenuate the blow as much as 50%. If this condition is noted, immediate correction is required. In some installations using magnetic drop hammers, there is a tendency for the hammer to magnetize and stick in the dropped position at low energy levels. Insertion of a flat aluminum plate or a shrink-fit stainless steel cap between the hammer and rapper rod will eliminate this problem. In no event should rubber pads be used.

For the rapper input force to perform the most work, it is essential that the collecting plate attachment to the casing be rigid and tight, and that no energy absorbers are built or allowed to creep into the system. Manufacturers of highly active and "cleanable" plates work hard at evolving designs that ring like a bell when rapped. If loose attachments are found, the manufacturer should be contacted for methods to correct the problem.

Doors. Access doors are usually gasketed to prevent air in-leakage to the precipitator. Overtightening can cut and eventually destroy the gasket material. If left uncorrected, local thermal gradients on the door can cause it to warp and the whole door will require replacement for lack of a properly maintained and inexpensive gasket. Doors with interlocks are the primary precipitator safety feature. Frequent inspection and powdered graphite lubrication are mandatory to maintain key interlocks in proper working order.

Alignment. One of the major maintenance requirements for a precipitator is the inspection for and correction of misalignment between electrode systems. Misalignment can be caused by mishandling of materials during construction, by improper construction techniques or quality checks, by overfull hoppers, or thermal excursions. Misalignment may result from mislocated electrode suspension systems or bending and bowing of collecting electrodes, discharge frames, or the discharge electrodes themselves. Manufacturers have design standards for alignment which are developed from the clean-air sparking-voltage characteristics of their design geometries. Maintenance alignment efforts should strive to obtain these tolerances.

Points of alignment measurement to ascertain compliance with design tolerances for various electrode systems are available. It should be noted, for example, if the discharge electrode system is properly installed but the elevation of the collecting electrode system is out of design tolerance, the upper end of the two systems could be within design tolerances, but the bottom ends could be significantly misaligned.

Gas Flow Distribution. The best operating condition for a precipitator will occur when the treatment velocity distribution is uniform. When significant maldistribution occurs, the higher velocity in one collecting plate area will decrease efficiency more than a lower velocity at another equal area will increase the efficiency of that area. Small particles tend to follow flow streamlines better than large particles, so a maldistribution in particle size distribution in areas of the precipitator will also occur. Gross flow maldistribution also contributes to re-entrainment losses and sneakage of raw gases around collecting plates. These concerns are addressed in a later problem.

Resistivity. Particle resistivity is a condition of the particle in the gas stream that can alter the actual collection efficiency of an ESP design. Resistivity is a term that describes the resistance of the collected dust layer to the flow of electrical current. By definition, the resistivity is the electrical resistance of a dust sample 1.0 cm^2 in cross-sectional area, 1.0 cm thick, and with units $\Omega \cdot cm$ (ohms × centimeters). It can also be described as the resistance to charge transfer by the dust. Dust resistivity values can be classified roughly into three groups:

1. Between 10^4 and 10^7 $\Omega \cdot cm$—low resistivity
2. Between 10^7 and 10^{10} $\Omega \cdot cm$—normal resistivity
3. Above 10^{10} $\Omega \cdot cm$—high resistivity

Particles that have low resistivity are difficult to collect since they are easily charged and lose their charge on arrival at the collection electrode. This happens rapidly, and the particles can take on the charge of the collection electrode. Particles can thus bounce off the plates and become re-entrained in the gas stream. Examples of low-resistivity dusts are unburned carbon in fly ash and carbon black. If the conductive particles are coarse, they can be removed upstream of the precipitator with another device such as a cyclone. Baffles are often installed on the collection plates to help eliminate this precipitation–repulsion phenomenon.

Particles that have normal resistivity do not rapidly lose their charge on arrival at the collection electrode. These particles slowly leak their charge to ground and are retained on the collection plates by intermolecular adhesive and cohesive forces. This allows a particulate layer to be built up, which is then dislodged into the hoppers. At this range of dust resistivity (between 10^7 and 10^{10} $\Omega \cdot cm$), fly ash can be collected most efficiently.

Particles that exhibit high resistivity are difficult to charge. Once they are finally charged, they do not readily give up the acquired negative charge on arrival at the collection electrode. As the dust layer builds up on the collection electrode, the layer and electrode form a high potential electric field. The surface of the dust layer is negatively charged, the interior is neutral, and the collection electrode is grounded. This causes a condition known as back corona. Under the influence of the corona discharge, the dust layer breaks down electrically, producing small holes or craters (in the layer) from which back corona discharges occur. Positive ions are generated within the dust layer and are accelerated toward the negative (discharge) electrode. The result of this event would be to counteract the ion generation of the charging electrode with a corresponding reduction in collection efficiency. Disruptions of the normal corona process greatly reduce precipitator collection efficiency which, in severe cases, may fall below 50%.

High resistivity can generally be reduced by adjusting the temperature and moisture content of the gas stream. Particle resistivity decreases for both high and low temperatures (see Figure 14.4). The moisture content of the gas stream also affects particle resistivity. Increasing the moisture content of the gas stream lowers the resistivity. This can be accomplished by spraying water or injecting steam into the ductwork preceding the ESP. In both temperature adjustment and moisture conditioning, one must maintain gas conditions above the dew point to prevent corrosion problems.

The presence of SO_3 in the gas stream has been shown to favor the electrostatic precipitation process when problems with high resistivity occur. Most of the sulfur content in the coal burned for combustion sources converts to SO_2. However, approximately 1% of the sulfur converts to SO_3 [8]. The amount of SO_3 in the flue gas normally increases with increasing sulfur content of the coal. The resistivity of the particles decreases as the sulfur content of the coal increases.

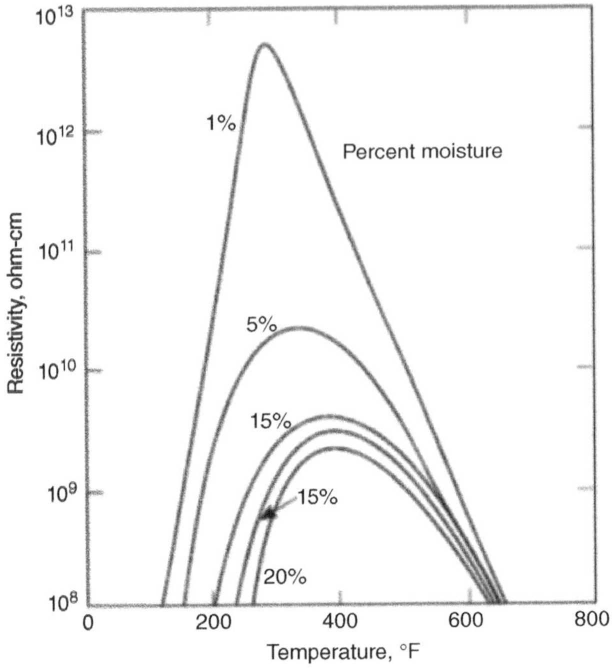

Figure 14.4 Effect on temperature and moisture content on apparent resistivity of precipitated cement dust.

The use of low-sulfur Western coal for boiler operations has caused fly ash resistivity problems for ESP operations. For coal fly ash dusts, the resistivity can be lowered below the critical level by the injection of as little as 10–20 ppm SO_3 into the gas stream. The SO_3 is injected into the ductwork preceding the precipitator.

Other conditioning agents such as sulfuric acid, ammonia, sodium chloride, and soda ash have also been used to reduce particle resistivity.

Two other methods to reduce particle resistivity include: increasing the collection surface area and by inletting the exhaust gas at higher temperatures. Increasing the collection area of the precipitator will increase the overall cost of the ESP. This may not be the most desirable method to reduce resistivity problems. Hot precipitators, which are usually located before the combustion air preheater section of the boiler, are also used to combat resistivity problems. The use of hot precipitators is not discussed in this chapter because of their limited application.

Electrical Sectionalization. Precipitator performance is dependent on the number of individual sections or fields installed (see Figure 14.5). The maximum voltage at which a given field can be maintained depends on the properties of the gas and dust being collected. These parameters may vary from one point to another in the unit. To keep each section of the precipitator working at high efficiency, a high degree of sectionalization is recommended. Multiple fields or stages are used to provide electrical sectionalization (see Figure 14.6). Each field has separate power supplies and controls to adjust for varying gas conditions in the unit.

Modem precipitators have voltage control devices that automatically limit precipitator power input. A well-designed automatic control system keeps the voltage level at approximately the value needed for optimum particle charging by the discharging electrodes. The voltage control

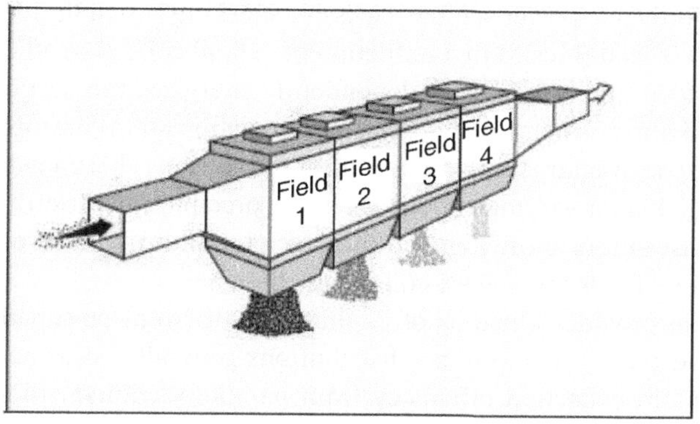

Figure 14.5 Stage or field sectionalization.

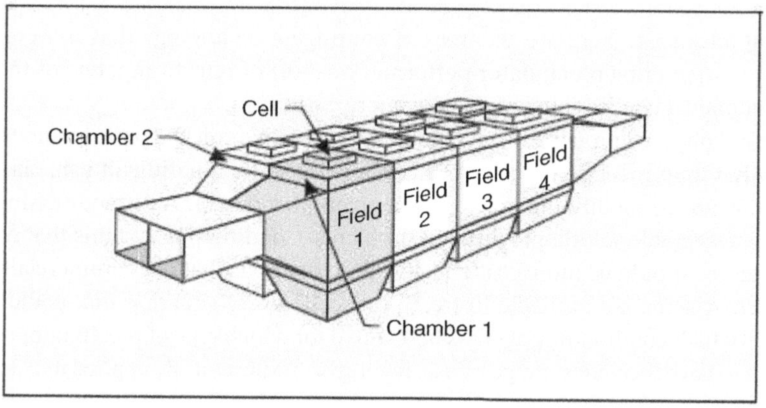

Figure 14.6 Parallel sectionalization.

device operates in the following manner. Increases in voltage will cause a greater spark rate between the discharge and collection electrodes. As noted, the occurrence of a spark counteracts high ESP performance since it causes an immediate, short-term collapse of the precipitator field. Consequently, useful power is not applied to capture particles. However, there is an optimal sparking rate where the gain in particle charging is just offset by corona current losses from sparkover. Measurements on commercial precipitators have determined that the optimal sparking rate is between 50 and 150 sparks per minute per electrical section. The objective in power control is to maintain corona power input at this optimal sparking rate. This can be accomplished by momentarily reducing precipitator power whenever excessive sparking occurs.

The need for separate fields arises mainly because power input requirements differ at various locations in a precipitator. The particulate matter concentration is generally higher at the inlet sections of the precipitator. High dust concentrations tend to suppress corona current. Therefore, a great deal of power is needed to generate a corona discharge for optimal particular charging.

In the downstream fields of a precipitator, the dust loading is usually lighter. Consequently, the corona current flows more freely in downstream fields. Particle charging will more likely be limited by excessive sparking in downstream fields than in the inlet fields. The power to the outlet sections must still be high in order to collect small particles, particularly if they exhibit high resistivity.

If the precipitator had only one power set, the excessive sparking would limit the power input to the entire precipitator. This would result in a reduction of overall collection efficiency.

The precipitator is divided into a series of independently energized bus sections or fields. Each bus section has individual transformer–rectifier sets, voltage stabilization controls, and high-voltage conductors that energize the discharge electrodes within the section. This allows greater flexibility for individual field energizing for varying conditions in the precipitator. Most ESP vendors recommend that there be at least four or more fields in the precipitator. It might be necessary to design the unit with seven or more fields for >99.9% collection efficiency.

Parallel sectionalization provides a means of coping with different power input needs due to uneven dust and gas distribution. Uneven gas distributions generally occur across the inlet face of the precipitator. Gains in collection efficiency from parallel sectionalization are likely to be small compared to field or stage sectionalization.

14.4.1 Recent Upgrades and Advances in ESP Technology

In terms of recent advances, there are six areas of contingent technology that have received the greatest attention in upgrading precipitator performance short of retrofitting; two of these are the aforementioned chemical conditioning and pulse energization.

Dust chemical and physical properties can combine to severely limit the performance capacity of a precipitator. Rather than increase the size of a precipitator to treat a difficult ash, chemical conditioning offers a means to modify the ash to suit a given precipitator. Ash conditioning has been used to provide an acceptable solution to difficult problems. Conditioning systems that make use of raw chemical materials should be more cost-effective than methods that use commercially prepared additives, and such systems are available to treat practically every situation that could arise. It is important to know which conditioning agent is best suited for which type of precipitator application. Ammonia appears to have the widest range of use, having been successfully applied to cold-side ESPs with both low- and high-sulfur coals, and on some hot-side units. The conditioning mechanism is not well known, but appears to enhance particle agglomeration and increase space charge. Sulfur trioxide has been most widely used to increase the surface conductivity of low-sulfur coal ashes. Since the mechanism depends on sorption, it cannot be expected to operate well above 190 °C. Sodium carbonate addition has seen the most limited use of these three. The charge-carrying capability of the sodium ions appears to predominate in this method, with the carbonate playing no known role.

Pulse energization can enhance the collection efficiency of precipitators where performance limitations are caused by poor energization, arising from either high resistivity or fine-particulate corona quenching. Pulse energization consists of electrical components that may superimpose high-voltage pulses of very short duration on an underlying, relatively constant electric potential, or the pulses may be solely applied and varied automatically with the pulsed parameters, such as voltage, frequency, and width, to tune the precipitator operation and minimize the debilitating effects of resistivity-induced sparking or back corona. The precipitator can then operate with the relatively high charging and collecting fields needed for more effective particulate removal in the precipitator. Pulse energization produces a profuse corona along the discharge electrodes, resulting in uniform corona distribution, which plays an important role in enhancing particulate collection.

As described earlier in this section, for increased performance and reliability, precipitators have been divided into a number of independently energized bus sections. Each bus section has its own transformer-rectifier, voltage stabilization controls, and high-voltage conductors that energize the discharge electrodes within that section. The main reasons for sectionalization are to offset the dampening effects on corona power input of heavy flue dust loadings and to reduce the effect of bus section failures. These heavy flue gas dust loadings occur mainly in the inlet sections of a precipitator. By sectionalization, corona power input and particle charging can be increased in the inlet sections, thereby raising the overall precipitator efficiency. This area continues to provide opportunities for improved performance in the future.

The final three developments with ESPs are:

1. Newly designed plates to prevent re-entrainment
2. Modification of rapping frequencies and intensity between bus sections
3. Renewed interest in tubular ESPs (reduced re-entrainment)

14.5 Applications

Three illustrative examples complement the information presented in this chapter of electrostatic precipitators.

14.5.1 Design Factors

Provide at least six design factors that require consideration for an ESP specification.

Solution:

1. Collection electrodes: Type, size (area), mounting, and mechanical and aerodynamic properties.
2. Discharge electrodes: Type, size, spacing, and method of support.
3. Shell: Dimensions, insulation requirements, and access.
4. Rectifier sets: Rating, automatic control system, number, instrumentation, and monitoring provisions.
5. Rappers for corona and collecting electrodes: Type, size, range of frequency and intensity settings, number, and arrangement.
6. Hoppers: Geometry, size, insulation requirements, number, and location.
7. Hopper dust removal system: Type, capacity, protection against air in leakage, and dust blowback.
8. Inlet and outlet gas duct arrangements, gas handling, and distribution system.
9. Degree of sectionalization.
10. Support insulators for high-tension frames: type, number, and reliability.

14.5.2 Effect of Particle Size Distribution

The following data are available for the proposed design of an ESP to operate at an efficiency of 97.5%:

- Air volumetric flow = 100,000 acfm
- Uniform flow distribution through 10 ducts
- Duct height = 30 ft with 12 inch plate-to-plate spacing
- Plate length = 36 ft
- Inlet loading = 14 gr/ft^3
- Outlet loading = 0.35 gr/ft^3

Table 14.2 Particle size distribution-drift velocity data.

d_p, μm	w,ft/s	x_i
0.1	0.27	0.01
0.25	0.15	0.01
0.5	0.012	0.01
1.0	0.11	0.01
1.5	0.15	0.016
2.0	0.20	0.016
2.5	0.26	0.016
5.0	0.50	0.016
10.0	0.60	0.016
25.0	0.70	0.016

In addition, the particle size distribution and drift velocity data listed in Table 14.2 have been provided in terms of the average particle size (d_p) of weight fraction (x_i) in a given size range and the corresponding drift velocity (w). Determine whether the proposed design will meet the desired efficiency. Also, prepare a graph of particle size vs. efficiency for the system and comment on the results.

Solution: Check if the average through put velocity is in an acceptable range. The average velocity should be 2–8 ft/s.

$$v = (100,000)/[(10)(30)(60)]$$
$$= 5.5 \text{ ft/s}$$

This is in the acceptable range.

The volumetric flow rate through each (one) duct is:

$$q = (100,000)/10$$
$$= 10,000 \text{ acfm}$$
$$= 167 \text{ acfs}$$

Calculate the plate area in square feet for each duct. Note once again that both plates contribute to the collection area:

$$\text{A} = 2(30)(36)$$
$$= 2,160 \text{ ft}^2$$

If the Deutsch–Anderson (DA) equation applies, then:

$$E = 1 - e^{-\left(\frac{wA}{q}\right)}$$
$$= 1 - e^{-\left(\frac{w(2,160)(60)}{10,000}\right)} = 1 - e^{-12.96w}$$

Calculate the efficiency for the 0.1 μm particle size:

$$E = 1 - e^{-12.96(0.27)}$$
$$= 0.9698 = 96.98\%$$

Calculate the collection efficiencies for all of the other particle size ranges. The results in Table 14.3 show that the overall efficiency is 95.30%. Since the desired efficiency is 97.5%, the proposed design is insufficient.

As is typical with particulate control, collection of large particles is highly efficient. A decline in efficiency is seen as particle size decreases. As noted earlier, when the particles become very small, diffusion effects occur that actually raise the collection efficiency for these particles. This is graphically illustrated in Figure 14.7. Refer to the literature for additional details [8].

Table 14.3 Overall efficiency results for illustrative example in Section 14.5.2.

d_p, µm	x_i	E_i	x_iE_i
0.1	0.01	0.9698	0.009698
0.25	0.01	0.8569	0.008569
0.5	0.01	0.7889	0.007889
1.0	0.01	0.7596	0.007596
1.5	0.016	0.8569	0.137104
2.0	0.016	0.9251	0.148016
2.5	0.016	0.9656	0.154496
5.0	0.016	0.9985	0.159760
10.0	0.016	0.9996	0.159936
25.0	0.016	0.9999	0.159984
		$E = \sum x_iE_i$	$E_i = 0.953048$

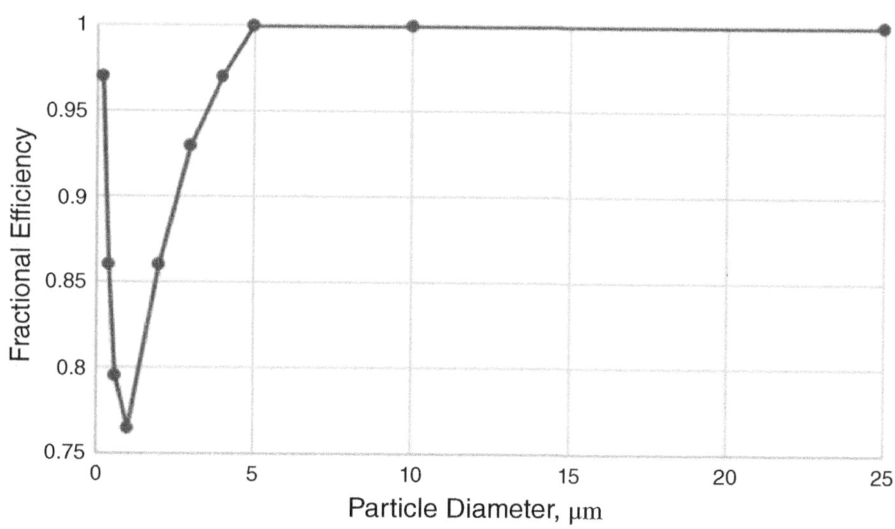

Figure 14.7 Effect of very small particles on collection efficiency.

14.5.3 Bus Section Failure

Bus section failure is one of the more important design operating and maintenance variables for ESPs. Detailed calculational procedures for estimating this effect are available in the literature. Two key studies addressing this issue are available in the literature [7–9].

A sectionalized electrostatic precipitator (ESP) consists of eight chambers (or parallel channels for airflow) and four independently energized electric fields per chamber. In effect, the unit contains 32 independent "chamber-fields" or "cells." As indicated earlier, ESPs are normally designed to allow for a certain number of cell failures before the unit has to be shut down for maintenance.

a. Assuming that the particulate penetration for all cells is 0.316* and that the incoming gas is perfectly distributed among all eight chambers, calculate the overall collection efficiency of the unit. In order to meet local particulate emission standards, the ESP must operate at a minimum collection efficiency of 95.0%.

b. Assuming that the collection efficiency of a "dead" or failed cell is 0% (or equivalently, that the cell penetration is 1.00), calculate the overall collection efficiency when four cells in the same chamber fail. Is the unit in compliance?

c. Calculate the overall collection efficiency when four cells, three in the same chamber and one in another chamber, fail. Is the unit in compliance?

d. Calculate the overall collection efficiency when four cells, two in the same chamber and two in two other different chambers, fail. Is the unit in compliance?

e. Calculate the overall collection efficiency when four cells, each in a different chamber, fail. Is the unit in compliance?

f. Comment on the results.

Solution:

a. The penetration for one chamber in the product of the cell penetrations.

Therefore,

$$\begin{aligned} \mathrm{P}_{\text{chamber}} &= (0.316)^4 \\ &= 0.00997 \end{aligned}$$

*Note: This is not normally the case. Since the larger particles are easier to collect and represent a disproportionate amount of the particulate mass, the upstream cells normally have a higher collection efficiency.

Since there is uniform flow, then

$$\begin{aligned} P_{\text{overall}} &= P_0 = 0.00997 = 0.997\% \\ &= E_0 = 0.9900 = 99.00\% \end{aligned}$$

b. When four cells in the same chamber fail, then

$$\begin{aligned} P_0 &= (7/8)(0.316)^4 + (1/8)(1)^4 \\ &= 0.1337 \\ E_0 &= 0.8663 = 86.63\% \text{ (as expected)} \end{aligned}$$

The unit is not in compliance.

c. When three cells fail in one chamber and one cell fails in another chamber, then

$$\begin{aligned} P_0 &= (6/8)(0.316)^4 + (1/8)(0.316)^3(1) + (1/8)(0.316)^1(1)^3 \\ &= 0.0509 \\ E_0 &= 0.9491 = 94.91\% \end{aligned}$$

The unit is marginally out of compliance.

d. For two failures in each of two other chambers

$$\begin{aligned} P_0 &= (5/8)(0.316)^4 + (2/8)(0.316)^3(1) + (1/8)(0.316)^2(1)^2 \\ &= 0.0328 \\ E_0 &= 0.9672 = 96.72\% \end{aligned}$$

The unit is in compliance.

e. With four failures in separate chambers

$$\begin{aligned} P_0 &= (4/8)(0.316)^4 + (4/8)(0.316)^3(1)^1 \\ &= 0.0208 \\ E_0 &= 0.9792 = 97.92\% \end{aligned}$$

In compliance!

f. The effect of failures is minimized when they are distributed and not "railroaded."

References

1. L. Theodore and A. J. Buonicore, *Control of Particulate Emissions*, EPA Manual for Course 413, RTP, NC, 1979.
2. L. Theodore, *Air Pollution Control Equipment Calculations*, John Wiley & Sons, Hoboken, NJ, 2008.
3. A. J. Buonicore, J. Reynolds and L. Theodore, *Assessment of Control Technology for Fine Particulate Emissions in Power Generation*, ANL/DOE Final Report No. 78-4249, 1980.
4. L. Theodore, *Self-Instructional Problem Workbook: Control of Gaseous and Particulate Emissions*. EPA Manual, SI: 412D, USEPA/APRI, RTP, NC, 1984.
5. L. Theodore, personal notes, East Williston, NY, 1975.
6. L. Theodore, personal notes, East Williston, NY, 1976.
7. P. L. Feldman, *Effects of Particle Size Distribution on the Performance of Electrostatic Precipitators*, 68th Annual Meeting of Air Pollution Control Association, Boston, June 15–20, 1975. Air Pollution Control Association, Pittsburgh, PA.
8. L. Theodore, personal notes, East Williston, NY, 2005.
9. L. Theodore, *Ask the Experts: Factors Affecting ESP Performance*, Chem. Eng. Prog., 20–21, New York City, NY, Oct 2006.
10. J. Reynolds, F. Karpowicz, L. Feeney and L. Theodore. *The Effect of Gas Velocity Profiles and Particle Size Distribution on the Collection Efficiency of Electrostatic Precipitators*, 128–134, Proceedings of the Fourth National Conference of Energy and the Environment, Cincinnati, OH, 1976.

11 J. Reynolds, F. Karpowicz, V. DePaola and L. Theodore. *Electrostatic Precipitator Modeling: Effect of Variations of Gas Velocity, Particle Concentration and Particle Size Distribution*, 128–134, Proceedings of the Sixth Annual Environmental and Science Conference, Louisville, KY, 1977.

12 J. Reynolds, F. Karpowicz, V. DePaola and L. Theodore. *The Influence of Length, Gas Velocity, Particle Concentration and Particle Size Distribution on Electrostatic Precipitator Behavior — A Theoretical Study*, Paper #77-51.4, Air Pollution Control Association Meeting, Toronto, 1977.

13 L. Theodore, Letter to the editor: ESP efficiency calculations, *J. Air Poll. Control Assoc.*, 29, 445–6, 1979.

14 L. Theodore and J. Reynolds, Some factors in minimizing precipitator costs, *J. Air Poll. Control Assoc.*, 7, 777–779, 1979.

15 L. Theodore and J. Reynolds, The effect on bus section failure on electrostatic precipitator performance, *J. Air Poll. Control Assoc.*, 33, 1202–5, 1983.

16 L. Theodore, J. Reynolds, F. Taylor and S. Errico, *Electrostatic Precipitator Bus Section Failure*, Paper #84-96.10, Air Poll. Control Assoc. Conference, San Francisco, CA, 1984.

17 L. Theodore, J. Reynolds, F. Taylor and S. Errico, *Electrostatic Precipitator Bus Section Failure: Operation and Maintenance*, Proceedings of the Fifth Symposium on the Transfer and Utilization of Particulate Control Technology, Kansas City, KS, 1984.

18 J. Reynolds, J. Tong, F. Taylor and L. Theodore, *Simplifying Bus Section Failure Calculations for Electrostatic Precipitators*, Paper #85-53.3, Air Poll. Control Assoc. Conference, Detroit, MI, 1985.

Suggested Readings

1 H. White, *Industrial Electrostatic Precipitation*. Addison-Wesley, Reading, MA, 1963.

2 W. Strauss, *Industrial Gas Cleaning*, Pergamon Press, New York City, NY, 1966.

3 S. Ogelsby and G. Nichols, *A Manual of Electrostatic Precipitator Technology*. Southern Research Institute, prepared for the USEPA, NTIS, PB-196, Birmingham, AB, 1970.

4 J. A. Danielson, *Air Pollution Engineering Manual*, EPA OAP Publication No. AP-40. Government Printing Office, Washington, D.C, 1967.

5 A. C. Stern, *Fundamentals of Air Pollution*, Academic Press, New York City, NY, 1974.

6 L. Theodore and A. J. Buonicore, *Industrial Control Equipment for Gaseous Pollutants*, Vols I and II, CRC Press, West Palm Beach, 1975.

7 P. L. Feldman, *Effects of Particle Size Distribution on the Performance of Electrostatic Precipitators*, 68th Annual Meeting of Air Pollution Control Association, Boston, June 15–20, 1975. Air Pollution Control Association, Pittsburgh, PA.

8 J. Katz, *The Art of Electrostatic Precipitation*, Precipitation Technology, Inc., Munhall, PA, 1979.

9 Air Pollution Control Association, ED, *Proceedings of the Specialty Conference on Operation and Maintenance of Electrostatic Precipitators*, Dearborn, Mich., April 10–12, 1978. Air Pollution Control Association, Pittsburgh, PA.

10 H. Hall, Solving problems in the electrical energization of electrostatic precipitators, *JAPCA*, 28(9), 871–878, 1978.

11 P. Bibbo and P. Aa, *Increasing Precipitator Reliability by Proper Logging and Interpretation of Operational Parameters-An Operator's Guide*, paper presented at the 2nd Annual Symposium of the Transfer and Utilization of Particulate Control Technology, sponsored by the USEPA, Denver, CO, 1979.

12 K. Darby, *Criteria for Designing Electrostatic Precipitators,* paper presented at the 2nd Annual Symposium of the Transfer and Utilization of Particulate Control Technology, sponsored by the USEPA, Denver, CO, 1979.

13 L. Theodore, *Air Pollution Control Equipment Calculations*, John Wiley & Sons, Hoboken, NJ, 2008.

14 L. Theodore, *Chemical Engineering: The Essential Reference*, McGraw Hill, New York City, NY, 2012.

15 L. Theodore, R. R. Dupont and K. Ganesan, *Unit Operations in Environmental Engineering*, John Wiley & Sons, Hoboken, NJ, 2017.

16 J. Richardson, R. R. Dupont, P. Farber, K. Ganesan and L. Theodore, *Air Pollution Control Equipment Revisited,* AWMA Conference, Calgary, CA, 2024.

17 J. Richardson and L. Theodore, *Optimizing Air Pollution Equipment Performance: Operation and Maintenance*, John Wiley & Sons, Hoboken, NJ, 2025.

15

Baghouses

15.1 Introduction

Filtration through fabric media is one of the oldest, simplest, and most efficient methods for removing solid particulate contaminants from gas streams. The fabric filter is capable of providing high collection efficiencies for particles as small as 0.5 μm and will remove a substantial quantity of those particles as small as 0.01 μm. In its simplest form, the industrial fabric filter consists of a woven or felted fabric through which dust-laden gases are forced. A combination of factors results in the collection of particles on the fabric fibers. When woven fabrics are used, a dust cake eventually forms; this, in turn, acts predominantly as a sieving mechanism. When felted fabrics are used, this dust cake is minimal or nonexistent. Instead, the primary filtering mechanisms are a combination of inertial forces, impingement, etc. related to individual particle collection on single fibers. These are essentially the same mechanisms which are applied to particle collection in wet scrubbers, wherein the collection media is in the form of liquid droplets rather than solid fibers.

As particles are collected, the pressure drop across the fabric filtering media increases. Due in part to fan limitations, the filter must be cleaned at predetermined intervals. Dust is removed from the fabric by gravity and/or mechanical means. The fabric filters or bags are usually tubular or flat. The structure in which the bags hang is frequently referred to as a baghouse. The number of bags in a baghouse may vary from a couple to several thousand. Quite often, when great numbers of bags are involved, the baghouse is compartmentalized so that one compartment may be cleaned while others are still in service.

The basic gas stream filtration may be conducted using many different types of fabric filters in which the physical arrangement of hardware and the method of removing collected material from the filter media will vary. The essential differences among these physical filtration techniques may be related, in general, to:

1. Type of fabric
2. Cleaning mechanism
3. Mode of operation

Gases to be cleaned can be either "pushed" or "pulled" through the baghouse. In the pressure system (push through), the gases may enter through the cleanout hopper in the bottom or through the top of the bags. In the suction type (pull through), the dirty gases are usually forced through the inside of the bag and exit through the outside.

Baghouse collectors are available for either intermittent or continuous operation. Intermittent operation is employed where the operational schedule of the dust-generating source permits halting the gas cleaning function at periodic intervals (regularly defined by time or by pressure differential) for removal of collected material from the filter media (cleaning). Collectors of this type are primarily utilized for particulate control from small-volume operations such as grinding and polishing, and so on, and for aerosols of a very coarse nature. For most air pollution control installations and major dust control problems, however, it is desirable to use collectors which allow for continuous operation. This is accomplished by arranging several filter areas in a parallel flow system and cleaning one area at a time according to some preset mode of operation.

Baghouses, are among the oldest and most widely applied particulate emission control devices. It is reported that there are currently over 200,000 fabric filters commercially operational in the United States. Different collector designs, fabric types, configurations, and a variety of cleaning methods provide numerous hardware combinations. A number of the basic designs emerged dominant in the marketplace. These designs and their maintenance features will be discussed subsequently.

There were once over 100 U.S. fabric-filter manufacturers. The larger companies usually offer more than one fabric-filter design type. The market is extremely competitive, and therefore, designs are constantly changed by the more aggressive companies. With so many manufacturers, product design possibilities, and a dynamic market, one realizes that hard and fast operating and maintenance rules are difficult to come by, and it is important for the user to establish where the fabric filter dust collector fits into the design and application picture.

15.2 Description of Baghouse Equipment

To better understand the baghouse, it is helpful to determine where it fits among the various types of fabric filters. When one tries to group fabric filters into a number of categories, it soon becomes obvious that the task is not simple. There appears to be an exception to each of the rules. The creation of certain categories, though they are not rigid, is yet very helpful.

One such approach is to group fabric-filter designs by cleaning method (Table 15.1). There are three major cleaning methods: shakers, reverse-air, and pulse jets. In addition to these three dominant cleaning methods, there exist a large number of other cleaning methods which are less often applied. Recently, combinations of the three primary methods have been occasionally employed. For example, reverse-air and shake have been used in combination, and reverse-air with a pulse assist. The three dominant cleaning methods are discussed below in the next three subsections.

Table 15.1 Fabric-filter cleaning methods.

Cleaning method	General characteristics associated with the cleaning method
Shaker	Inside bag collection
	Manual and automatic cleaning
	Woven filter media
	Filtration by the filter cake
	Generally intermittent—small operation
	Large and small gas volumes
	Often single compartment design

Table 15.1 (Continued)

Cleaning method	General characteristics associated with the cleaning method
Reverse air	Inside bag collection
	Automatic cleaning
	Woven filter media
	Filtration by the filter cake
	Continuous operation
	Medium and large gas volumes
	Multiple compartment design
Pulse jet	Outside bag collection
	Automatic cleaning
	Felted filter media
	Media filtration
	Continuous operation
	Small, medium, and large gas volumes
	Can be multiple compartment design

Another approach to grouping fabric filters is by capacity (Table 15.2). Generally, the groupings are small volumes (i.e., below 10,000 acfm), medium volumes (i.e., 10,000 to 100,000 acfm), and large volumes (i.e., >100,000 to the multimillion acfm level).

The filter-media type and temperature capabilities provide two other ways to categorize and view fabric filters. Woven vs. felted media are the media-type categories and high-temperature (>400 °F), medium-temperature (200 to 400 °F), and low-temperature applications (<200 °F) are useful temperature groupings. Some of the fibers used for baghouse filter media are shown in Table 15.3.

Two additional ways to group filters are intermittent vs. continuous service and inside vs. outside bag collection.

One should be aware of these distinctions and attempt to find where the collector in question fits; thus, when considering operating or troubleshooting recommendations, one only applies recommendations that are suited to the type of collector being used.

15.2.1 Cleaning Methods

15.2.1.1 Shakers

Simply stated, a shaker cleaning system removes the collected dust from the surface of the bag by mechanically shaking the bag. This can be done manually in the case of small dust collectors. Generally, however, it is accomplished by the use of a motor driving an eccentric, which in turn has a rod connected to the bags. It is probably the oldest bag cleaning technique and is currently employed over the full range of baghouse capacities from very small off the-shelf units to extremely large structural design units. The bag is generally open at the bottom and closed at the top, being fixed in the tube sheet at the bottom and attached to the shaking mechanism at the top.

Table 15.2 Fabric-filter design difference based on gas volume capacity.

General capacity	General characteristics associated with the design gas capacity
Low capacity	"Off the Shelf" small units
	Treating gas volumes of hundreds to a few thousands actual cubic feet (of gas) per minute (acfm)
	Little or no assembly
	Bags installed at the factory (i.e., prebagged)
	Manual cleaning systems available with some types
	Cleaning may be shaker, cartridge, or pulse-jet
Medium capacity	Modular designs
	Treating gas volumes of a few thousands to approximately less than a 100,000 acfm
	Little field assembly required
	Generally prebagged
	Pulse-jet—high gas-to-cloth ratio
High capacity	Shippable, modular design—50,000 to over 1 million acfm
	Moderate field assembly
	Bags installed in the field
	Reverse air, shake, and pulse-jet
	Structural considerations in assembly
	Shake and reverse air—low gas-to-cloth ratio
	Pulse jet—high gas-to-cloth ratio

With this configuration the dust is collected on the inside of the bag. The bag normally contains no rings or cages. The flow of dirty gas to the bags is stopped during the cleaning process. Table 15.4 shows some of the typical characteristics of a shaker cleaned baghouse.

In the United States, shakers normally employ woven cloth at gas/cloth ratios below 4:1. Attempts thus far to use shake cleaning in combination with domestic felts have led either to ineffective cleaning and thus high pressure drop or to the filter medium being shaken apart. The important parameters affecting the efficiency of cleaning are frequency, oscillation, and amplitude.

To achieve a continuous shake-clean baghouse, the unit needs to be subdivided into a number of isolatable compartments, such that one compartment at a time may be taken off-stream for cleaning. The normal procedure for the intermittent-shake cleaning process is to first close the inlet flow with a damper and, if necessary, the outlet flow of dirty gas. This should almost immediately result in zero pressure drop across the bags. Once this condition is achieved, shaking of the bags can be initiated. Shaking continues for as many as 100 shake cycles. The cleaning is then stopped, and the compartment is momentarily kept dormant until the dust settles. The dampers are then opened and the filtration process resumed. If reverse-air is employed to assist the cleaning process, the cleaning cycle may consist of a series of repeated shakes, then reverse-air cycles before the unit is brought on-stream.

Table 15.3 Characteristics of filter bag fibers.

Generic name	Fiber	Maximum temperature				Acid resistance	Alkali resistance	Flex abrasion resistance	Relative cost
		Continuous		Surges					
		°F	°C	°F	°C				
Natural fiber cellulose	Cotton	180	82	225	107	Poor	Excellent	Average	0.4
Polyolefin	Polypropylene	190	88	200	93	Excellent	Excellent	Good	0.5
Natural fiber protein	Wool	200	93	250	121	Good	Poor	Average	0.8
Polyamide	Nylon	200	93	250	121	Poor to fair	Excellent	Excellent	0.6
Acrylic	Orlon*	240	116	260	127	Very good	Fair	Average	0.7
Polyester	Dacron*	275	135	325	163	Good	Fair	Excellent	0.5
Aromatic polyamide	Nomex*	400	204	425	218	Fair	Very good	Very good	2.0
Fluoro-carbon	Teflon*	450	232	500	260	Excellent (except poor if fluorine is present	Excellent except poor in the presence of trifluoride, chlorine, and molten alkali metals	Fair	6.7
Glass	Fiberglass* or Glass	500	260	550	288	Good	Poor	Poor to fair	1.0
Polymer	P84*	450	232	500	260	Good	Fair	Fair	2.5
Polymer	Ryton*	375	191	450	232	Excellent	Excellent	Good	2.5–4.0
Nextel	Ceramic	2192	1200	2372	1300	Excellent	Excellent	Fair	10

Source: [1, 2].

Table 15.4 Shaker baghouse parameters.

Item	Parameter
Frequency	Usually several cycles/minute; adjustable
Motion type	Simple harmonic or sinusoidal
Peak acceleration	1–10 gs
Amplitude	Fraction of an inch to several inches
Mode	Offline (compartment isolated) usually
Duration	10–100 cycles, 30 seconds to few minutes
Common bag diameters	4–12 inches

Table 15.5 Reverse-air filter parameters.

Frequency	One compartment cleaned at a time, sequencing one compartment after another. Cleaning may be initiated either automatically by system maximum pressure drop or continuously.
Motion	Gentle collapse of the bag concave inward upon deflation. Compartment slowly depressurized after completion of a back-flush. NOTE that anticollapse rings are sewn into the bag at intervals along the length to ensure that the bag is reinflated after cleaning.
Mode	Compartments are isolated off-line during cleaning
Duration	1–2 minutes, including valve opening and closing and time for dust settling periods. Reverse-air flow itself is normally 10–30 seconds
Common bag dimensions	8–12″ diameter, up to 30 feet long
Bag tension	50–75 psi typical; optimal values are determined after initial on-line operation

15.2.2 Reverse-air/Reverse-flow Cleaning

Reverse-air cleaning is perhaps the gentlest cleaning method. In this method, the dust is removed from the bags by back-flushing with low-pressure (a few inches water gauge) reversed flow. In the case of high-temperature applications, the just-cleaned hot gas is employed to back-flush rather than ambient-temperature air. Woven-filter media are generally employed in conjunction with reverse-air cleaning. The cleaning is motivated by a separate cleaning fan which is normally much smaller than the main system fan since only a fraction of the total system is cleaned at any one time. In the case of a negative pressure system, one often can clean without a reverse flow fan. The flowrate of cleaning gas is normally about equal to that of the dirty gas (see Table 15.5).

Most often, reverse-flow systems are comprised of a number of isolatable compartments. The gas/cloth ratios usually employed are less than 4:1. Normally, the dust is collected on the inside of the bag, the bag being open at the bottom and closed at the top. The bag will contain rings to keep it from collapsing completely during flow reversal. Complete collapse would, of course, prohibit cleaning since the dust particles would be unable to fall down within the bag to the hopper.

Cleaning is accomplished both with and without flexing (partial collapsing) of the bag. Reverse flow without flexing can be, employed when the dust is very easily dislodged from the bag surface. Figure 15.1 shows a typical cleaning cycle. In the on-stream-gas filtering mode, the compartment and outlet dampers are open and the dirty gas enters the bag at the bottom. The dust is collected on

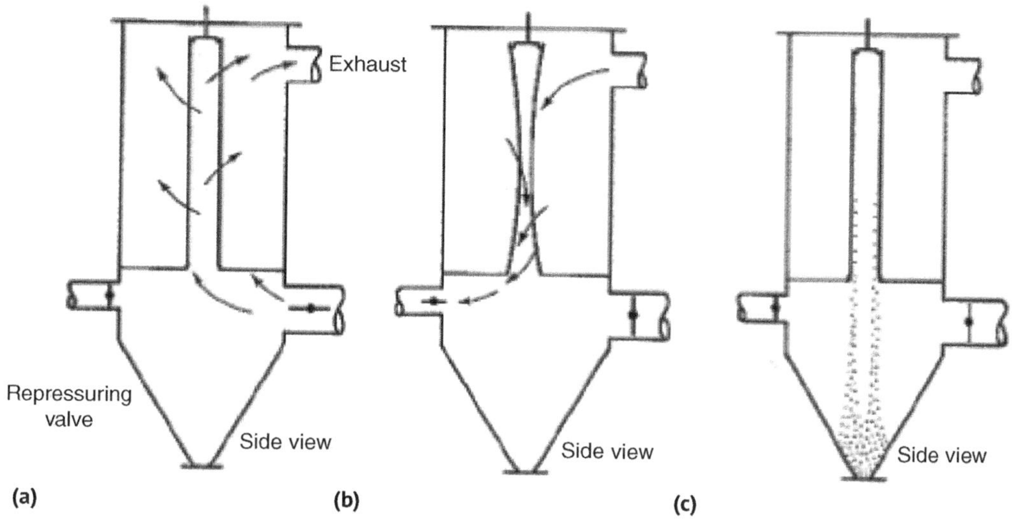

Figure 15.1 Reverse-air cleaning: (a) filtering; (b) collapsing; (c) cleaning.

the inside of the bag and the cleaned gas exits the outlet damper. During the bag cleaning cycle, the flow is reversed by closing the outlet plenum and opening a third damper, which allows cleaned gas to enter the compartment on the clean side of the bags, thus back-flushing the bags and exiting the compartment through the inlet damper. This now-dirty gas progresses to the balance of compartments which are on-stream. It should be noted that this process increases the system gas/cloth ratio by adding to the total gas volume, the volume of gas employed in the back-flushing process. The system described here, although generally applicable, is normally associated with large gas volumes. Low- and medium-volume reverse-air systems can be found; however, these often employ felted media with a moving reverse-flow plenum.

15.2.3 Pulse-jet Cleaning

Pulse-jet cleaning Pulse jet cleaning is a method that employes high-pressure (60 to 120 psi) compressed air, with or without a venturi, to back-flush the bags vigorously. This method creates a shock wave that travels down the bag, knocking the dust away from the filter medium. Normally, this method is employed in conjunction with felted-filter media and the gas/cloth ratio is generally higher than that of the shake and reverse-air cleaning methods. The duration of cleaning is lower than that of the other two methods; generally, the pulse lasts only a fraction of a second. The baghouse is often not subdivided when pulse-jet cleaning is employed.

The usual configuration has the bag closed at the bottom and open at the top, A metal cage is employed within the bag to keep the bag from collapsing. In the normal mode of operation, the dirty gas enters the hopper and proceeds to the bags. The dust is collected on the outside of the bags and the cleaned gas exits through the top of the bags and baghouse. Usually, a row of bags is cleaned simultaneously by introducing compressed air briefly at the top of each bag. The shock wave thereby created drives the dust off the outside of the bag and down into the hopper. Continuous discharge of the dust from the hopper is often employed. The importance of the venturi to provide an onrush of secondary air for cleaning is now being questioned, and some pulse systems operate with only the compressed-air manifold and without venturis over each bag. Pulse-jet bag parameters are given in Table 15.6. Notable is that this system has no internal moving

Table 15.6 Pulse-jet filter parameters.

Frequency	Usually one row of bags at a time in the compartment being cleaned. Sequenced one row after another. Cleaning can be controlled such that every other row is cleaned. Cleaning may be continuous or activated by maximum baghouse pressure drop.
Motion	Shock wave of cleaning pulse travels down the bag, momentarily distending the bag and releasing the dust cake.
Mode	Generally bags are cleaned on-line. Off-line cleaning may be used with difficult to remove filter cakes or to thoroughly clean bags prior to compartment maintenance
Duration	Compressed air pulse generally approximately 0.1 seconds. The row of bags being cleaned is effectively "off-line" during the pulse.
Common bag dimensions	5–8″ diameter, bag length up to 20 feet.

parts and allows for removal of the bags from the clean side of the baghouse compartment, since the bags are usually connected only at the top.

15.2.4 Miscellaneous Cleaning Methods

In addition to the three most common cleaning methods, a number of other methods have been employed. These methods were either made obsolete by one or more of the three major methods or else never achieved the commercial significance of the three major cleaning methods.

One method that achieved some degree of commercial success is "reverse jet" cleaning, sometimes referred to as "blow ring" cleaning. This method employed a jet ring surrounding each filter tube; the rings were carried up and down the bags on a carriage driven by chains and sprockets or a cable system. Today, this method is found only in very limited number of applications.

In addition to the reverse jet, high-frequency agitation has been employed. One such approach is to achieve fluttering of the cloth by vibration and rapping. This is normally associated with envelope filters. In the case of tubular bags, fluttering has been achieved by the use of an "air shake."

Even-higher-frequency agitation is achieved by the use of sonic and ultrasonic cleaning. Sonic cleaning has also been employed in combination with reverse flow and pulse-jet filters. In this method, sonic horns may be installed in baghouse compartments and activated during the compartment cleaning cycle. These sonic horns are often a "postoperation" installation when it is found out that there is a hard-to-remove dust from the bags. In this way, the sonic horns provide extra cleaning, when necessary.

One of the simplest cleaning methods used in the case of very small collectors is manual cleaning, where the bags are struck, shaken, or even removed and turned inside out.

A variation of the pulse jet which is commercial today but not as common as the pulse jet is the plenum pulse method of cleaning. Like the pulse jet, this method employes compressed air, but instead of delivering the air to each bag via a manifold, the compressed air is delivered to a compartment of bags via a compressed air plenum and damper system, cleaning the entire compartment simultaneously.

15.2.5 Cleaning Method Versus Capacity

As shown in Tables 15.1 and 15.2, fabric filters can be grouped by both cleaning method and gas capacity. There is a loose correlation between the two groupings. The extremely small collectors are generally found to be shakers. This is especially true if an intermittent collector is suitable.

Recently, compact cartridge filters (similar to truck air filters) employing simple pulse cleaning have been finding success in the small-volume marketplace. The small and especially the medium-size collectors, where continuous operation is required, are most often found to be pulse jet collectors.

The large continuous collectors of recent vintage are most often pulse-jet or reverse-air collectors, although large shakers are found as well. In all cases, continuous operation is achieved by subdividing the unit into a number of compartments so that when one compartment is off-line for bag cleaning, the total process gas volume can be handled by the balance of the baghouse system.

15.2.6 Bags and Cages

As previously mentioned, each fabric-filter dust collector consists of a number of fabric-filter elements. These elements normally consist of textile material and are constructed as tubes, envelopes, or most recently, cartridges. The textile can be either felted or woven with a bag material (Table 15.3) chosen due to the chemical and thermal conditions of the gas stream being filtered as well as the chemical and physical characteristics of the dust being filtered.

If woven textiles are used, there are numerous constructions that can be employed, such as sateen, 2 × 2 twill, or crowfoot. In addition to various yarn constructions, the filter element can be varied further by the finishing process applied. The most commonly employed finishing steps include glazed, silicone graphite, Teflon, napping, calendaring, and singeing. Details of these finishes are provided in Table 15.7. Finally, when reviewing bag construction, one must also consider stitching and seam types.

The most commonly employed filter element design is that of a tubular shaped "bag." This is employed to collect dust either on the outside or the inside of the tube. The envelope-constructed filter element is not as common in the United States as the tubular-shaped. The flat-enveloped baghouse normally collects on the outside of the filter element and provides more filtering area per volume of housing. Finally, the cartridge design, which is a pleated construction, provides maximum filtering area, but at this time seems to have temperature and filtering velocities limitations more restrictive than the other two constructions.

When reviewing the filter element and its construction, one must also consider the filter element support structure. In the case of shakers, as mentioned, the filter element normally has no support other than attachment at the top and bottom. Reverse-air filter elements normally have support rings sewn into the bag. In the case of pulse-jet cleaning systems, the bag is usually supported by a cage placed within the bag. The standard construction has been 6 to 20 vertical wires (stringers) spot-welded to a number of rings that are generally located about 6 to 12 inches apart.

Table 15.7 Fabric finishing.

Calendaring	High-pressure pressing of the fabric; pushes the surface fibers down onto the body of the fabric
Napping	A scraping of the fabric surface that raises the surface fibers
Singeing	Passing of the filter medium over an open flame, thereby removing the independent surface fibers
Glazing	High-pressure pressing of the fabric at elevated temperatures which fuses the surface fibers to the body of the fabric
Coating	Immerses the fabric into a solution that will provide the fabric with a coating that will lubricate and thereby reduce self-abrasion; in the case of many woven fabrics the preferred coating is Teflon and/or silicon-graphite

The use of polymer and glass bags in conjunction with pulse-jet cleaning has led to the use of close-mesh cages with either a 2 × 1 inch or a 1/ 2 × 1/ 2 inch mesh to provide greater fabric support. Variations in cage design beyond mesh considerations include differences in the bottom cage pan, top collar, and top flanges. Usually, these variations are related to bag–cage attachment schemes.

Finally, for baghouses as well as most air pollution control installations with major dust control problems, it is desirable to use collectors which allow for continuous operation. This is accomplished by arranging several filter areas in a parallel flow system and cleaning one area at a time according to some preset mode of operation.

15.3 Process Design Considerations

The number of variables necessary to design a fabric filter is very large. Since fundamentals cannot treat all of these factors in the design and/or prediction of performance of a filter, this determination is basically left up to the experience and judgment of the design engineer. In addition, there is no one formula that can determine whether a fabric filter application is feasible. A qualitative description of the filtration process is possible, although quantitatively the theories are far less successful. Theory, coupled with some experimental data, can help predict the performance and design of the unit.

The state of the art of engineering process design is the selection of filter medium, superficial velocity, and cleaning method that will yield the best economic compromise. Industry relies on certain simple guidelines and calculations, which are usually considered proprietary information, to achieve this. Despite the progress in developing pure filtration theory, and in view of the complexity of the phenomena, the most common methods of correlation are based on predicting a form of a final equation that can be verified by experiment.

The gas-to-cloth (G/C) ratio is a measure of the amount of gas passed through each square foot of fabric in the baghouse. It is given in terms of the number of cubic feet of gas per minute passing through one square foot of cloth. In other words, the G/C ratio = gas volume rate/cloth area. Also note that this velocity is not the actual velocity through the openings in the fabric, but rather the apparent velocity of the gas approaching the cloth. As the G/C ratio increases, pressure drop (AP) also increases. In the United States, pressure drop in baghouse applications is generally measured by inches of water column (in H_2O or ″WC).

Despite several sophisticated formulas that have been developed, there is no satisfactory set of published equations that allows a designer to accurately calculate the efficiency of a prospective baghouse. However, there are three more heuristic formulas worth mentioning that can help baghouse designers:

$$\frac{\text{Gross gas}}{\text{cloth ratio}} = \frac{\text{total inlet volume rate}}{\text{total filter cloth area in collector}} \tag{15.1}$$

$$\frac{\text{Net gas}}{\text{cloth ratio}} = \frac{\text{total inlet gas volume rate + cleaning volume rate}}{\text{on stream cloth}} \tag{15.2}$$

$$\frac{\text{Units gas}}{\text{cloth ratio}} = \frac{\text{gas volume rate}}{\text{cloth area}} = \frac{\frac{\text{ft}^3}{\text{min}}}{\text{ft}^2} = \text{ft/min} \tag{15.3}$$

As mentioned earlier, baghouse design is still very much an artform and nowhere is this more evident than in the selection of G/C ratios. Factors influencing the G/C include the cleaning method, filter media, dust size, dust density, dust loading, and other factors unique to each situation. Because of their variability, it has never been able to satisfactorily quantify all of these factors for every application.

There are two basic design approaches that are routinely employed throughout the industry. One approach is to collect all empirical data available for the source in question. If there is no data for the industry at hand, then one could check with a similar industry which is using a baghouse and determine the G/C range successfully employed in that industry and conservatively apply it. There is a risk that the subtle but essential differences between the two industries may prohibit the G/C transition. An alternative approach is to run a pilot unit on a slip stream and vary the G/C ratio, and then select and design around the G/C ratio which appears to provide the best results. The pilot plant should run for a few months in order to obtain data representative of a long-term operation.

Once the gas/cloth ratio (G/C) has been selected, the size of the baghouse is nearly defined. Variations also occur in the number of walkways, hopper slope, and other design considerations. These other variations will influence the total size of the baghouse to a limited degree, once the basic needs have been identified by the G/C.

The G/C and cleaning method are tied to each other, as shown in Table 15.8. Once the G/C ratio and cleaning method have been selected, the next major decision is the selection of the filter medium. At last, selection of the filter medium is more science than art! Temperature limitations and chemical resistance often determine filter medium selection. As the operating temperature rises, options decrease.

Table 15.8 Gas-to-cloth ratios. Generally safe design values—application requires consideration of particle size and grain loading; ft/min.

	Shaker/Woven; reverse-air/woven	Pulse-jet/Feld; reverse-air/felt
Alumina	2.5	8
Asbestos	3	10
Bauxite	2.5	8
Carbon black	1.5	5
Coal	2.5	8
Cocoa, chocolate	2.8	12
Clay	2.5	9
Cement	2	8
Cosmetics	1.5	10
Enamel frit	2.5	9
Feeds, grain	3.5	14
Feldspar	2.2	9
Fertilizer	3	9
Flour	3	12
Fly ash	2.5	5
Graphite	2	5

(Continued)

Table 15.8 (Continued)

	Shaker/Woven; reverse-air/woven	Pulse-jet/Feld; reverse-air/felt
Gypsum	2	10
Iron ore	3	11
Iron oxide	2.5	7
Iron sulfate	2	6
Lead oxide	2	6
Leather dust	3.5	12
Lime	2.5	10
Limestone	2.7	8
Mica	2.7	9
Paint pigments	2.5	7
Paper	3.5	10
Plastics	2.5	7
Quartz	2.8	9
Rock dust	3	9
Sand	2.5	10
Sawdust (wood)	3.5	12
Silica	2.5	7
Slate	3.5	12
Soap, detergents	2	5
Spices	2	10
Starch	3	8
Sugar	2	7
Talc	2.5	10
Tobacco	3.5	13
Zinc oxide	3	5

The maximum temperature for economical and commercially available filter media is approximately 550 °F. If the gas stream is hotter than this level, it is important to cool the gas to within safe parameters before it reaches the filter medium. Important fiber characteristics are:

1. Temperature. The fiber must be able to perform without failure at temperatures higher than those expected during operation. The fiber must also be able to handle temporary heat surges (temperature excursions).
2. Corrosiveness. The fiber must be able to resist degradation from exposure to specific acids, alkalis, solvents, or oxidizing agents found in the dust-laden gas stream.
3. Hydrolysis. Humidity levels must be accounted for.
4. Dimensional Stability. If the fiber is expected to shrink or stretch within the process environment, those effects must be controlled.
5. Cost. The least costly selection that will satisfy overall requirements is usually preferred.

Table 15.9 Fabric selection chart.

Fabric	Maximum temp, °F	Acid resistance	Fluoride resistance	Alkali resistance	Flex abrasion resistance
Cotton	180	Poor	Poor	Good	Very good
Polypropylene	200	Excellent	Poor	Excellent	Very good
Polyester	275	Good	Poor to fair	Good	Very good
Nomex	400	Poor to fair	Good	Excellent	Excellent
Teflon	450	Excellent	Poor to fair	Excellent	Fair
Fiberglass	500	Fair to good	Poor	Fair to good	Fair

Table 15.9 compares the characteristics of the more widely used filtration media. Most bag manufacturers, as well as baghouse vendors, will provide media selection charts similar to Table 15.9. Once the critical parameters of a specific application have been identified, such a chart can be used to make an initial selection of media possibilities. A more detailed evaluation of the media possibilities, including cost, availability, and previous application history, should then be made.

Once a preliminary selection of filtering fabric has been made, the media supplier can usually provide additional information that should be considered prior to finalizing the fabric choice. The information below gives basic information that may be used along with the media selection charts [3, 4].

Cotton—Low temperature capability; good flex abrasion resistance; still used in some low temperature applications.

Polypropylene—Very sleek fabric; demonstrates good cake release and resistance to blinding. Does not readily absorb moisture and would be a good selection for a low temperature, high moisture condition, e.g., dye production.

Polyester—Very sturdy material; good resistance to acids and alkalis, slightly higher temperature capability than polypropylene. Costs are about the same as polypropylene. Used in most routine low temperature applications including quarry, woodworking, and sand handling operations.

Nomex—Extremely sturdy material with respect to flex abrasion. Superior to fiberglass in resistance to fluorides and abrasion. Good temperature capability. Poor acid resistance; not used in gas streams containing SO_2 and SO_3. Costs about 2.5 times that of polypropylene and polyester. Used in asphalt, steel, carbon black, and cement industries.

Teflon—Generally chemically inert and for that reason useful in severe environments. Very expensive, but the cost is usually justified because of superior bag life. Used in the carbon black industry, lead smelting, coal-fired boilers, and various unusual applications. Costs about 10 times that of polyester for the same size and weight bag.

Fiberglass—Normally used in high temperature applications. Improvements in finishes and techniques of fabrication, installation, and operation have paid dividends in extended bag life. Fiberglass is the primary fabric used in the boiler market. Cost is between polyester and Nomex.

Overall system component selection can be simple, as in the case of a single-point pickup for an ambient application—or it can be complex, as in the case of a high temperature, acid gas, dust control system involving system bypass, auxiliary heat, numerous dampers, temperature and pressure alarms, automatic shutdown, and extra modules for maintenance and off-line cleaning.

However, once a system is designed, it is important to have criteria for performance and reliability. Also, local ordinances as well as state and federal regulations must be adhered to. Flange-to-flange requirements, such as temperature limitations and volume and dew point restrictions, must be considered in advance. Importantly, the successful baghouse must be built as a partner with the production process, not as a restrictor of output. Finally, all extreme ranges must be anticipated and accounted for within the design.

A variation on Darcy's formula for the flow of through a porous bed has been developed for the flow of gases through a filter medium [5]. The basic Darcy equation can be used to predict the pressure drop for an operating fabric filter with accumulated dust cake:

$$\Delta P = S_E v + K_2 c_1 v^2 t \tag{15.4}$$

where
ΔP = pressure drop, in H_2O
S_E = effective residual drag, in H_2O
v = velocity, fpm
K_2 = specific cake coefficient

An equation that can be used for determining the collection efficiency of a baghouse is [6–11]:

$$E = 1 - e^{-(\psi L + \phi t)} \tag{15.5}$$

where
ψ = constant based on fabric ft^{-1}
ϕ = constant based on cake, s^{-1}
t = time of operation to develop the cake thickness, s
L = fabric thickness, ft
E = collection efficiency (dimensionless)

The exit concentration (we) for the combined resistance system (the fiber and the cake) is:

$$w_e = w_i e^{-(\psi L + \phi t)} \tag{15.6}$$

The reader should note that ϕ and ψ in Equations 15.5 and 15.6 are modified inertial impaction numbers based on the cake and fabric, respectively. Unfortunately, this key equation has been totally ignored by responsible EPA individuals in this field. Taxpayers' dollars continue (for over a dozen years) to be provided to contractors whose research efforts have produced little, if any, usable results in developing quantitative equations to describe collection efficiency.

The effect of bag failure on baghouse efficiency can be described by the following equation:

$$\begin{aligned} P_t^* &= P_t + P_{tc} \\ P_{tc} &= 0.582(\Delta P)^{\frac{1}{2}}/\phi \\ \phi &= q/\left[LD^2(T+460)^{\frac{1}{2}}\right] \end{aligned} \tag{15.7}$$

where
P_t^* = penetration after bag failure
P_t = penetration before bag failure
P_{tc} = penetration correction term; contribution of broken bags to P_t^*
ΔP = pressure drop, in H_2O

ϕ = dimensional parameter
q = volumetric flow rate of contaminated gas, acfm
L = number of broken bags
D = bag diameter, inches
T = temperature, °F

A detailed development of Equation 15.7 is available in the literature [12].

When designing a baghouse, two major areas of design need to be considered in conjunction with each other. The first is that of the total dust collection system design. This includes all ductwork, fans, heater, dampers, dust conveying equipment, and controls that will directly affect and are intimately associated with the dust collector itself. The second area of concentration is that of the flange-to-flange dust collector, its cleaning method, bag type, materials selection, and so on. The flange-to-flange collector should not be selected without concurrent specification of the total dust collection system, and this in turn requires full consideration of the dust source operation (i.e., normally the production process). It can be that such an overview will lead to some modification of the production process or process controls in order to attain trouble-free baghouse operation.

The specification of the dust collector and its auxiliaries must begin with a definition of the emission problem. The better and more complete the problem definition, the better are the chances that the dust collection system will function in a trouble-free, reliable manner. The dust properties of primary interest are density, particle size, and dust loading. The gas properties essential to proper problem definition are volume, temperature, moisture content, and acid gas concentration. A general description of the source or production operation is helpful, particularly in defining whether the collector is to be continuous or intermittent. Additional information, as requested in Figure 15.2 can be of great help. It is not uncommon to find a dust collector selected simply on the basis of estimated gas volume, temperature, and a brief description of the production operation; however, one probably enhances the risk of selecting a baghouse system requiring constant maintenance and bag replacement.

The size of the fabric-filter dust collector is tied directly to parameters known as the air/cloth (A/C) or gas/cloth ratio (G/C). This parameter is the apparent velocity of the dust-laden gas stream as it approaches the filter elements. It can be calculated by dividing the total gas volume, usually in ft^3/min, by the total area in ft^2 of cloth or filter medium available for filtration.

Baghouse design is still very much an art form rather than an exact science, and nowhere is this more evident than in the selection of the gas/cloth ratio. Factors influencing the gas/cloth selection include the cleaning method, filter media, dust size, dust density, dust loading, and other factors. Quantifying the general relationship between these parameters has been attempted with very limited success. See Table 15.10.

The overall system component selection and system design can be straightforward and relatively simple, as in the case of a single point pickup for an ambient application, thus involving a single hood, a small amount of ductwork, and a system fan on stack, or it can be fairly complex, as in the case of a high-temperature, acid gas, dust control system involving system bypass, auxiliary heat, numerous dampers, temperature and pressure alarms, automatic shutdown, extra modules for maintenance, and off-line cleaning. The system design must start with a definition of system objectives. Consideration must be given to local ordinances (e.g., bypass restrictions, height restrictions, etc.). The production system requirements must be entered into the analysis (e.g., impact of back pressure and baghouse upset conditions). Last, but certainly of equal importance,

Design Data Sheet

Company Name ____________________ Phone Number ____________________

Address ____________________ City ____________________

State __________ Zip Code __________ Contact ____________________

Essential Information

1. Description of Application ____________________

2. Gas inlet volume ____________________ ACFM Gas inlet temp ____________________ °F
3. Description of particulates ____________________

Other Information Required

1. Space available for equipment (attach sketch or drawings) ____________________

2. Auxiliary equipment required ____________________

3. Particulate pollutions: Type of dust: __________ Particulate Size __________

   Abrasive? __________ Explosive? __________ Inflammable? __________

   Sticky? __________ Moisture Content __________ Density __________

   Specific Gravity __________ Inlet Loading __________ grains/SCF

   Allowable Outlet Loading __________ grains/SCF
4. Gaseous pollutants: Type(s) and concentration(s) ____________________

5. Additional comments (i.e., location, etc.) ____________________

6. Type of duty: Intermittent __________ Continuous __________
7. Operating Pressure __________ inches of WC
8. Existing electrical and compressed air capabilities ____________________

Figure 15.2 Customer-inquiry design data sheet.

are the flange-to-flange baghouse requirements (e.g., temperature limitations and volume and dew-point restrictions). Full consideration of the interplay of baghouse and production operations must be made since the baghouse can restrict production and the production process can restrict or shut down the baghouse. The designer must not only consider typical operating conditions but must also review and consider the extreme ranges of the production process and their impact on the dust control system.

Table 15.10 Fiber selection chart.

		Fiber properties					Recommended Operating temp (°F)	
Fiber	Name	Tensile strength	Abrasion resistance	Chemical acids	Resistance alkalis	Support combustion	Cont.	Surges
Cotton	Natural fiber cellulose	Good	Average	Poor	Excellent	Yes	180	225
		Comment: Excellent selection in ventilation-type collector						
Polypropylene	Polyolefin	Excellent	Good	Excellent	Excellent	Yes	190	190
		Comment: strong fiber, low moisture absorption and possesses excellent chemical resistance						
Glass	Glass	Excellent	Poor	Good	Poor	No	500	550
		Comment: All properties are highly dependent on fabric treatment. Can be used at high temperature and has high tensile strength						
Nylon	Polyamide	Excellent	Excellent	Poor	Excellent	Yes	200	250
		Comment: rugged fiber with excellent resistance to abrasion and alkalis						
Dacron[a]	Polyester	Excellent	Excellent	Good	Fair	Yes	275[b]	325[b]
		Comment: High tensile strength, good dimensional stability, excellent heat resistance. Susceptible to moist-heat hydrolysis						
Orlon[a]	Acrylic	Average	Average	Very Good	Fair	Yes	240	260
		Comment: Good at elevated temperatures and in acid conditions. Microtain fabrics possess excellent dimensional stability						
Microtain[c]	Acrylic	Average	Average	Very Good	Fair	Yes	260	280
		Comment: Good at elevated temperatures and in acid conditions. Microtain fabrics possess excellent dimensional stability						
Wool	Natural fiber protein	Poor	Average	Fair	Poor	No	200	250
		Comment: Good filterability						
Nomex[a]	Aromatic polyamide	Very Good	Very Good	Fair	Very Good	No	400	425
		Comment: Outstanding heat resistance and good resistance to abrasion						
Teflon[a]	Fluorocarbon	Average	Below Average	Excellent	Excellent	No	450	500
		Comment: Can be used at elevated temperatures and possesses excellent chemical resistance						

a. E.I. du Pont registered trademark.
b. Dry heat.
c. Registered trademark for Globe Albany.

15.4 Equipment Design Considerations

A dozen key baghouse equipment and equipment-related considerations are presented below:

1. Perhaps the most important equipment design consideration for baghouses is *bag failure*. The effect of bag failure on baghouse can be described by Equation 15.7.
2. *Uniform air and dust distribution to all filters*. Duct design, turning vanes, and deflection plates all assist in obtaining this. They often arrive loose and are field-installed. If improperly installed, they can induce high-airflow regions that will abrade the duct or bag filters or cause re-entrainment, high-dust-concentration regions that can produce uneven hopper loading and uneven filter bag dust cake.
3. *Total seal of system from dust pickup to stack outlet*. In-leakage at flanges or collector access points either adds additional airflow to be processed or short-circuits the process gases. In-leakage to a high-temperature system is extremely damaging, as it creates cold spots and can lead to dew-point excursions and corrosion. If severe, it can cause the entire process gas temperature to pass through the dewpoint and result in condensate on the bags. Early bag failure and high pressure drop will generally result. The best check for leaks is to inspect the walls from inside the system during daylight. Light penetration from outside isolates the problem areas. It is particularly important to seal the dust discharge points in negative systems. In-leakage here will result in incomplete or no discharge, which can lead to re-entrainment problems, yielding high-pressure drop and hopper fires.
4. *Effective coatings and paint*. Most systems are painted on the exterior surfaces only. Take extra care to touch up damaged areas with a good primer and if equipment is not delivered finish-painted, apply it as soon as possible following erection. Unprotected primers will soon allow corrosion and require sandblasting plus many dollars to repair. If the system has been internally protected with a coating, thoroughly inspect for cracks or chips, particularly in corners, and repair before operating. A poor interior coating can be worse than none at all because it will trap corrosive elements between the coating and the surface it was intended to protect.
5. *Properly installed filter bags*. The filter bags are the heart of any fabric filter collection system. Improper installation can result in early bag failure, loss of cleaning effectiveness, and thus high pressure drop and operating costs or increased stack emission. Each manufacturer provides instructions on the proper filter bag installation and tensioning (where required). These must be explicitly followed. Very often, early bag failures can be traced to improper installation. Remember, it is much easier to check and recheck bag connections, tensioning, locations, and so on, in a clean, cool, dry collector than it will be one day after start-up. Bag maintenance usually accounts for 70% of annual maintenance time and money. Extra efforts in this area during installation can have a significant effect.
6. *Proper insulation installation*. High-temperature collector systems require special consideration in order to prevent O&M problems. When handling high-temperature gases, it is important to maintain the temperature of the gas and all collector components coming in contact with it above the gas dew point. Insulation is usually employed. Much of the time, all or a part of the insulation is field installed. Check to see that all surfaces and areas of potential heat loss are adequately covered. In particular, check to see that field flashing also has insulation beneath it. Cold spots cause local corrosion. Gross heat loss may cause excessive warm-up time or lower the gas temperature below the dew point.
7. *Total seal between dirty side and clean side of collector*. Remember, the primary purpose of the dust collector is to separate the particulate from the gas by means of fabric filtration. This

means that all the gas must pass through the fabric. Any leaks bypassing the fabric filters will directly emit dust to the stack and therefore reduce the collection efficiency of the system. The time to inspect "bypass leaks" is before start-up, when everything is clean and accessible. The best technique is to utilize a bright light on one side of the plenum and visually observe for light penetration on the other. This is the most effective in total darkness. Take that extra time to check this important area. Tracking down stack emissions not associated with bag failures can be extremely difficult after start-up.

8. *Properly installed and operating dampers.* Most systems employ several dampers to isolate areas of the system or control the volume of air flow. Proper alignment is important both in the case of internal blades and the operating linkage. In high-temperature applications, special care must be taken to allow for proper operation and sealing at the operating temperatures. Some dampers may require readjusting after reaching high-temperature operation. In modular systems, single modules are normally isolated for bag cleaning and maintenance. Leakage through these isolation dampers can cause improper bag cleaning. It will also create a very poor ambient condition for maintenance workers to work in. This, in some applications, can pose a health hazard, and in all applications results in lower quality workmanship or incomplete maintenance.
9. *Properly operating mechanical components.* Most mechanical components are designed with a normal operating direction. Cylinder rod location, motor rotation, and so on, must be checked. Remember, when hooking up an ac motor, one has a 50% chance of being correct on the first try. Not only will a backward-moving conveyor produce no discharge, but it can pack material so that it bends the screw. Left uncorrected, a reversed screw conveyor will result in a full hopper. The industry abounds with horror stories where full hoppers have led to burned bags, or dust set up, requiring jackhammers to remove it.
10. *Smoothly running fans.* Fans must be checked for proper rotation, drive component alignments, and vibration. Fans should be securely mounted to sufficient mass to eliminate excessive vibration.
11. *Clean, dry compressed air.* Most systems employ compressed air to operate dampers, controls, instruments, and so on. Probably more systems suffer shutdowns and maintenance problems due to poor-quality compressed air than for any other reason. Clean, dry air is necessary to maintain proper operation of the pneumatic components. In installations where the ambient temperature drops below 32 °F, a desiccant dryer system is generally employed. Sometimes, insulation of air lines and pneumatic components will be required. Often, these considerations are not included in the dust collector system, with "clean, dry compressed air to be supplied by owner." Remember, the air must be clean and dry when it reaches the pneumatic component.
12. *Bag Cleaning.* A comparison of the various equipment cleaning methods is presented in Table 15.11.

The major equipment concerns are generally the filter bags. All bag sets have a finite lifetime which will vary by application, installation, operating parameters, fabric type, and so on. Typical causes of bag failure are as follows:

- High G/C abrasion
- Metal-to-cloth abrasion
- Chemical attack
- Bag-to-bag abrasion

Table 15.11 Comparison of bag cleaning methods.

Cleaning method	Uniformity of cleaning	Bag attrition	Equipment ruggedness	Type of fabric	Filter velocity	Apparatus cost	Power cost	Dust loading	Submicron efficiency
Shake	Average	Average	Average	Woven	Average	Average	Low	Average	Good
Reverse flow, no flexing	Good	Low	Good	Woven	Average	Average	Medium	Average	Good
Reverse flow, with collapse	Average	High	Good	Woven	Average	Average	Medium	Average	Good
Pulse-compartment	Good	Low	Good	Felt, woven	High	High	Medium	High	High
Pulse-bags	Average	Average	Good	Felt, woven	High	High	High	Very High	High
Pulse-jet	Very Good	Very Good	Low	Felt, woven	Very High	High	High	High	Very High
Vibration, rapping	Good	Average	Low	Woven	Average	Average	Medium	Average	Good
Sonic Assist	Average	Low	Low	Woven	Average	Average	Medium	-	Good
Manual flexing	Good	High	-	Felt, woven	Average	Low	-	Low	Good

- Inlet velocity abrasion (on inside-out cleaning)
- Accidents
- Upset conditions (e.g., temperature)
- Thread mismatched cuff mismatched

In addition, each bag in a set may have a different life as a result of fabric quality, bag manufacturing tolerances, location in the collector, variation in the bag cleaning mechanism, and so on. Any one or a combination of these factors can cause bags to fail. This means that a baghouse will experience a series of intermittent bag failures until the failure rate requires total bag replacement. Typically, a few bags will fail initially or after a short period of operation due to installation damage or manufacturing defects. The failure rate should then remain very low until the operating life of the bags is reached, unless a unique failure mode is present within the system.

The importance of when to correct a broken bag will depend on the type of collector and the resultant effect on outlet emissions. In "inside bag collection" types of collectors, it is very important that dust leaks be stopped as quickly as possible to prevent adjacent bags from being abraded by jet streams of dust emitting from the broken bag. This is called the "domino effect" of bag failure. "Outside bag collection" systems do not have this problem, and the speed of repair is determined by whether the outlet opacity has exceeded its limits. Often, it will take several broken bags to create an opacity problem and a convenient maintenance schedule can be employed instead of emergency maintenance.

In either type of collector, the location of the broken bag or bags has to be determined and corrective action taken. In a noncompartmentalized unit, this requires system shutdown and visual inspection. In inside collectors, bags often fail close to the bottoms, near the tube sheet. Look for accumulation of dust on the tube sheets, the holes themselves, or unusual dust patterns on the outside of the bags. Other probable bag failure locations in reverse-air bags are near anticollapse rings or below the top cuff. In shaker bags, also inspect the area below the top attachment. Improper tension can cause early failure.

In outside collectors, which are normally top-access systems, inspection of the bag itself is difficult; however, location of the broken bag or bags can normally be found by looking for dust accumulation on top of the tube sheet, on the underside of the top-access door or on a blow pipe.

If the system is compartmentalized, the search for broken bags can be narrowed by monitoring the stack while isolating one known compartment at a time. Through an opacity monitor, and sometimes visual observation of the stack, the compartment containing broken bags can be identified, since the emissions will be reduced when a compartment containing broken bags is isolated from the system. Once the compartment is identified, find the specific broken bag as detailed above.

One technique for locating failed bags is the use of fluorescent or phosphorescent powder and an ultraviolet light. The powder is injected into the inlet gas stream and the ultraviolet light used to scan the clean-air side of the collector. Very small leaks can be detected from the glow of the powder under the ultraviolet light where it has penetrated the clean air plenum.

Once a broken bag is located, it must be corrected. Individual bags within a compartment should not be replaced with clean bags, as they will be forced to filter a much higher volume of air than the older dust-laden bags, due to its lower pressure drop. This higher filtering velocity could cause permanent "blinding" and high pressure drop or early failure through dust abrasion. Instead, the broken bag should be tied off or plugged up. In inside collection bags where the failure has not occurred too close to the tube sheet, the bag should be cut, tied, and stuffed into its hole. For outside collection bags, an inside collection bag with holes too close to tie, a hole plug should be used. The collector manufacturer can typically supply hole plugs or they can be made from steel plate with proper gasketing or heavy sand bags made from the same material as the filter bags.

Further action will be dictated by increased pressure drop. As the number of bags taken out of service rises, the pressure will drop. At some point, the broken bags will need to be replaced. When this occurs, it is recommended that all broken bags are replaced with used bags from one designated compartment. Change out all bags in the designated compartment with new bags and damper the flow into that compartment until a sufficient dust cake has been developed. Store extra "used" bags for future replacement.

When inspecting for plugging or changing out broken bags, care must be taken for the safety of the maintenance personnel. Protection from dust, temperature, and flue gas contaminants must be provided. This is especially important for inside collectors, where the maintenance personnel must enter the bag plenum.

Care should also be taken not to damage adjacent bags and to properly reinstall the new bag. This operation should not be rushed and as good a working environment as possible should be provided. *Errors here will only create more problems tomorrow.*

One final equipment note is directed toward the total production process. A review of operating cost will indicate that utilities are of ever-growing concern and the dust collection system fan is usually the major utility controller of the dust collection system costs. Reduction of the system fan operating cost requires reduction of the system pressure drop. This can be accomplished by optimization of the duct run design, inlet and outlet damper designs, the bag cleaning system, filter media type, or reduction of the gas volume processed. Often, the fan speed can be adjusted to an improved fan curve operating point. It is important that constant vigilance be maintained to ensure that the process volume is no greater than required (e.g., boiler excess air control) and that process gas inleaks are eliminated.

The manufacturer's operating manual is at best only a guide to proper operation. The optimization of operation and minimization of operating costs are left in the user's hands. Awareness and sensitivity to the operational cost parameters and a semiannual review of the operational data collected are the route to improved operation, performance, and costs. Reynolds and Theodore provide a detailed survey study of operation and maintenance data and information [13].

15.5 Applications

Three illustrative examples complement the presentation for this chapter on baghouses.

15.5.1 Filter Bag Fabric Selection

It is proposed to install a pulse-jet fabric filter system to clean an air stream containing particulate pollutants. You are asked to select the most appropriate filter bag fabric considering performance and cost. Pertinent design and operating data, as well as fabric information, are given below (see also Table 15.12).

- Volumetric flow rate of polluted air stream = 10,000 scfm (60 °F, 1 atm)
- Operating temperature = 250 °F
- Concentration of pollutants = 4.00 gr/ft^3
- Average, ACR (G/C) = 2.5 cfm/ft^2 cloth
- Collection efficiency requirement = 99%

Table 15.12 Fabric information.

Filter bag	A	B	C	D
Tensile strength	Excellent	Above average	Fair	Excellent
Recommended max temp, °F	260	275	260	220
Resistance factor	0.9	1.0	0.5	0.9
Cost per bag, $	26	38	10	20
Standard size	8 in × 16 ft	10 in × 16 ft	1 ft × 16 ft	1 ft × 20 ft

Note: Not bag has an advantage from the standpoint of durability under the operating conditions for which the bag was designed.

Solution:

Consider the bag types listed in Table 15.12. Bag D is eliminated since its recommended maximum temperature (220 °F) is below the operating temperature of 250 °F. Bag C is also eliminated since a pulse-jet fabric filter system requires the tensile strength of the bag to be at least above average. Consider the economics for the two remaining choices. The actual gas flow rate and filtration velocity are (applying Charles' law):

$$q = 10{,}000\left(\frac{250 + 460}{60 + 460}\right)$$
$$= 13{,}354 \text{ acfm}$$
$$\text{v} = 2.5 \text{ cfm/ft}^2 \text{ cloth}$$
$$= 2.5 \text{ ft/min}$$

The filtering (bag) area is then

$$\text{A} = \text{q/v}$$
$$= 13{,}654/2.5$$
$$= 5{,}462 \text{ ft}^2$$

For bag A, the area, a, and number, N, of bags are

$$\text{a} = \pi\text{DH}$$
$$= \pi\left(\frac{8}{12}\right)(16)$$
$$= 33.5\text{ft}^2$$
$$\text{N} = \text{A/a} = 5{,}462/33.5 = 163$$

For bag B:

$$a = \pi\left(\frac{10}{12}\right)(16)$$
$$= 41.9 \text{ ft}^2$$
$$\text{N} = 5{,}462/41.9 = 130$$

The cost per bag is \$26.00 for A and \$38.00 for B. The total cost (TC) for each bag is as follows:

For bag A:

$$
\begin{aligned}
\mathrm{TC_A} &= \mathrm{N(cost\ per\ bag)} \\
&= (163)(26.00) \\
&= \$4{,}238
\end{aligned}
$$

For bag B:

$$
\begin{aligned}
\mathrm{TC_B} &= (130)(38.00) \\
&= \$4{,}940
\end{aligned}
$$

Since the total cost for bag A is less than bag B, bag A is preferred.

15.5.2 Maximum Allowable Bag Failure

A baghouse system consists of one compartment with 100 bags. The bags are 4 inches in diameter, and the pressure drop across the system is 7.0 in H_2O. The operating temperature and pressure are 70 °F and 11 atm, respectively. The inlet load to the baghouse is 4.0 gr/ft^3, and the system is 99.5% efficient, assuming that all bags are completely functional. The filtering area is 5,500 ft^2 and the filtering velocity is 420 ft/hr.

a. Calculate the efficiency assuming that 3 bags fail.

b. What is the maximum number of bag failures that can be tolerated to ensure a minimum collection efficiency of 91.50%?

Solution:

a. Relate efficiency to penetration and employ Equation 15.7

$$
\begin{aligned}
P_t &= 1 - E \\
&= 1 - 0.995 \\
&= 0.005 \text{ (before failure)}
\end{aligned}
$$

$$
\begin{aligned}
q &= (A)(v) = (5{,}500)\left(\frac{420}{60}\right) \\
&= 38{,}500 \text{ ft}^3/\text{min}
\end{aligned}
$$

$$
\begin{aligned}
\phi &= a/LD^2(T+460)^{0.5} \\
&= 38{,}500/(3)(4)^2(70+460)^{0.5} \\
&= 34.84
\end{aligned}
$$

$$
\begin{aligned}
P_{tc} &= (0.582)(\Delta P)^{0.5}/\phi \\
&= (0.582)(7.0)^{0.5}/34.84 \\
&= 0.0442
\end{aligned}
$$

and

$$
\begin{aligned}
P_t^* &= P_t + P_{tc} \\
&= 0.005 + 00442 \\
&= 0.0492
\end{aligned}
$$

$$
E^* = 1 - P_t^*
$$

$$
E^* = 0.915 = 91.5\%
$$

b. For

$$E^* = 0.915 = 91.5\%$$
$$P^* = 1 - 0.915 = 0.085$$

The calculation in part (a) is now reversed.

$$\begin{aligned} P_{tc} &= P_t^* - P_t \\ &= 0.085 - 0.005 \\ &= 0.08 \end{aligned}$$

$$\phi = \frac{(0.582)(\Delta P)^{0.5}}{P_{tc}} = \frac{(0.582)(7)^{0.5}}{0.08} = 19.25$$

One can now solve directly for the number of bag failures:

$$19.25 = 38{,}500/(L)(4)^2(70 + 460)^{0.5}$$
$$L = 5.43$$

Therefore, a maximum of five bags can fail before the efficiency drops below 91.5%. The reader should note that there is a major inconsistency in the information provided in the problem statement. Indicate what you believe to be wrong.

H = bag height one may write:

$$(100)(H)(\pi)\left(\frac{4}{12}\right) = 5{,}500$$
$$\mathrm{H} = 52.5\ \mathrm{ft}$$

This is an unreasonable height. Generally, the recommended maximum height is 36 ft. Therefore, 52.5 ft is not a reasonable height. Once a solution is determined with the equations, it is imperative to look deeper and ensure the solution makes sense.

15.5.3 Normally Distributed Bag Failures

The lifetime T of a bag employed in a baghouse is normally distributed with a mean of 2,500 days. What is the largest lifetime standard deviation that the installed bags can have if 95% of them need to last at least 365 days?

Solution:

For this application, refer to the literature [14].

$$P(T > 365) = 0.95$$

Normalizing gives:

$$P\left(\left[\frac{T-\mu}{\sigma}\right] > \left[\frac{365 - 2{,}500}{\sigma}\right]\right) = 0.95$$
$$P\left(Z > -\frac{2{,}135}{\sigma}\right) = -1.645$$

The following equation must apply for this condition:

$$-\frac{2,135}{\sigma} = -1.645$$

Solving for the standard deviation, one obtains:

$$\sigma = 1,298\ days$$

References

1 L. Theodore and A. J. Buonicore, *Control of Particulate Emissions*, USEPA/APTI Manual for Course 413, RTP, NC, 1979.
2 L. Theodore, *Air Pollution Control Equipment Calculations*, John Wiley & Sons, Hoboken, NJ, 2008.
3 L. Theodore, *Air Pollution Control and Waste Incineration for Hospitals and Other Medical Facilities*, Van Nostrand Reinhold, New York City, NY, 1990.
4 L Theodore and P. Feldman, *Air Pollution Control Equipment for Particulates*, a Theodore Tutorial, Research-Cottrell, NJ, 1991.
5 J. Nowak, M. Mangone, J. Reynolds and L. Theodore, *A Theoretical Model to Predict Pressure Drop for Flow Through Fabric Filters*, 141–50, Proceedings of the Fourth National Conference on Energy and the Environment, Cincinnati, 1976.
6 L. Theodore, *Personal Notes*, East Williston, NY, 1980.
7 L. Theodore, J. Reynolds and P. Farber, *Performance Model for Fabric Filtration: Particle Capture and Collection Efficiency*, J. Air Poll. Control Assoc., Kansas City, 1981.
8 P. Farber, J. Reynolds and L. Theodore, *A Simplified Model for Fabric Filter Cleaning and Initial Cake Buildup*, Fine Particle Society-13th Annual Meeting, Chicago, 1982.
9 P. Barber, J. Reynolds and L. Theodore, *A Model for Fabric Filtration and Multiple Compartment Effects*, J. Air Poll. Control Assoc., 1982.
10 L. Theodore, and J. Reynolds, Some factors in minimizing precipitator costs, *J. Air Poll. Control Assoc.*, 29(7), 777–779, 1979.
11 L. Theodore and J. Reynolds, Effect of bag failure on baghouse outlet loading, *J. Air Poll. Control Assoc.*, 29(8), 870–872, 1979.
12 L. Theodore, *Engineering Calculations: Baghouse Specification and Operation Simplified*, Chem. Eng. Progress, p. 22, New York City, NY, June 2000.
13 L. Theodore and J. Reynolds, Analysis of an APCA baghouse operation and maintenance survey, *J. Air Poll. Control Assoc.*, 30(7), 1218–1223, 1980.
14 S. Shaefer and L. Theodore, *Probability and Statistics Applications for Environmental Science*, CRC Press/Taylor & Francis Group, Boca Raton, FL, 2007.

Suggested Readings

1 W. Strauss, *Industrial Gas Cleaning*, Pergamon Press, New York City, NY, 1966.
2 J. A. Danielson, *Air Pollution Engineering Manual*, EPA OPA Publication No. AP-40, Government Printing Office, Washington, D.C., 1973.
3 Stern et al., *Fundamental of Air Pollution*, Academic Press, New York City, NY, 277–280, 1973.

4 A. J. Buonicore and L. Theodore, *Industrial Control Equipment for Gaseous Pollutants*, Volume I, CRC Press, Inc. Boca Raton, FL, 1975.
5 L. Theodore and A. J. Buonicore, *Vapor Control by Condensation: Performance Equations and Design Procedures*, paper#75-23.2, Air Pollution Control Association Meeting, Boston, 1975.
6 D. Richard, et al., *Filtration Model for Coal Fly Ash with Glass Fabrics*. EPA-600/7-77-084, 1977.
7 J. Nowak, M. Mangone, J. Reynolds and L. Theodore, "*A Theoretical Model to Predict Pressure Drop for Flow Through Fabric Filters*", 141–150, Proceedings of the Fourth National Conference on Energy and the Environment, Cincinnati, 1976.
8 L. Theodore and J. Reynolds, Effect of bag failure on baghouse outlet loading, *J. Air Poll. Control Assoc.*, 29(8), 870–872, 1979.
9 A. J. Buonicore, J. Reynolds and L. Theodore, Control of fly ash from conventional coal-fired utility boilers, *J. Air Poll. Control Assoc.*, 30, 602–611, 1980.
10 L. Theodore and J. Reynolds, *Analysis of an APCA Baghouse Operation and Maintenance Survey*, *J. Air Poll. Control Assoc.*, 30(7), 1218–1223, 1980.
11 L. Theodore, *Engineering Calculations: Baghouse Specification and Operation Simplified*, Chem. Eng. Progress, p. 22, New York City, NY, June 2000.
12 L. Theodore, Letter to the Editor, *J. Air Poll. Control Assoc.*, 56, 1367, 2006.
13 L. Theodore, *Air Pollution Control Equipment Calculations*, John Wiley & Sons, Hoboken, NJ, 2008.
14 L. Theodore, *Chemical Engineering: The Essential Reference*, McGraw Hill, New York City, NY, 2012.
15 L. Theodore, R. R. Dupont and K. Ganesan, *Unit Operations in Environmental Engineering*, John Wiley & Sons, Hoboken, NJ, 2017.
16 J. Richardson, R. R. Dupont, P. Farber, K. Ganesan and L. Theodore, *Air Pollution Control Equipment Revisited*, AWMA Conference, Calgary, CA, 2024.
17 J. Richardson and L. Theodore, *Optimizing Air Pollution Equipment Performance: Operation and Maintenance*, John Wiley & Sons, Hoboken, NJ, 2025.

Part IV

Design of Other Air Pollution Control Equipment

Part IV of this book examines what the author has come to define as "other" air pollution control equipment. These are devices and/or processes that are either outside the domain of the traditional control equipment or have recently taken on greater significance. The devices include:

- Chapter 16: Hybrid Systems
- Chapter 17: Flue Gas Desulfurization Systems
- Chapter 18: Controlling the Oxides of Nitrogen
- Chapter 19: Carbon Capture and Storage
- Chapter 20: Stacks

Some of the material in these chapters has been drawn from the authors' earlier work: *Optimizing Pollution Control Equipment Performance: Optimization and Performance.*

16

Hybrid Systems

16.1 Introduction to Hybrid Systems

The basic design of air pollution control equipment has remained relatively unchanged since first used in the early part of the 20th century. Some modest equipment changes and new types of devices have appeared in the first 25 years of this century, but all have essentially employed the same capture mechanism of the past.

Hybrid systems are one area that has recently received some attention. Hybrid systems have shown to operate at higher efficiencies or more economically than conventional devices in some cases. Regulations continue to tighten which has placed emphasis on developing new technologies and creating synergies from existing technologies.

This chapter is concerned with hybrid systems, as they apply to air pollution control equipment. The term hybrid system has come to mean different things to different people. The term is defined here as:

1. Two or more different air pollution control devices connected in series, e.g., baghouse followed by an absorber.
2. An air pollution control unit that utilizes two or more collection mechanisms to enhance pollutant capture.

With regard to item (1), equipment connected in series are usually treated, from an engineering and calculational point of view, sequentially, with the outlet of one device becoming the inlet of the next control device. As such, this topic does not require detailed additional treatment other than that provided earlier in this book. It should be noted that this approach has found application in the treatment of metal particulates. A dry scrubber followed by a wet scrubber has been employed in this country to improve the collection of fine particulate metals in hazardous waste incinerators; the dry scrubber captures metal that condenses at the operating temperature of the unit, and the wet scrubber captures residual metals (particularly mercury) and dioxins/furan compounds. Another recent application in Europe involves the injection of powdered activated carbon into a flue gas stream from a hazardous waste incinerator at a location between the spray drier (dry scrubber) and the baghouse (or electrostatic precipitator [ESP]). The powdered carbon mixing with the lime particles from the dry scrubbing system and the gas stream itself adsorb the mercury vapors and residual dioxins/furnace compounds and then are separated from the gas stream by a particulate control device.

This chapter will focus on item (2). Although there are numerous hybrid systems available in the marketplace, the discussion that follows will examine only the following control devices:

1. Dry scrubbers—DRYS
2. Ionizing wet scrubbers—IWS
3. Wet ESP—WEP
4. Elektrofil ESP—EESP
5. Electrostatic stimulation of fabric filtration—ESFF

These have been chosen because of their widespread use and/or potential application in the future, or both. Tighter regulations and a greater concern for environmental control by society have placed increased emphasis on the development and application of these systems.

Each system above is treated separately in individual sections and includes (where applicable) a brief introduction, a description of the system, some operational considerations, and applications. Some of the material has been drawn from company literature, reports, and/or bulletins. (These have been acknowledged in the reference section.)

16.2 Dry Scrubbers (DRYS)

As environmental concerns continue to become more mainstream in everyday life, new emissions regulations have become tighter. One of the most studied and publicized effects of industrial pollution is that of climate change and acid rain. Hydrochloric acid (HCI), sulfuric acid (H_2SO_4), and sulfur dioxide (SO_2) are typical byproducts of common combustion processes and are major contributors to climate change and acid rain. To operate responsibly and within environmental compliance, emissions of HCl and SO_2 need to be controlled and monitored. Since the 1990s, a common method of control has been to chemically react pollutant acid gases with an alkaline material to generate a relatively benign waste product; this process is known as dry scrubbing or dry sorbent injection (DSI).

The DSI industry in this century has expanded to include other pollutant treatment targets such as hydrofluoric acid (HF), stack opacity improvement, particulate matter (PM) collection efficiency increases (on ESP applications), and metal leachate stabilization for process byproducts. Dry scrubbing and in particular DSI is now used as a complete solution to solve many issues arising from post combustion emissions and byproducts, rather than a complement to other scrubbing technologies in the past. When DSI is applied (with proper particulate collection devices like a baghouse, cyclone, or ESP), the resulting air pollution control system becomes robust and flexible in terms of pollution capture efficiencies.

DSI processes involve pneumatic introduction (injection) of a dry, powdery alkaline material (usually a calcium or sodium-based sorbent) into the flue gas stream with subsequent fabric-filter (or other PM control equipment) collection downstream of the injection point. The injection point in such processes can vary from the boiler/furnace section all the way to the flue gas entrance at the particulate collection device. The injection points are determined by process operating conditions and design criteria.

The method of operation of dry scrubbing via DSI is relatively simple, requiring only two major equipment items; (1) a pneumatic conveyor and storage system (the DSI system), and (2) a particulate control system (baghouse, cyclone or ESP) to collect the fly ash or entrained flue gas PM and the sorbent PM. Baghouses have become most common, because baghouses are generally designed to "cake" or generate a layer of sorbent material on their bags to increase sorbent interaction with pollutants. The solid-bearing gas passes through the particulate control devices where the dry product is collected, and where a percentage of the unreacted alkali reacts with acid gases for further removal. Usually, the cleaned gas is then discharged through the particulate control device's plenum to an induced draft (ID) fan and the stack.

As noted earlier, fabric filters are very successful in removing fine particles from flue gas streams which has encouraged the combination of dry scrubbing and fabric filter systems for the purpose of simultaneously removing both particulates and acid gases. Dry scrubber processes such as DSI offer potential advantages over their wet counterparts, especially in the areas of energy savings and capital costs; a dry system can cost up to 90% more than a wet system with comparable performance. Furthermore, the dry-scrubbing process design is relatively simple for construction and operating purposes in the future, especially in existing facilities. These dry scrubbing systems generate a solid byproduct waste at the PM control locations. Waste removal and disposal are often an overlooked, yet expensive and critical component of these air pollution control designs. A dry scrubber generates a dry powder waste product that is more easily and economically disposed by the users, rather than the more difficult and expensive wet sludge of a wet system.

Among the inherent advantages that the DSI enjoys over wet scrubbers are as follows:

1. Lower capital cost
2. Lower draft losses
3. Reduced auxiliary power
4. Reduced water consumption
5. Lower waste disposal cost

DSI is evolving from its roots, first commercially initiated in the coal-fired power utility market in the 1990s for sulfur dioxide and sulfuric acid treatment. The application is now finding use in new markets such as hydrogen production, glass, cement, and refining or any other industrial application where acid gas or metal pollutants are generated and in need of capture.

DSI can take on numerous different applications with varying different sorbent supplies. DSI can utilize the following different injection materials for distinct types of pollution control or remediation:

1. Hydrated lime $Ca(OH)_2$
2. Sodium bicarbonate $NaHCO_3$
3. Sodium sesquicarbonate $Na_3H(CO_3)_2$
4. Activated carbon C
5. Magnesium hydroxide $Mg(OH)_2$
6. Or various blends of the above materials

Sorbent injection systems now are not only used to treat flue gas acid pollutants but also to treat process byproducts. In many cases, the process byproducts or PM that are generated from the process are metals that exceed local and federal guidelines for traditional waste classification and need to be disposed of by hazardous waste guidelines. In a patented process, facilities and users are instituting specialized sorbent injection methods solely for the treatment of the waste byproduct solids within the flue gas stream. In a sense, these dry sorbent injection systems are preventing the forming hazardous waste within the process. There can be many combinations of these hazardous waste stabilizers and dry sorbent injection configurations used on the same process and treatment system. Dual Resource Partners, an air pollution control technology provider outside of Chicago, IL, initiated the first hazardous metals waste stabilization program in 2023 [1]. The DSI system and methodology transformed a previously hazardous waste generating process into a stable byproduct solution, meeting nonhazardous waste criterion. Additional details on the chemical reactions and pneumatic convey systems are available in the literature [2].

16.2.1 DSI Expansion

There are new applications within the carbon sequestration and capture market that will need the dry sorbent application for improving their systems. Carbon capture systems utilize amine solutions to strip or remove carbon dioxide from the gas stream. Amine solutions are usually enormously expensive proprietary products that have been created over vast amounts of research and funding. Amine solutions are incredibly sensitive to acid gases like sulfur dioxide, prolonged exposure to acid gases will destroy amines that could otherwise be recycled and reused for further carbon dioxide stripping. The sensitivity to sulfur dioxide necessitates the use of intense and powerful dry scrubbing like a DSI system to remove the acid gases prior to the introduction of the amines.

16.3 Ionizing Wet Scrubber (IWS)

The term wet scrubber was first used by the Ceilcote Co., located in Berea, OH. The presentation to follow will key on the Ceilcote system (Patent Nos.: 3,874,858 and 3,958,958), with information provided by the vendor [3, 4]. This system has found wide application in the air pollution control field, particularly in controlling discharges from hazardous waste incinerators. A short description of MikroPul's short-lived Elektrofil ESP is also included.

The Ceilcote IWS system is a proven technique for the removal of pollutant from industrial gas streams. The IWS combines electrostatic particle charging, inertial impaction, image force attraction, and gas absorption to simultaneously collect submicron solid, liquid, and gases submicron in size. The IWS system requires relatively little energy and its collection efficiency is high for both submicron and micron-size particles.

The IWS utilizes high-voltage ionization to electrostatically charge particulates in the gas stream before the particles enter a Tellerette-packed scrubber section where they are removed by attraction of the charged particles to neutral surfaces. Particles in the 3 to 5 μm range are collected through inertial impaction. Smaller particles continue to flow through the scrubber. The electrostatic charge on the particles causes them to be attracted to the neutral surfaces of the Tellerettes by image force attraction. Contaminant gases are absorbed and reacted in the exit liquor which washes out all collected particles.

Tellerette is the primary packing material used in the IWS systems, as well as the scrubbing liquor itself. Particles of any size or composition are collected by the IWS. IWS systems are highly efficient at collecting both fine particles (0.05–2 μm) and coarse particles (2 μm or larger). Composition has little effect as well, with high collection efficiencies on organic or inorganic particles with either high or low resistivity. Particle collection efficiency remains consistently high over long-term service as well.

The IWS system acts as a fractional collector. The percentage of particulates removed remains constant over wide ranges of load and particle size distribution. The IWS system also simultaneously absorbs gases. Contaminant gases are removed through both physical absorption and absorption. The pressure drop through a single-stage IWS is moderate at only 0.5–1.5 in H_2O. The energy for particles charging is low—approximately 0.2–0.4 kVA/1,000 CFM. The components of the system are commonly fabricated with FRP (fiberglass reinforced plastic) and thermoplastic materials. The predominance of plastic construction assures corrosion-free operation in the presence of acid gases such as HCl, HF, Cl NH_3, SO_2, and SO_3. Metallic construction is also available for noncorrosive applications. Factory assembled modules, available in standard capacities from 900 to 54,000 ACFM, can also be grouped together to handle virtually any gas volume [5, 6].

In IWS systems, particles are initially electrostatically charged within an ionizer section powered by a high-voltage DC power source. Discharge electrodes have negative polarity and wetted "mini-plates" serve as grounded electrodes. High-voltage DC power for the ionizing section is provided by a high-voltage transformer/rectifier. This is a sealed, watertight, oil-filled unit. Rectification is accomplished by silicon diodes which are suitably protected against transient voltages for efficient and long-term, trouble-free service. There is typically a prewired central control panel that controls all control power functions. The maximum voltage on the interconnecting wiring is 480 V. Voltage control for the high voltage is achieved by thyristor control through phase-angle variation. An ultrafast response, solid state, plug-in type standard control module automatically controls high-voltage levels and spark rate for optimum performance. A manual control function is also provided for backup and for troubleshooting.

Operation is designed to be simple. Interlocks and process parameters are annunciated with indicator lights to monitor the operation. Protective interlocks and circuits safeguard the equipment against overloads or malfunctions. A continuous flow of water (or other suitable liquid) is utilized to flush the grounded electrode plate surfaces of the IWS. The water washes away any resistive solids or residues that could reduce performance. A second flushing system is also provided which periodically deluge flushes all internal areas of the ionizer section. The deluge flushing is an automatic operation typically set for every eight hours or as required by the experience. Water for the periodic deluge flush is recycled liquid from the scrubber recycle system. The IWS incorporates a cross-flow scrubber section following the ionization section. The scrubber section typically uses Tellerette packing to collect gases, liquid, and solid particles from the gas stream. The contaminated gas stream moves through the packing bed material that is flushed by scrubbing liquid through the packing.

The cross-flow design has the inherent advantages of high operating reliability, low pressure drop (with high airflow), and low operating cost. A sump pump recirculates the scrubbing liquid after it passes through the packed-bed section. Use of an integral sump reduces cost and minimizes floor space requirements. The recirculation is provided for continuous wetting of the ionizer plates, deluge flushing of the ionizer section, and continuous irrigation of the cross-flow scrubber. The method described above can be adjusted based on the customer's requirements and site specific restrictions.

As described earlier, the IWS shell and internals are typically fabricated of fiberglass plastic (FRP) and other polymeric materials. The use of plastic assures corrosion-free operation in the presence of gases such as HCl, HF, NH_3, SO_2, and SO_3. The light weight aids in structural requirements as well. Plastics are used in the IWS because they have proven for long service lifespans. The only metal required is for the electrical components employed within the ionizing section. These metallic components are typically alloys chosen specifically for each application to provide maximum corrosion resistance. The IWS can also be constructed from metal in specific applications where there are no corrosive issues and the use of FRP materials is not economical or practical.

Results from earlier pilot tests and more than 50 actual full-size operating systems verify that particles of virtually any size and composition are collected by the IWS. Submicron solid/liquid particles as minute as 0.05 μm, coarse particles having organic or inorganic structures, and conductive or highly resistive particles are collected at high efficiency.

The predominant influence acting within the IWS to collect submicron-size particulate is the mechanism of "image force attraction." Image force attraction occurs whenever an electrostatically charged particle comes within the boundary layer (usually within 1 mm) of a neutral surface. As the charged particles come close to the neutral surface, an electrostatic charge of opposite polarity is induced at the neutral surface. The force of attraction is similar to that which exists between a charged particle and an imaginary particle of equal and opposite polarity located at an equal distance behind the surface. It is as though the charged particle was seeing its "mirror image [7]."

16.4 Elektrofil ESP

MikroPul's Elektrofil ESP utilizes a continuous flow of finely divided liquid droplets, which become electrically charged and are attracted to the collecting plates. The precipitated water droplets create an evenly distributed water film over the entire collector surface. The solid particulates are charged and simultaneously migrate to the collecting surfaces and are captured by the continuously downward-moving film of water. The water film on each of the collecting plates and discharge electrodes flows by gravity into the slurry trough for disposal.

The operation of the Elektrofil system is somewhat unique. The dirty gas stream is preconditioned to achieve saturation. It then enters at the inlet and loses much of its velocity in the inlet expansion chamber. Strategically located spray nozzles in the inlet section ensure complete saturation and maintain a steady water flow down the housing sides and baffles to remove any particulates collected by impingement. The water sprays in the inlet expansion chamber help break up the airstream, and transverse baffles, double or triple staggered rows of channels, assure even distribution of the gas flow over the entire cross section of the precipitator. After passing through the transverse baffles, the dirty gas stream then passes into the electrostatic field. Water sprays located in front of and behind the transverse baffles, and above the electrostatic field section, introduce the proper amount of water droplets to the gas stream. The solid particulates and the water droplets in the electrostatic field pick up a charge and migrate to the collecting plates or to the discharge electrodes. The water droplets form a continuous downward moving film over all the collecting surfaces, and help to entrap the droplets and particulates as they migrate to the electrostatic elements. The water film and collected particulates slowly flow down the collecting elements and fall to the discharge trough below. The collected slurry is then withdrawn through a drain and piped to a treatment system [8, 9].

Where efficiency specifications require the use of more than one field, a row of transverse baffles between each field redistributes the gas flow. Each field has its own spray system which introduces the proper amount of water to ensure effective cleaning. After passing through all the electrostatic fields, the gas stream again is subjected to another set of transverse baffles where any water droplets or dust particles not deposited on the plates are eliminated. All baffle systems are arranged so that a walkway runs across the front or back of each of the electrostatic fields. The discharge electrode frames are mounted on collar type high-voltage support insulators. Insulator compartments are heated and pressurized to prevent moisture and particulate leakage.

16.5 Wet Electrostatic Precipitators (WEPs)

The two major types of high-voltage single-stage ESP configurations are plate and tubular. As the term single stage implies, particles are both charged and collected in a single stage. The plate-type configuration was examined in detail in Chapter 14. This section examines the wet ESP—a unit that has a tubular configuration and is normally operated wet. This unit has appropriately been called a wet ESP. It should be noted that its use has declined in recent years. In addition, plate ESPs are used more often than tubular ESPs in industrial applications.

Tubular precipitators consist of cylindrical collection electrodes with discharge electrodes located in the center of the cylinder. Dirty gas flows into the cylinder where precipitation occurs. The negatively charged particles migrate to and are collected on grounded collecting tubes. The collected dust or liquid is removed by washing the tubes with water sprays located directly above the tubes. These precipitators have also been referred to as water-walled ESPs. Tubular precipitators are generally used for collecting mists or fogs. Tube diameters typically vary from 0.5 to 1 ft (0.15–0.31 m), with lengths ranging from 6 to 15 ft (1.85–4.6 m). The gas to be cleaned flows around the outside of the cylindrical electrodes and up through the inside of the cylinders where precipitation takes place. The collected dust or liquid is removed from the bottom of the chamber. On occasion, hexagonal tubes (enabling closer packing) may be used to increase the amount of available collecting surface in a given volume, a problem treated in the literature [8].

16.6 Electrostatic Stimulation of Fabric Filtration (ESFF)

The use of fabric filters to remove particles from gas streams is a well-established industrial practice and its importance is growing for electrical utility boilers. The fabric filter is roughly equivalent in price to other utility control devices and provides increased particle removal efficiency. However, fabric filter operating experience is limited, and enough questions have been raised because of unexpectedly high baghouse pressure drops to cause concern in the industry.

A concept to improve fabric filtration through the use of electric fields ESFF was developed on a laboratory scale by researchers at the Triangle Research Institute (TRI), Princeton, NJ, whose work was partially supported by EPA [10, 11]. TRI showed that this concept could significantly improve filter performance by reducing filter drag and increasing filter efficiency. In work supported by EPA, the Research Triangle Institute (RTI), Research Triangle Park, NC, and ETS, Inc., Roanoke, VA, successfully transferred the ESFF technology to the field environment by successfully operating a pulse-cleaned ESFF baghouse on a slipstream from an industrial boiler [8].

The ESFF effect can best be understood as the net effect of fundamental forces on particles—either charged or uncharged—in the presence of an electric field. While only one primary electric field is applied to the filter, that field interacts with the filter fibers and collected particles to form many localized gradients in the electric field. A particle approaching the filter is thus exposed to a complex field before it is collected. The electric field continues to have an effect on the particle after it is collected.

There are several changes in the pattern of dust deposition on a filter that might account for the reduced pressure drop effects observed with ESFF. These include the following:

1. An increased fraction of the collected dust being deposited on or near the upstream filter surface.
2. Changes in the pattern of collection on a filter fiber as more dendritic collection and bridging occur.
3. Formation of a highly nonuniform dust deposit—at the scale of the electrode spacing—on the filter surface.

An increased fraction of the dust being deposited near the upstream surface of the filter has been observed in laboratory studies of ESFF. This observation is important because the upstream side of a fabric filter has a lower packing density than the body of the fabric. Collection in the less dense surface region, which is enhanced by the electric field, leads to reduced pressure drop for a given quantity of dust and reduces the amount of dust that penetrates the fabric. Reduced penetration into the depth of the fabric accounts for the reduced residual pressure drop observed.

Changes in the pattern of dust deposition on fibers have been observed by a number of researchers. In general, how the presence of an electric field leads to increased dendritic deposition on filter drag is not clear. If the dendrites bridge between fibers and pack tightly, the resistance should increase; if they remain in an open structure, the flow resistance should be reduced. Visual observation within a filter is difficult, and the effects of particle deposition on fibers remain uncertain.

If electrostatic forces cause the formation of a dust deposit with a nonuniform areal mass density, a reduction in pressure drop should result. This will be true, for instance, if Coulombic forces cause most of the dust to collect in the immediate region of the electrodes, leaving reduced deposits on the balance of the filter [12]. Highly nonuniform deposits have been observed with redispersed fly ash in a laboratory ESFF filter apparatus [10].

Stabilization of the deposited dust layer by the electric field has also been studied as a possible ESFF mechanism. The theory was developed in conjunction with fluidized beds and adhesion forces within electrostatic precipitator dust layers, but its application to the ESFF situation is fairly direct. It is postulated that the initial dust cake formation in an ESFF filter is quite porous due to the electrical gradient forces, and that the electric field prevents the collapse of this porous cake as the pressure drop increases. The electrical stabilization preserves the initial high porosity, resulting in prolonged operation at lower drag than in the more compressible conventional operation [13].

16.7 Recent Advances in Control Equipment Technology

Recent advances in this field have been primarily involved in the treatment and control of metals. A dry scrubber followed by a wet scrubber has been employed in the United States to improve the collection of fine particulate metals in hazardous-waste incinerators; the dry scrubber captures metals that condense at the operating temperature of the unit and the wet scrubber captures residue metals (particularly mercury) and dioxin/furan compounds. Wastes from the wet scrubber are recycled into the dry scrubber resulting in a no liquid discharge system.

Another recent application in Europe and the United States involves the injection of powdered activated carbon into a flue gas stream from a hazardous waste incinerator at a location between the spray dryer (the dry scrubber) and the baghouse (or ESP). The carbon, mixing with the lime particulates from the dry-scrubbing system and the gas stream itself, adsorbs the mercury vapors and residual dioxin/furan compounds and is separated from the gas stream by a particulate control device.

16.8 Conclusion

The two most prevalent definitions employed today for hybrid systems are (1) two or more different air pollution control devices connected in series, e.g., a baghouse followed by an absorber, and (2) an air pollution control unit that utilizes two or more collection mechanisms to enhance pollutant capture.

There are two major types of so-called dry scrubber systems: spray drying and dry injection. The first process is often referred to as a wet-dry system. When compared to a conventional wet scrubber, it uses significantly less scrubbing liquid. The second process may be referred to as a dry–dry system since no liquid scrubbing is involved. The spray drying system is predominantly used in utility and industrial applications.

The IWS incorporates a cross-flow scrubber section following an ionization section. The scrubber section employs Tellerette packing to absorb gases and collect both liquid and solid particles from the gas stream.

Wet tubular ESPs consist of cylindrical collection electrodes with discharge electrodes located in the center of the cylinder. Dirty gas flows into the cylinder where precipitation occurs. The negatively charged particles migrate to and are collected on grounded collecting tubes.

A concept to improve fabric filtration through the use of electric fields ESFF was developed. The ESFF effect can best be understood as the net effect of fundamental forces on particles—either charged or uncharged—in the presence of an electric field. While only one primary electric field is applied to the filter, that field interacts with the filter fibers and collected particles to form many localized gradients in the electric field.

References

1 S. Dooley, Personal Notes, Dual Resource Partners, Chicago, IL, 2023.

2 J. Richardson and L. Theodore, *Optimizing Air Pollution Equipment Performance: Operation and Maintenance*, John Wiley & Sons, Hoboken, NJ, 2025.

3 Ceilcote Co. Bulletin 12-19, *IWS System*, Berea, OH, not dated.

4 J. MacDonald, Private Communication to L. Theodore, Ceilcote Co., Berea, OH, 1985.

5 L. Theodore, *Air Pollution Control Equipment Calculations*, John Wiley & Sons, Hoboken, NJ, 2008.

6 L. Theodore and R. Allen, *Air Pollution Control Equipment*, A Theodore Tutorial, Theodore Tutorials, East Williston, NY, 1993 (Originally published by USEPA/APTI, RTP, NC).

7 J. Mycock, J. McKenna and L. Theodore, *Air Pollution Control Engineering and Technology*, CRC Press Taylor & Francis Group, Boca Raton, FL, 2010.

8 C. Wang and C. Ho, *Inertial and Electrostatic Deposition of Solid Particles on Isolated Fibers*, Invited paper presented to the Fiber Society and the Filtration Society Joint Symp. Fibers, Electrostatics, and Filtration, Princeton, NJ, November 14–15, 1979.

9 MikroPul Bulletin, *The Elektrofil Electrostatic Precipitator*, NJ, 1972.

10 D. Van Osdell and D. Furlong, *Electrostatic Augmentation of Fabric Filtration*, USEPA Report No. 68-02-3186, E.S. EPA, Research Triangle Park, NC, 1986.

11 G. Lamb and P. Costanza, A low-energy electrified filter system, *Filtr. Sep.*, 17, 319–322, 1980.

12 J. Louie Kovach, Private Communication, President and CEO, Nucon, Columbus, OH, 1980.

13 C. Wang, C. Ho, H. Makino and K. Iinoya, Effect of electrostatic fields on accumulation of solid particles on single cylinders, *AICHE J.*, 26(4), 680–685, 1980.

17

Flue Gas Desulfurization Systems

Note: Parts of this chapter on flue gas desulfurization systems are an edited and updated version of an earlier chapter written by T.W. Devitt for a 1981 Prentice-Hall publication titled "Air Pollution Control Equipment: Selection, Design, Operation, and Maintenance," which was edited by L. Theodore and A.J. Buonicore (the rights of the book were later transferred to Theodore Tutorials during the 1985–1986 time period), and the 2025 John Wiley & Sons book by J. Richardson and L. Theodore, "Optimizing Air Pollution Control Equipment Performance: Operation and Maintenance."

17.1 Introduction

Flue gas desulfurization (FGD) is the process of removing sulfur oxides, primarily SO_2 from combustion gases. In wet process FGD systems, flue gases are contacted with an absorbent in a vessel called either an absorber or a scrubber. The SO_2 reacts with the absorbent or dissolves into the solution to produce a slurry or liquid that contains dissolved or solidified sulfur compounds. In most operating systems, water is used to dissolve or suspend the reacting chemicals, although a number of dry absorption systems are used in commercial applications. The advantages of dry scrubbing include potential cost and energy savings and relative ease of operation compared with wet FGD systems. Dry scrubbing, however, requires higher chemical consumption and may be restricted to low-sulfur coals. Because most of the commercial dry scrubbing installations are lesser in number, there is a lack of data on full-scale systems. Therefore, this chapter is limited to wet FGD systems.

17.2 Description of Control Device

Wet FGD processes are grouped into two general categories—regenerable and nonregenerable—depending on whether the sulfur is separated from the absorbent as waste. Nonregenerable processes produce a sludge that requires disposal in an environmentally sound manner. Regenerable processes include additional steps to convert the sulfur into by-products such as liquid sulfur dioxide, sulfuric acid, or elemental sulfur.

Six basic wet FGD systems are used in this country, primarily to treat flue gas from large, coal-fired boilers. Of these, the lime scrubbing, limestone scrubbing, sodium carbonate scrubbing, and dual alkali systems are classified as nonregenerable. The magnesium oxide scrubbing and Wellman–Lord systems are classified as regenerable processes.

Because most operating FGD systems are nonregenerable (a trend that is likely to continue because of their generally lower capital and operating costs), only the nonregenerable systems are reviewed here. Furthermore, only the wet lime- and limestone-based systems are discussed, although many of the design, operation, and maintenance considerations are similar for other systems.

17.2.1 The Lime Scrubbing Process

The first step of the process is the preparation of a lime slurry. Lime (CaO) is reacted with water in a slaker to produce a slurry of calcium hydroxide ($Ca[OH]_2$) and water, according to the following reaction: $CaO + H_2O \rightarrow Ca(OH)_2$ + heat. This material is diluted with recycled water to produce a scrubber slurry containing about 15–20% solids.

The lime slurry is added at a regulated rate to a recycle tank that contains the scrubbing slurry. Large pumps circulate the scrubbing slurry through the scrubber or absorber vessel. In the absorber, the slurry droplets are brought into contact with the flue gas containing SO_2, and the SO_2 passes from the gas into the droplets driven by a concentration gradient driving force. The dissolved SO_2 reacts with calcium from the lime according to the following reactions:

$$Ca(OH)_2(s) \leftrightarrow Ca(OH)_2(aq) \leftrightarrow Ca^{2+} + 2OH^- \quad (17.1)$$

$$2OH^- + CO_2 \leftrightarrow CO_3^{2-} + H_2O \quad (17.2)$$

$$CO_3^{2-} + CO_2 + H_2O \leftrightarrow 2HCO_3^- \quad (17.3)$$

$$Ca^{2+} + CO_3^{2-} \leftrightarrow CaCO_3 \quad (17.4)$$

$$CaCO_3(s) \leftrightarrow CaCO_3(aq) \leftrightarrow Ca^{2+} + CO_3^{2-} \quad (17.5)$$

$$SO_2(g) \leftrightarrow SO_2(aq) + H_2O \leftrightarrow H^+HSO_3^- \quad (17.6)$$

$$HSO_3^- \leftrightarrow H^+ + SO_3^{2-} \quad (17.7)$$

$$Ca^{2+} + SO_3^{2-} \leftrightarrow CaSO_3(aq) \quad (17.8)$$

$$Ca^{2+} + SO_3^{2-} + \frac{1}{2}H_2O \leftrightarrow CaSO_3 \cdot \frac{1}{2}H_2O(s) \quad (17.9)$$

$$CaSO_3 + H^+ \leftrightarrow Ca^{2+} + HSO_3^- \quad (17.10)$$

$$H^+ + HCO_3^- \leftrightarrow H_2CO_3 \leftrightarrow CO_2(g) + H_2O \quad (17.11)$$

$$CaSO_3 \cdot \frac{1}{2}H_2O + \frac{3}{2}H_2O + \frac{1}{2}O_2 \leftrightarrow CaSO_4 \cdot 1H_2O \quad (17.12)$$

These reactions take place to some extent in every lime scrubbing system; similar reactions occur in limestone systems. Factors such as ionic concentrations, pH, temperature, and retention time in the reaction tank influence the speed and completeness of the various reactions. The net effect, however, is the removal of SO_2 from the flue gas, depletion of calcium hydroxide in the slurry, and generation of insoluble calcium sulfite and calcium sulfate. The calcium ions used in the reactions are replenished by the dissociation of calcium hydroxide. The use of an insoluble alkaline reagent is the primary difference between the lime and limestone slurry scrubbing systems and the so-called clear solution scrubbers (e.g., double alkali scrubbers), where the predominant reactive alkali (sodium) is more soluble.

Forcing most of the sulfite/sulfate precipitation to occur in the reaction tank requires a certain amount of retention time to permit the slurry to reach chemical equilibrium before it is recirculated to the scrubber. For this purpose, a reaction tank, sometimes an integral part of the scrubber vessel, is usually sized with a retention time of between 5 and 10 minutes.

The calcium sulfite and calcium sulfate that are formed in the reaction tank must be removed from the system. This is accomplished by bleeding a stream from the recirculating slurry to a thickener or clarifier. The calcium sulfite and sulfate settle out and are removed from the underflow of the clarifier in a slurry of about 30% solids. The overflow, a clear liquid of less than 1% solids, is brought back to the system for reuse in the process. The sludge from the underflow is disposed of by one of several means.

One method of disposal is to pump the sludge into a pond. In this case, the sludge is pumped from the thickener underflow (or occasionally directly from the reaction tank) to a large pond, where the solids settle. Clear water is recycled from the pond, as necessary, to keep the pond from overflowing. Return of all this clear liquor from the pond to the process is called a "closed-loop" operation.

Another method of disposal is to dewater the sludge in a vacuum filter, chemically stabilize it using generic or proprietary processes, and dispose of it in a landfill. When vacuum filtration is used, the filtrate is returned to the system to create a closed-loop operation [1–9].

17.2.2 Limestone Scrubbing Process

Limestone slurry preparation is the only step that is significantly different from the lime scrubbing process. Lump-size limestone is pulverized to ~200 to ~300 mesh in a ball mill and slurried with water. The fine size is needed because the dissolution of $CaCO_3$ is a slow reaction, the rate being directly related to the particle surface area and therefore to the particle size. For further limestone dissolution, the slurry is maintained at an acidic pH, in contrast to lime-based systems, in which scrubbing slurry is slightly alkaline. The overall process reactions are the same as for the lime slurry scrubbing process, except the reactant species is $CaCO_3$, not $Ca(OH)_2$.

Figure 17.1 illustrates a typical limestone process. The lime process is similar, except a lime slaker replaces the ball mill of the limestone process.

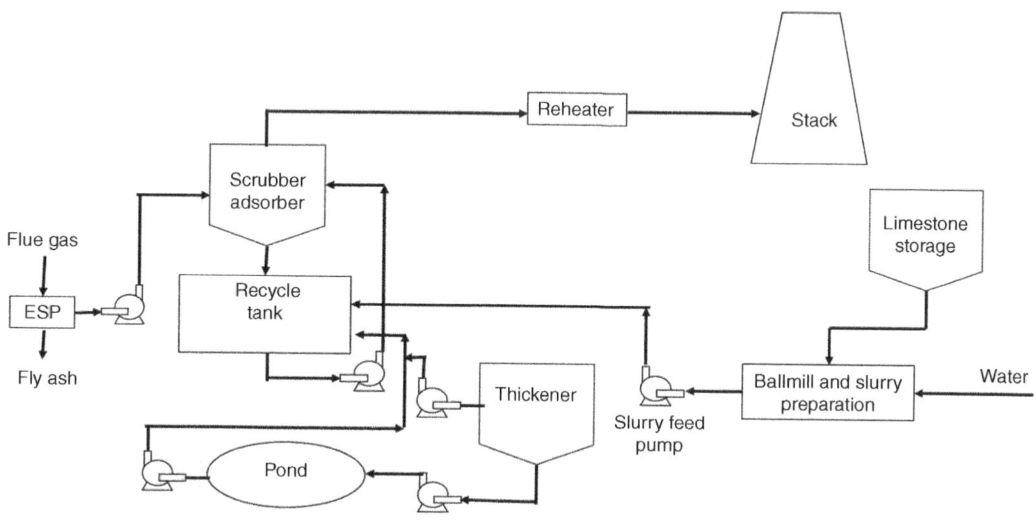

Figure 17.1 Flow diagram for a limestone scrubbing system.

Other wet process FGD systems operate similarly. The SO_2 is scrubbed from the gases in a contactor, and reaction products are removed in a separate vessel. Several excellent texts and reports are available that describe the various systems, their chemistry, and the status of commercialization [1–8].

17.3 Process Design Considerations

Proper system design is crucial to the successful operation of an FGD system. Although this is true for all air pollution control systems, it is especially important for FGD systems. In contrast to most other air pollution control systems, there are two design considerations; in addition to adequate mechanical design, an FGD system must receive careful chemical process design. Many operating problems have developed in these systems as a result of inadequate chemical design.

Table 17.1 presents a partial list of the factors outside the scrubbing system battery limits that influence system design, and Table 17.2 lists the mechanically and chemically related design parameters that must be considered within the scrubber battery limits. A significant body of knowledge has been established concerning proper mechanical design, but the factors bearing on chemical process design and their interrelationships with the mechanical equipment design have only begun to receive adequate consideration.

Table 17.1 Factors influencing scrubber system design outside the battery limits.

Coal properties	Sulfur content Ash content Fly ash composition/particle size Chloride content Heating value Moisture content Composition variability
Boiler design	Type of boiler Size of boiler Age of boiler Flue gas flow Loading characteristics
Reagent properties	Percent inert Ca, Mg contents Reactivity Size
Size considerations	Land availability Soil permeability Rainfall Climate
Regulations	SO_2 emission/ambient standards Particulate standards Plume visibility standards PSD/nonattainment considerations Water quality standards

Makeup water	Source Chemical composition
Flue gas	Temperature Flow Dew point Particulate loading Particulate alkalinity SO_2 content O_2 level

Table 17.2 Factors influencing scrubber system design inside the battery limits.

Absorber type
Waste slurry disposal scheme
Degree of redundancy
Reheat type and amount
Degree of instrumentation
Mist eliminator
Raw material reactivity
Process layout
Materials of construction
Makeup water distribution
Process chemistry
pH gradient
Sulfite oxidation
Chloride balance
Liquid/gas ratio
Point of fresh water addition
Reagent utilization

Chemically related problems, primarily scaling and corrosion, are due to improper consideration and/or inadequate control in the operational phase of process chemistry. Mechanically related problems refer to the malfunction or failure of mechanical components in the scrubber and auxiliary systems, such as pumps, fans, and agitators. Failure of field instruments, expansion joints, and piping may sometimes be caused by interrelated mechanical and chemical conditions. Indeed, the most persistent and significant problems with FGD systems have both chemically and mechanically related contributing factors. The design practices being used to minimize these problems are discussed below.

17.3.1 Control of Scale Formation

Scale-free operation can be accomplished by operating in either of two basic modes: coprecipitation or control of supersaturation. Coprecipitation involves the removal of calcium sulfate from the system as part of a calcium sulfite/sulfate solid solution. If a system is operated so that the maximum oxidation in the slurry circuit is about 16% (i.e., 16% of the SO_2, on a molar basis, is oxidized completely to the SO_4^{2-}), the scrubbing liquor can remain subsaturated with respect to calcium sulfate (gypsum). Above this level, more calcium sulfate is formed in the slurry circuit than can leave the system in a coprecipitated form. This causes the sulfate concentration to increase and the system to operate supersaturated with respect to gypsum. When the relative saturation levels approach a critical value of 140%, a hard scale can form within the system.

Supersaturation can be controlled by providing nucleation sites (seed crystals) for the precipitation of gypsum and calcium sulfite/sulfate solid solution. This maintains the system below the critical (140%) level of saturation with respect to gypsum.

Alternatively, the slurry in a forced-oxidation system contains predominantly gypsum crystals, and saturation levels are maintained at 100–110% with respect to gypsum. The use of forced-oxidation systems has been especially beneficial for limestone systems treating flue gas with low-sulfur dioxide concentrations. For these applications, the low-sulfur dioxide levels, coupled with the long liquid retention times, result in high "natural" oxidation levels. This makes forced oxidation a particularly attractive method for improving sludge quality, minimizing scaling, increasing sulfur dioxide removal, and improving reagent utilization.

Another method for controlling scaling involves increasing the magnesium-ion concentration. This increases the liquid-phase alkalinity of the scrubbing slurry and the amount of sulfite and sulfate the scrubbing slurry can hold without exceeding solubility limits. The overall effect is a subsaturated operation that produces higher sulfur dioxide removal efficiencies and higher utilization. Organic acids, such as citric, benzoic, and (more recently) adipic, are also being evaluated for use in increasing the total alkalinity of the scrubbing liquor.

17.3.2 Mechanical System Design

17.3.2.1 Balanced Draft or Booster Fans

In addition to placing these fans upstream of the FGD system, another development is the use of variable-pitch, axial-flow fans. The main advantage of this design is its consistently higher efficiency (versus centrifugal fans) over the entire boiler operating range, which produces substantial energy savings. Other advantages are better flow control with variable frequency motor drives, easy access and maintenance, less severe construction requirements, and increased reliability.

17.3.2.2 Dampers

Bypass and isolation dampers are used to regulate the flow of flue gas into and around the FGD system. The primary purpose of isolation is continuance of unit operation while the scrubber modules are under maintenance. Common damper designs include slide-gate (guillotine), single-blade butterfly, and multiblade parallel (louver). Corrosion and erosion of the various types of dampers and damper seals have been common. In some cases, dampers have failed entirely or have been so inefficient that the modules could not be maintained during bypass situations. The current trend is toward two-stage louver dampers having a pressurized seal-air system that maintains a positive pressure between them. Pressurized seal air increases the energy demand of the system, but it contributes significantly to successful damper operation.

17.3.2.3 Scrubbers

The trend in the design of lime/limestone slurry FGD systems is toward the use of simple, relatively open units, such as spray towers, or combinations of venturis and spray towers. Spray towers have few internal components in the gas–liquid contact zone and thereby offer fewer sites for the deposition of solids, collected fly ash and unused reagents, and for the formation of scale. Spray tower operation has been very successful, but SO_2 removal efficiency may be limited when high-sulfur coal is fired. In such circumstances, the use of more reactive absorbents such as lime or magnesium-promoted lime or limestone, or the use of combinations of scrubbers such as a venturi in series with a spray tower, may be required.

17.3.2.4 Reaction Tanks

Coinciding with gas-side staging via a venturi/spray tower or similar combination is liquid-side staging, in which the hold tanks are arranged in series to approach plug flow reactor design [10].

17.3.2.5 Mist Eliminators

Chevron and baffle-type mist eliminators [10] continue to be the predominant designs used in U.S. utility FGD systems. The popularity of these separators is due to design simplicity and flexibility, adequate collection efficiency for medium- to large-size drops, relatively low-pressure drops, open construction, easy access for maintenance, and relatively low cost.

17.3.2.6 Reheaters

A pronounced preference for stack gas reheat is still evident for systems in service and committed for future operation. In several systems, the use of wet stacks has been abandoned in favor of reheat because of problems with stack corrosion, plume dispersion, and localized "rain" caused by plume condensation. Stack corrosion is more pronounced where high-sulfur coal is burned because sulfur dioxide concentrations are higher.

17.3.2.7 Solids Separation (Sludge Dewatering)

The major development in this area is the increased emphasis on dewatering and stabilization of FGD wastes for managed landfill disposal and the corresponding decreased emphasis on ponding. Formerly, a pond was expected to fulfill three functions: clarification, dewatering, and temporary or final sludge storage. The increased emphasis on disposal in landfills and structural fills and on attaining closed water-loop operations further stimulated the use of clarifiers, centrifuges, and vacuum filters. In addition to using these techniques, several installations are using forced-oxidation strategies to enhance solids settling and filtration properties, to improve process chemistry and waste sludge quality, and to decrease land requirements for sludge disposal.

17.3.2.8 Process Control and Instrumentation

Virtually all full-scale operating systems regulate the reagent feed rate by controlling slurry pH. Dip-type pH sensors have been more successful than in-line sensors because they are easier to clean and calibrate; the in-line, flow-through sensors are subject to more wear and abrasion and generally require much more frequent maintenance. Other systems for control of reagent feed have also been evaluated on full-scale systems. One type involves feed-forward reagent control based on the inlet flue gas flow rate and sulfur dioxide concentration, with trim provided by slurry pH. Another type involves control of reagent feed rate by using the outlet sulfur dioxide as the control variable. Limited success has been reported for both systems, primarily because of the difficulty in obtaining accurate and consistent readings from sulfur dioxide gas analyzers; this has been especially true on high-sulfur coal applications. Nonetheless, this type of process control should enable much better control of process chemistry, particularly with limestone systems and variable boiler loading, and should continue to be developed.

Other critical instruments are those that measure and control the density of recycled slurry and the level of reaction tank liquid. In general, the more sophisticated instruments based on nuclear absorption or ultrasonic principles have proved more satisfactory than mechanical bubble-tube devices.

17.3.2.9 Construction Materials

The construction materials generally preferred for use in critical areas of the FGD system are 316 and 316L stainless steels. However, some designers prefer carbon steel with a surface lining or coating that physically shields the bare metal from the corrosive environment. These linings are usually resins applied in liquid or semiliquid form, by spray or trowel, and allowed to cure. The use of lined or coated carbon steel offers the potential benefits of being able to withstand low pH/high-chloride environments much better than the 316 stainless steels and being considerably less expensive. Operating experience, however, has shown that these materials are susceptible to damage by high-temperature excursions. In addition, improper application of linings has been the suspected cause of lining breakdown and subsequent base metal corrosion, especially in reapplications, where proper preparation of the metal surface is more difficult.

The use of natural rubber, neoprene, polyvinyl chloride (PVC), fiber-reinforced plastic (FRP), and flaked-glass polyester generally predominates in the liquid and thickener circuits of the FGD system (tanks, pumps, agitators, piping, and thickeners) where the metal parts are 316 stainless steel or Alloy 20. Neoprene, polypropylene, FRP, and Alloy 20 are generally used in filtration systems.

Many systems are also being constructed of more resistant alloys in trouble spots (e.g., a wet/dry, high-temperature, high-chloride environment such as a prescrubber or presaturator). Alloys such as Hastelloy C-276, Hastelloy G, Alloy 20, Inconel 625, Incoloy 825, 317 low-carbon stainless steel, 904 low-carbon stainless steel, Jessop JS-700, and E-Brite 26-1 are being selected for special problem areas.

The use of stack lining or coating materials has been troublesome, especially where high-sulfur eastern coal is burned. Lining failures have also been reported in systems that included an apparently adequate degree of reheat. Developments in this area indicate that elastomer (Coldbran's CXL-2000) has been applied with apparent success on the flue gases at the Bruce Mansfield station, which burns high-sulfur eastern coal without reheat. The use of acid brick also appears to be successful in a similar application. At installations burning low-sulfur coal, adequate linings have not been as great a problem; many installations can get by with unprotected carbon steel flues.

17.4 Equipment Design Considerations

The most serious design errors result when an item of equipment that should be expected to require eventual replacement or major repair is buried deeply within the assembly. In particular, the rotating machinery, such as pumps, fans, reagent conveyors, and vessel agitators, should be installed with sufficient free space to permit their removal without an extensive disassembly of other parts of the plant. Large dampers also may require eventual replacement. Some compromise of this provision is usually necessary because open space in a process unit greatly increases the original construction cost. Design arrangements can usually be devised, however, in which those items most likely to require major service are located near the outside of the assembly.

Instruments present a special installation problem because the cost of maintaining field instrumentation has been especially high in some FGD systems. Since the instruments are often the most delicate items of equipment in the plant, they will usually require much more frequent maintenance than will the larger pieces of equipment. These instruments can usually be serviced easily and quickly, however, and the time required to service an instrument is often much less than that required to gain access to it and to remove it from the process line or vessel. It is especially important that the installation of such higher-maintenance instruments as SO_2 analyzers and pH electrodes permit simplified access and removal.

Installation procedures that simplify FGD system maintenance can usually also simplify operations. The system design must be reviewed to ensure that the test connections provided for routine operational use are conveniently accessible to the operators.

The continuing maintenance cost of an FGD system can be adversely affected by errors that occur during checkout or start-up of a new system. Failure to flush piping thoroughly before the unit is placed in operation can cause damage to pumps or vessel linings that cannot be field-repaired to its original condition and therefore create continued maintenance problems. Improper start-up can begin the process of corrosion or scale formation, which will show up later as a premature equipment failure. The first few days in the operation of a new system may form the basis for excessive maintenance that may require years to correct.

17.5 Installation Procedures

The installation of an FGD system is a major construction project, and the details that affect operation and maintenance are primarily a part of the engineering design. Unless considerations of operation and maintenance are included in the design, the system will be expensive to operate and maintain.

17.6 Operation

Two areas of FGD system operation are addressed here, in which the problems of operation are quite different. Start-up of an FGD system involves not only a thorough checklist of the unit, but also a rapid approach to the conditions of normal operation. On completion of the start-up phase, the slower-paced period of normal operation begins, in which a fixed routine of activities must be maintained to keep the system performing properly. The following paragraphs first describe the considerations that apply during normal operation and then discuss the activities associated with start-up.

17.6.1 Normal Operation

The routine program conducted during the normal operation of an FGD system has two objectives. First, it is necessary to determine the conditions under which the system performs best and to hold the system to those conditions. Second, an organized procedure must be devised and executed to continuously assess the mechanical condition of the unit and each unit component. Although problems and minor emergencies are always changing during normal operation, the routine of operation is never-ending.

Especially with high-sulfur coal, operation of the reagent feed equipment necessitates close coordination with the operation of the power-generation equipment to permit the FGD system operator to anticipate a change in boiler loading. In such systems, it may be necessary to manually override the automatic instrumentation during excessive load changes to prevent an out-of-limits excursion in outlet SO_2 concentration.

Another parameter that must be optimized is the concentration of solids in the scrubbing slurry. The recycle of previously precipitated calcium sulfite and sulfate is necessary to supply nucleation sites and thereby to minimize accumulation of crystals on scrubber surfaces as scale. A high concentration of solids in scrubbing slurry, however, not only minimizes scale formation, but also increases erosive wear in equipment and increases carryover of solids into the mist eliminators. The fresh or clarified water used to wash solids from the mist eliminators becomes mixed with the scrubbing slurry, and thereby reduces solids concentration. In each FGD system, therefore, an optimum balance must be made between the tendency to create scale and the tendency to plug the mist eliminators and exacerbate equipment erosion. This balance is expressed in a concentration of solids in the slurry and is measured by the density of the slurry. Optimum solids concentration is around 8% solids in lime-based FGD systems and 12–15% in limestone-based systems. Equivalent densities range between 1.06 and 1.12 g/cm^3.

Within limits, some FGD system operators also consider the cost of reagent in defining optimum conditions of operation. Especially in limestone-based systems, lower PH and higher solids content provide a greater percentage of utilization of the reagent added. However, the quality of the reagent in regard to the thoroughness of lime slaking or fineness of limestone grind probably has a great effect on reagent utilization.

The optimum pH and slurry density for a specific FGD system must be established, at least in part, by trial and error. Chemical analyses of a scrubbing slurry can establish such characteristics as scaling potential and degree of reagent utilization to speed up the stepwise procedure of defining optimum parameters. It is not yet possible to obtain a complete prediction of these values from theoretical calculations because data that define the precise effect of combinations of various operational and design options have not been fully developed.

17.6.2 System Health

The other facet of normal operation is the routine program of checks and examinations that are conducted to spot potential problems before they increase in complexity and result in costly correction. Following is a list of diagnostic checks that constitute a typical program of this type.

1. *Scrubber/absorber (slurry recycle).* Check the recirculation slurry header pressure at the branch take-offs to the scrubber. If the header pressure has increased at one or more of the branches, it is indicative of nozzle pluggage. If it has decreased, it indicates nozzle wear. The pressure should be checked and recorded once per shift by using handheld gauges (static pressure sensor taps frequently become plugged).
2. *Scrubber/absorber.* Check the differential pressure across the absorber section. Slow buildup can indicate scale formation, although scale formation can be quite rapid if control of the process chemistry is lost. Suggested frequency is two to four times per shift. If a low-pressure drop is detected, further checks should be made to determine whether there is low scrubbing liquor flow or low flue gas flow due to a partially closed damper, inoperable or damaged variable venturi throat, where applicable, or some similar constriction. A sudden high-pressure drop probably indicates a sudden increase in gas flow rate, possibly due to an improper fan damper setting. If the fan amperage is too high, the gas flow should be restricted immediately to avoid fan-motor overheating and serious damage.
3. *Scrubber/absorber.* Check the SO_2 removal efficiency by using data from the continuous SO_2 monitors. Decreased removal efficiency could indicate a chemical imbalance or poor gas flow distribution. The gas velocity could be exceeding design limits in one or more of the sections.

The SO_2 removal efficiency information coupled with pressure drop and slurry chemical composition data can be used to determine whether poor gas flow distribution may be the problem. Suggested frequency of checking SO_2 removal efficiency is twice per shift.

4. *Scrubber/absorber*. Another indication of poor gas flow distribution is entrained solids or mist carryover. If inspection ports are available, the amount of carryover from the scrubber to the mist eliminator as well as downstream of the mist eliminator can be assessed. In severe cases, entrained solids and mist carryover can be observed. Another indication of carryover would be excessive fan vibration due to slurry deposits on the fan blades. Suggested frequency of inspection and notation in the operating log is once per day.
5. *Mist eliminator*. The differential pressure across the mist-eliminator section should be checked and visually inspected (through portholes) the appearance of blade surfaces. A pressure increase indicates scale buildup or pluggage, which can be verified by visual inspection. Excessive moisture in the scrubber outlet duct is another indication of mist-eliminator malfunction. Suggested frequency is once per shift.
6. *Mist eliminator*. The composition of the spray liquor should be checked, especially for suspended solids, alkalinity, and pH. Often, the spray liquor source is the overflow from the clarifier, although many systems use the mist-eliminator wash as the source of freshwater makeup; thus, checking the composition of the spray liquor often ties in with assessing the performance of the clarifier. High solids content in the spray liquor can lead to erosion; more significantly, however, high pH or dissolved alkalinity can lead to removal of additional SO_2 in the mist eliminators and scale formation. If the potential for scaling exists, more freshwater makeup should be used, consistent with overall water-balance limitations. The composition of the wash water should be checked and noted several times per week.
7. *Mist eliminator*. The check timing sequence and valve operation should be checked. Flow in deluge-type systems can be checked for proper timing by observing valve actuators and by feeling the headers for flow. A significant number of deluge-type systems have been found to have leaks in the automatic shut-off valves which lead to upsets in the overall water balance. Leaks can be detected by observing the pressure in the supply line with flow to all points from that line shut off and comparing that pressure with the pressure in the line with the supply open to the mist-eliminator header and flow from the mist eliminator curtailed only by the automatic shut-off valves. The ability to perform this check and the frequency of such checks depend upon the overall piping configuration and the degree of concern about the water balance.
8. *Slurry recirculation pumps*. A slow loss of discharge pressure indicates probable impeller erosion. High pressure can indicate restricted flow. Intake restrictions to the pumps can be caused by plugging of the intake lines or screens, which can lead to cavitation. Slurry flow can be checked by touching the pipe; a warm pipe (about 120 °F) indicates flow. The discharge pressure should be noted at least once per shift.
9. *Slurry recirculation pumps*. Seal-water consumption should be individually monitored for each pump. Excessive water consumption indicates seal wear, which is significant because it may necessitate restricting the freshwater supply to the mist eliminators in order to maintain the overall water balance. Seal-water consumption should be noted once per shift.
10. *Scrubber slurry*. The density meter controlling the bleed rate to the thickener should be calibrated every shift by using a simple pulp tester or comparable device. This will permit control of the solids concentration in the spray liquor. The scrubber slurry also should be manually checked for pH once per shift to verify the in-place pH instruments. In addition, on limestone systems, the available limestone in the slurry should be checked, because pH is a poor control indicator in such systems.

11. *Thickener underflow*. The solids concentration in the underflow should be checked. The sludge blanket depth should be maintained in the 3- to 7-ft range and blown down only at 4-ft intervals; continuous blowdown is to be avoided. The blanket depth can be checked from the top of the thickener with a dipstick. Too high or low a concentration can be controlled by adjusting blowdown frequency, the solids levels of incoming system bleed streams, the amount of flocculant addition, and rake operation. In severe instances, the sludge blanket will float so that it cannot be pumped out by the underflow pumps. This may be due to too much flocculant addition or too high a rake speed (both compounded by inattentive operation). The incoming bleed stream must be shut off, as well as the rake. Where possible, the rake should be moved up and operated in an on–off fashion to try to promote settling (e.g., 30 seconds on every 3 minutes). If it is possible to introduce the system bleed at another point, this may also help initiate settling. Proper thickener operation is critical for overall system operation and should be carefully monitored by a full-time operator.
12. *Thickener overflow*. The solids concentration in the thickener overflow should be measured periodically to determine system performance. Visual inspection of the overflow clarity is usually sufficient for assessing performance on a routine basis and should be done at least twice per shift.

17.7 Start-up

The initial checkout and operation of a new FGD system is a critical phase in the life of the unit. An FGD system is a complex process unit, and no two are exactly alike. Unexpected problems develop during start-up, and the available personnel are often unable to handle all the problems at once. Incidents of severe damage to FGD systems have occurred during the first days of operation.

Three considerations are of primary importance during the start-up phase:

1. *Full and complete mechanical checkout of every component in the system*. There is usually strong economic pressure to get the unit on line, but extended delays and greatly increased costs can result from attempts to operate the plant before construction and checkout are completed.
2. *Prevention of system overheating*. Loss of recycle slurry flow will cause hot flue gas to pass through the scrubber/absorber, which can severely damage rubber linings, plastic mist eliminators, and metal components not designed for any temperature excursion. Loss of recycle flow can occur through pump or electrical failure, severe plugging from inadequately mixed slurry, pipeline failure, loss of liquid level in the recycle tank, and many other reasons. At all times during operation, but especially during start-up, the flue gas must be immediately prevented from passing through the absorber vessel if the recycle slurry flow stops. Various systems permit this to be accomplished in a number of ways, ranging from opening a bypass damper to shutting down the boiler.
3. *Prevention of scaling*. When the process chemistry is badly out of balance, as it often is during start-up, severe scaling can occur in a matter of a few hours. Speed in approaching normal pH and solids content during start-up is of prime importance.

Actual start-up is not difficult. It consists of first establishing slurry recycle, then starting the flow of flue gas. The slurry should be as close to optimum conditions as possible. If another FGD system is operating at the station, underflow from the second thickener can be used to prepare a slurry for the new unit. If a second unit is unavailable, best practice includes the preparation of a synthetic slurry using purchased gypsum. In either instance, a rapid adjustment of pH is important, and the pH controller should be placed in the automatic mode as soon as the proper pH value can be attained.

Many FGD systems include a start-up line to permit the return of thickener underflow slurry to the recycle tank. The objective is to allow a more rapid approach to operating slurry density. If the necessary recycle line has not been provided, the use of a portable pump and a hose line may be indicated. If a sludge filter is used in the sludge disposal area, manual return of the filter cake to the recycle tank is another option. Still, another technique is the use of reagent slurry that is more concentrated than normal.

Especially for use prior to a planned shutdown in which the recycle tank will be drained, special sumps with the necessary piping are designed into some FGD systems to store slurry and thereby simplify restart. Accumulation of these solids in the thickener is also possible unless the thickener is common to several modules.

17.8 Maintenance

In a large, complex, and continuously operating process unit, programs of maintenance have evolved almost into a science, which cannot be discussed fully in this chapter. Two general areas of maintenance are recognized: emergency and preventive. Emergency maintenance follows an unexpected failure of a plant component, which often disrupts plant operation, sometimes causes a major shutdown, and almost always results in an abnormally high repair expense. Preventive maintenance is scheduled so as not to disrupt operations, and it does not involve overtime use of maintenance personnel and the high costs of emergency procurement of parts or tools. The purpose of preventive maintenance is to minimize emergency maintenance, and a judicious program of preventive maintenance is vital to the continued satisfactory performance of an FGD system.

The program is based on an inventory of every component of the plant, categorized with respect to its potential for failure and its availability for service. Some components, such as rotating machinery, slurry nozzles, and control valves, will wear more rapidly than major process vessels or utility piping. Preventive maintenance programs must assign a frequency of attention to all these items; although the components least likely to fail will be examined less frequently, they must not be ignored entirely.

Availability for service defines when each component will be examined. Components will be classified into four groups. A program of routine renovation may not be applied to a few noncritical or expendable items, which can be quickly repaired or replaced by shift personnel. Electrodes for pH measurement may be in this category. Three schedules of availability will be prepared, however, to cover all other equipment items.

A program of regular renovation is practiced while the system continues to operate on those equipment items where redundant equipment was included as part of the original design. In some FGD systems, this may be limited to pumps and fans; in other systems, it may include entire scrubbing modules. A second program of renovation covers items that, through minor modifications to operation, can be removed temporarily from service for cleaning, replacement of parts, or total replacement with a spare unit. This program includes such items as ball mills, sludge filters, and most instruments. Finally, a third program of renovation covers those equipment items that cannot be removed from service without shutting down the plant. This may be a major part of some systems, whereas in others it may only involve low-maintenance devices such as clarifier vessels and major sections of ductwork and piping. An effort must be made to refurbish this class of equipment during scheduled shutdowns, which also usually involve the boiler and the power-generation equipment.

A well-conceived maintenance program replete with appropriate spare equipment and trained personnel, coupled with a systematic approach to monitoring and control of process operations, will pay significant dividends.

17.9 Improving Operation and Performance

Optimum conditions are expressed primarily in the chemical and physical characteristics of the scrubbing slurry. Of all the measurable characteristics, slurry pH is the most important. If all other variables are constant, a consistent pH ensures that the amount of reagent fed to the scrubber/absorber is in the proper ratio to the amount of SO_2 being absorbed from the flue gas. If the pH varies, the variation will be reflected in other parameters throughout the system. If pH is too low, less SO_2 will be absorbed and more SO_2 will pass through into the stack. If pH is too high, an excess of reagent will be present in the slurry and formation of scale may result.

In general, most lime-based FGD systems operate best at a pH between 8.0 and 8.5, and limestone-based systems operate best at a pH between 5.5 and 6.0. The optimum pH will be different for different systems, however, depending on the location within the system where pH is measured, the reactivity of the reagent, and details of the unit design. The degree to which pH can be permitted to vary from optimum conditions also differs for different systems. Units designed for low- and medium sulfur coal can usually tolerate a rather wide variation in the percentage of total SO_2 that is absorbed from the flue gas and still remain within emission limits. Since high-sulfur coal produces a flue gas richer in SO_2, an equivalent percentage variation cannot usually be tolerated, and closer control of pH is necessary.

17.10 Conclusion

FGD continues to be a very important technology for the control of SO_2 emissions from combustion sources. The selection of an FGD process for a specific plant generally includes preliminary screening of applicable processes and final evaluation of the selected processes. The preliminary screening would consider performance requirements, site-specific constraints, raw material requirements, and desired end products. The final evaluation should consider operational experience with similar systems, control efficiency, energy needs, and costs.

As noted earlier, the most advanced FGD processes from a commercialization standpoint are wet lime and limestone scrubbing. Proper system design is crucial to the successful operation of any FGD system. Most of the operational problems of FGD systems are related to two primary design considerations: mechanical and chemical. The mechanical design considerations consist of materials of construction, process control, scrubber and recycle tank sizing and configuration, component redundancy, gas and liquid flow distribution, and accessibility for operation and maintenance. The chemical design considerations include recycle slurry pH, liquid/gas ratio, reagent stoichiometry, degree of oxidation, and homogeneous precipitation of solids in the recycle tank.

Inadequate mechanical design will lead to plugging and erosion of mechanical components in the scrubber and auxiliary equipment. Inadequate chemical design, or loss of chemical control, will lead to scaling (deposition of calcium sulfite/sulfate/carbonate solids) and corrosion in the FGD system. Many of the problems are interrelated, and it is essential to recognize the key variables in the design phase.

Control of calcium sulfite scale formation is achieved by controlling the slurry pH and reagent stoichiometry below a certain level and operating at a high liquid/gas ratio and recycle tank residence time. Control of calcium sulfate formation is accomplished either by coprecipitation, when the oxidation level is below 16% and the scrubbing liquor is subsaturated with gypsum, or by circulating gypsum seed crystals for supersaturation, when the oxidation level is above 16%. Supersaturation can also be controlled by operating in a forced-oxidized mode. Magnesium additives (MgO or $MgSO_4$) or buffer additives (adipic or benzoic acid) have also been used to control scaling and improve SO_2 removal efficiency and reagent utilization.

Specification of proper materials of construction is another important consideration. The materials generally preferred for the scrubber vessel are 316 and 316L stainless steel, although rubber or resin lining is also widely used. The capital-cost advantage of lining should be weighed against the potential failure due to improper application or contact with hot flue gases during upset conditions. Higher-grade alloys are used for special areas in the scrubber. The use of natural rubber, PVC, and FRP predominates in the liquid circuits of the FGD system. The materials for stack lining when high-sulfur, high-chloride coal is burned require careful consideration, and no single material has appeared as the predominantly successful choice.

The process control systems using integration of automatic and manual control techniques have proven to be much more successful than fully automatic systems. The primary variables that require process control include gas flow, recycle slurry pH, reagent feed rate, recycle slurry density, and overall water balance.

An FGD system is essentially a chemical processing plant; thus, the installation philosophy used in the chemical industry should be followed. The rotating machinery, such as fans, pumps, and agitators, should be installed with sufficient free space for maintenance. Easy accessibility is also essential for high-maintenance instruments such as SO_2 analyzers and pH electrodes. The initial checkout and start-up is a critical phase for an FGD system. The most important considerations in system start-up are complete mechanical checkout of every component, prevention of hot gas flow through the scrubber, and rapid attainment of stable system chemistry.

During initial normal operation, the optimum values of process parameters should be identified and organized operating and maintenance procedures established to hold the system close to the optimum conditions. As noted, the optimum pH at the scrubber inlet varies between 8.0 and 8.5 for lime, and 5.5 and 6.0 for limestone systems. The optimum recycle slurry solids content is 8–10% and 12–15% for the lime and limestone systems, respectively. Typical manual diagnostic checks include pressure in the recycle header, pressure drop across the absorber and the mist eliminator, SO_2 removal efficiency, pH and composition of recycle slurry and mist-eliminator wash water, and slurry solids content for the reagent feed, recycle, and thickener underflow streams.

It is essential that personnel be assigned exclusively to the FGD operation and maintenance team. These operators should be well trained in emergency and preventive maintenance programs. It is only through proper training of operator and maintenance personnel that careful attention paid to design, construction, and initial operation phases will yield the long-term dividend.

References

1 T. Devitt et al., *Flue Gas Desulfurization System Capabilities for Coal-Fired Steam Generators*, 2 vols. EPA-600/7-78-032, RTP, NC, March 1978.
2 Electric Power Research Institute, *Lime FGD Systems Data Book*. Final report, FP-1030, Research Project 982-1. Palo Alto, California, May 1979.
3 Electric Power Research Institute, *Lime/Limestone Scrubber Operation and Control Study*. Final report, RP-630-2. Palo Alto, California, April 1978.
4 H. Head and S. Wang, *EPA Alkali Scrubbing Test Facility: Advanced Program – Fourth Progress Report*, 2 vols. EPA-600/7-79-244, RTP, NC, November 1979.
5 B. Laseke and T. Devitt, Status of flue gas desulfurization, *Chem. Eng. Prog.*, 75(2), 37–50, New York City, NY, 1979.
6 Proceedings of the Symposium on Flue Gas Desulfurization, Las Vegas, Nevada, March 1979, 2 vols. EPA-600/7-79-167, RTP, NC, July 1979.
7 A. Slack and G. Holliden, *Sulfur Dioxide Removal from Waste Gases*. Noyes Data Corp., Park Ridge, NJ, 1975.
8 J. Smith et al., *EPA Utility FGD Survey: December 1978 - January 1979*. Updated through July—September 1979. EPA-600/7-79-022, RTP, NC, May–October 1979.
9 L. Theodore, *Chemical Reactor Analysis and Applications for the Practicing Engineer*, John Wiley & Sons, Hoboken, NJ, 2012.
10 L. Theodore, *Air Pollution Control Equipment Calculations*, John Wiley & Sons, Hoboken, NJ, 2008.

18

Controlling the Oxides of Nitrogen

Note: Parts of this chapter on the oxides of nitrogen are an edited and updated version of an earlier chapter in the J. Richardson and L. Theodore publication titled, *Optimizing Air Pollution Control Equipment Performance: Operation and Maintenance.*

18.1 Introduction

The oxides of nitrogen consist of the elements nitrogen and oxygen, both of which are briefly described below.

1. *Oxygen.* Oxygen is a gaseous element with the symbol O and located in group VI of the periodic table, with atomic number 8, and atomic weight of 16. Oxygen is the most abundant of all of the elements on planet Earth. It represents 21% by volume (23% by mass) of the atmosphere; 85.8% of the oceans; and as a constituent of most rocks and minerals, 46.7% of the solid crust of the Earth. Oxygen comprises 60% of the human body. It is a constituent of all living tissues; almost all plants and animals, including all humans, require oxygen in the free or combined state to maintain life.

   Oxygen is a colorless, odorless, and tasteless nontoxic gas. It can be transformed into a pale blue liquid by compression. Three allotropic forms of oxygen are known to exist: ordinary oxygen containing two atoms per molecule (O_2), ozone containing three atoms per molecule (O_3), and a pale blue nonmagnetic form (O_4) containing four atoms per molecule which readily breaks down again.
2. *Nitrogen.* Nitrogen is a gaseous element with the symbol N and located in group V of the periodic table, with atomizing number 7, and atomic weight of 14. The free gas represents about 78% by volume of the atmosphere. The element occurs in the combined state in several minerals.

   Nitrogen is a colorless, odorless, tasteless, nontoxic gas. It can be transformed into a colorless liquid, which can be compressed into a colorless, crystalline solid. Nitrogen exists in two natural isotopic forms and four radioactive isotopes.

18.2 The Oxides of Nitrogen

Molecular nitrogen (N_2) is a relatively inert gas. However, the element nitrogen can be reactive as a single atom. Thus, nitrogen can form several different oxides. The stable gaseous oxides of nitrogen include N_2O (nitrous oxide), NO (nitric oxide), N_2O_2 (dinitrogen dioxide), N_2O_3 (nitrogen trioxide), NO_2 (nitrogen dioxide), and N_2O_5 (nitrogen pentoxide). NO_3 also exists but is unstable. Of these, the only ones present in the atmosphere in any significant amount are N_2O, NO, and NO_2. Thus, these are potential contributors to air pollution and are referred to as NO_x.

1. Nitrous oxide (N_2O) is an inert gas with anesthetic characteristics. Its ambient concentration is approximately 0.5 ppm, which is considerably below the threshold for biological effects. In addition, it has a balanced environmental cycle which is independent of the other oxides of nitrogen.
2. Nitric oxide (NO) is a colorless gas, and its ambient concentration is usually significantly less than 0.5 ppm. At these concentrations, its biological toxicity in terms of human health is insignificant. However, nitric oxide readily oxidizes in the atmosphere to the form of nitrogen dioxide (NO_2) and is an active participant in photochemical smog formation as well. It initiates reactions that produce air pollutants.
3. Nitrogen dioxide (NO_2) is a reddish-brown gas and is quite visible in sufficient amounts. A concentration of 1 ppm of NO_2 is detectable to the naked eye. The NO_2 ambient air quality standard in California was initially set at 0.25 ppm (on an hourly average), primarily on the basis of visibility concerns, but toxicology and epidemiological effects of NO_2 on human beings are not completely known.

Over 90% of all man-made NO_x that enter the atmosphere are produced by the combustion of various fossil fuels. In the United States, approximately one-half of the NO_x is from stationary sources, while the remainder is from mobile sources. Over 95% of NO_x formed from industrial and utility combustion is in the form of NO, with the remaining 5% as NO_2 and a trace amount of N_2O. Environmental regulations typically consider all NO_x as NO_2 because all the NO in the atmosphere is eventually converted to NO_2. N_2O is considered a Greenhouse Gas (GHG) as it can remain in the atmosphere for hundreds of years. By comparison, NO_x only lasts a matter of days. Because of its long lifespan, N_2O will eventually reach the upper atmosphere where it decomposes under UV light and the products of this reaction remove ozone from the upper atmosphere. The man-made emissions rate of NO_x is of increasing concern as NO_x reacts with moisture and other compounds to form nitric acid and create acid rain. To form smog, NO_x undergoes a photochemical reaction with volatile organic compounds (VOCs) and sunlight that result in airborne particles and ground level ozone.

NO_x formation from a specific stationary combustion source (boiler, turbine, incinerator, etc.) is determined by the interaction of chemical and physical processes occurring within the boiler furnace. There are three principal chemical processes for NO_x formation:

A. Thermal NO_x, which is the oxidation of atmospheric nitrogen from the combustion air supplied to the burners and furnace.
B. Prompt NO_x, which is formed by chemical reactions between hydrocarbon fragments and atmospheric nitrogen from the combustion air. Prompt NO_x tends to form quickly, often at the lower temperature stages of combustion.
C. Fuel NO_x is formed from chemical reactions involving nitrogen atoms chemically bound within the fuel. Fuel NO_x is generally a more active pathway with solid fuels (such as coal and heavy oil) due to the Nitrogen present in the material.

18.2.1 Thermal Oxide of Nitrogen Formation

Thermal NO_x results from the oxidation of atmospheric nitrogen from the combustion air in the high-temperature postflame region of the boiler. During combustion, oxygen radicals are formed and attack atmospheric nitrogen molecules to start the reactions that comprise the thermal NO_x Zeldovich formation mechanism:

$$N_2 + O \leftrightarrow NO + N \tag{18.1}$$

$$N + O_2 \leftrightarrow NO + O \tag{18.2}$$

$$N + OH \leftrightarrow NO + H \tag{18.3}$$

The first reaction is generally assumed to determine the rate of thermal NO_x formation because of its high activation energy of 76.5 kcal/mol. Because of this reaction's high activation energy, NO_x formation is slower than other combustion reactions causing large amounts of NO to form after combustion is "complete" [1–3].

The major parameters that influence thermal NO_x formation are: (a) temperature; (b) oxygen concentrations; (c) nitrogen concentrations; and (d) residence time. If temperature, oxygen concentrations, or nitrogen concentrations can be reduced quickly after combustion, thermal NO_x formation is suppressed. Of these four factors, temperature is the most important, and as such, suppression of NO_x formation is generally referred to as "quenching."

Thermal NO_x formation is an exponential function of temperature and related to flame temperature by the Arrhenius rate relationship [1–3]. Generation begins at 2,200 °F and increases exponentially with an estimated doubling of NO_x for every 100 °F of flame temperature increase (see Figure 18.1).

One of the fundamental parameters affecting temperature is the local equivalence ratio, ϕ, within the combustion zone. Flame temperature peaks at equivalence ratios near 1.0 as shown in Figure 18.2. If localized areas within the boiler furnace are considered fuel-rich, then there is insufficient oxygen to burn all the fuel, and peak temperatures decreases. If localized areas within the combustion zone are considered fuel-lean, this will result in decreasing peak temperatures. A nonpremixed flame, or a flame that exists where the reactants must diffuse into each other during chemical reaction, will generally react near an equivalence ratio of one, causing high peak temperatures. Thus, reducing peak flame temperature through combustion modifications can significantly reduce thermal NO_x formation.

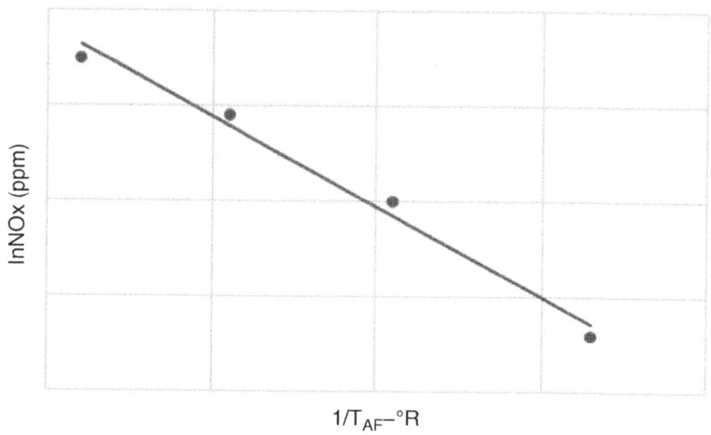

Figure 18.1 Relationship of adiabatic flame temperature (AFT) to thermal oxides of nitrogen production.

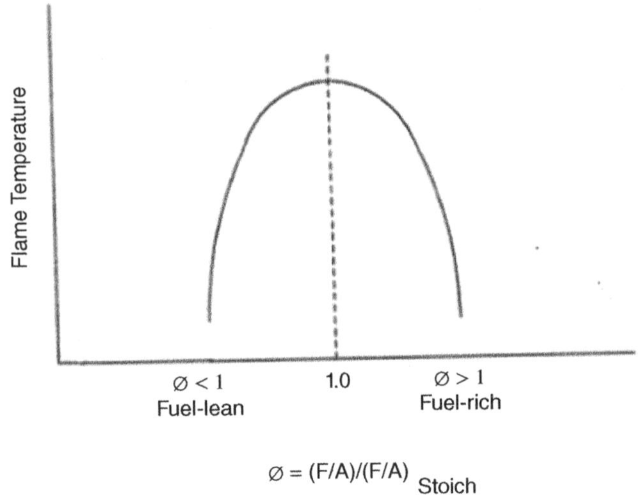

Figure 18.2 Variation of flame temperature with equivalence ratio.

18.2.2 Prompt Oxide of Nitrogen Formation

Prompt NO_x formation occurs in the boiler furnace through the reactions of hydrocarbon fragments and atmospheric nitrogen from the combustion air. As opposed to slower thermal NO_x formation, prompt NO_x formation is rapid and on a time scale comparable to the energy release reactions (i.e., within the flame). It's not possible to "quench" prompt NO_x formation by decreasing the temperature as is the case with thermal NO_x formation. However, the contribution of prompt NO_x to the total NO_x emissions of a system is rarely significant and accounts for a relatively small percentage of the total NO_x production (typically <10%). Although there is some uncertainty in the detailed mechanisms for prompt NO_x formation, it is generally believed that the principal product of the initial reactions is hydrogen cyanide (HCN) or CN radicals, and that the presence of hydrocarbon species is essential for the reactions to take place.

The HCN radical is further reacted to form NO_x. Measured levels of prompt NO_x for a number of hydrocarbon compounds in a premixed flame show that the maximum prompt NO_x is reached on the fuel-rich side of stoichiometry. On the fuel-lean side of stoichiometry, few hydrocarbon fragments are free to react with atmospheric nitrogen from the combustion air to form HCN, the precursor to prompt NO_x. With increasingly fuel-rich conditions, an increasing amount of HCN is formed, creating more NO_x. However, above an equivalence ratio of approximately 1.4, there are not enough O_2 radicals present to react with HCN and form NO, so NO levels decrease, and as such is the preferred methodology employed by many low NO_x burners or combustion staging technologies to reduce prompt NO_x formation.

18.2.3 Fuel-bound Oxide of Nitrogen Formation

The oxidation of fuel-bound nitrogen (FBN) is the principal source of NO_x emissions in combustion of coal and some oils. All indications are that the oxidation of FBN compounds to NO is rapid and occurs on a time scale comparable to the energy release reactions during combustion. Thus, as with prompt NO_x, the reaction system cannot be quenched as it can be for thermal NO_x.

Although some details of the kinetic mechanism for conversion of fuel nitrogen to NO_x are unresolved at the present time, the sequence of kinetic processes is believed to be a rapid thermal decomposition of the parent fuel-nitrogen species to low molecular weight compounds such as HCN, and subsequent decay of these intermediates to NO or nitrogen (N_2).

In stoichiometric or fuel-lean situations, the intermediates will generally react to form NO over N_2, whereas in fuel-rich systems, there is evidence that the formation of N_2 is competitive with the formation of NO. This may, in part, be the cause of high NO_x emissions in fuel-lean and stoichiometric mixtures and lower NO_x emissions in fuel-rich systems.

For high nitrogen fuels such as No. 6 oil, fuel NO_x may be as high as 70% of the total NO_x emissions. For low nitrogen fuels such as natural gas, fuel NO_x is negligible.

18.3 NO_x Control Methods

Factors that affect the formation of NO_x in combustion processes are described in this and the next two sections. A brief discussion of each factor is also included. Much of this material is driven from the principles associated with the 3 Ts of combustion: time, temperature, and turbulence [1–3].

There are essentially six principles of NO_x control:

1. Reducing peak flame temperature
   a. Flue gas recirculation (FGR)
   b. Fuel reburning
   c. Low or ultra-low NO_x burners (LNB/ULNB)
   d. Combustion optimization
   e. Burners out of service (BOOS)
   f. Reducing excess air
   g. Water/Steam injection
   h. Over-fired air
2. Reducing residence time at peak temperature
   a. Air injection
   b. Fuel injection
   c. Steam injection
3. Chemical reduction of NO_x
   a. Selective catalytic reduction (SCR)
   b. Selective noncatalytic reduction (SNCR)
4. Removal of nitrogen
   a. Using oxygen instead of air
   b. Low nitrogen fuels
5. Use of a sorbent
6. Combination of multiple methods

Details on each method follow:

1. Reducing peak flame temperatures often involved avoiding the stoichiometric ratio (the exact ratio of chemicals that enter into reaction). Essentially, this technique dilutes the reaction with excess air, flue gas, or steam. Combustion controls use different forms of this technique and are different for fuels with high and low nitrogen content. The basic technique is to reduce the temperature of the combustion products. This method keeps the vast majority of nitrogen from reacting to form NO and is limited by the kinetics at the temperature in question.

2. Reducing residence time at high combustion temperature can be accomplished by ignition or injection timing with internal combustion engines. It can also be in boilers by restricting the flame to a short region in which the combustion air becomes flue gas. This is immediately followed by injection of fuel, steam, air, or recirculating flue gas. This residence time at peak temperatures keeps the vast majority of nitrogen from reacting and NO formation is limited by the kinetics at the operating temperature.
3. Chemical reduction of NO_x provides a chemically reducing (for example: reversal of oxidization) substance to remove oxygen from nitrogen oxides. Examples include selective catalytic reduction (SCR) which uses ammonia, selective noncatalytic reduction (SNCR) which uses ammonia or urea, fuel reburning (FR), nonthermal plasma, and other emerging technology.
4. Removal of nitrogen from combustion is accomplished by removing nitrogen as a reactant either by using oxygen instead of air in the combustion process or using low-nitrogen content fuel to form less fuel NO_x. Eliminating nitrogen by using oxygen tends to produce a rather intense flame that requires special refractory chambers and must be subsequently and suitably diluted.
5. Treatment of flue gas by injection of sorbents (such as ammonia, powdered limestone, aluminum oxide, or carbon) can remove NO_x and other pollutants (principally sulfur). There have been successful efforts to make sorption products a marketable commodity. This kind of treatment has been applied in the combustion chamber, flue, and particulate control devices. This method uses either absorption or adsorption followed by filtration and/or electrostatic precipitation to remove the resultant sorbent.
6. Many of these methods can be combined to achieve a lower NO_x concentration than can be achieved alone by any one method.

18.4 Reducing NO_x Generation via Pollution Prevention

Some information on NO_x pollution prevention methods are described below.

1. Reducing Excess Air
   Excess air flow for combustion has been correlated to the amount of NO_x generated. Limiting the net excess air flow to under 2% can significantly limit the NO_x content of the flue gas. Although there are fuel-rich and fuel-lean zones in the combustion region, the overall net excess air is limited when using this approach.
2. Burners Out of Service (BOOS)
   Multiple-burner equipment can have part of an array of burners with some "burners out of service" (not feeding fuel but supplying air or flue gas). This allows the burners around them to supply fuel and air to the air or flue gas flowing from the BOOS. The result is combustion in stages with the temperature always lower than when all burners are in service [4].
3. Over-fired Air (OFA)
   The over-fired air approach removes a portion of the air from the burner and reinjects it downstream. Similar to fuel staging, OFA allows combustion to operate outside of the peak flame temperature zone and operate rich at the burner front and lean when OFA is introduced downstream. The amount and location of OFA is designed to ensure complete burnout of the fuel before exiting the boiler.

   The supply for the over-fired air is typically taken from the burner wind box or supply ducting where sufficient pressure is available (see Figure 18.3). In general, the higher the

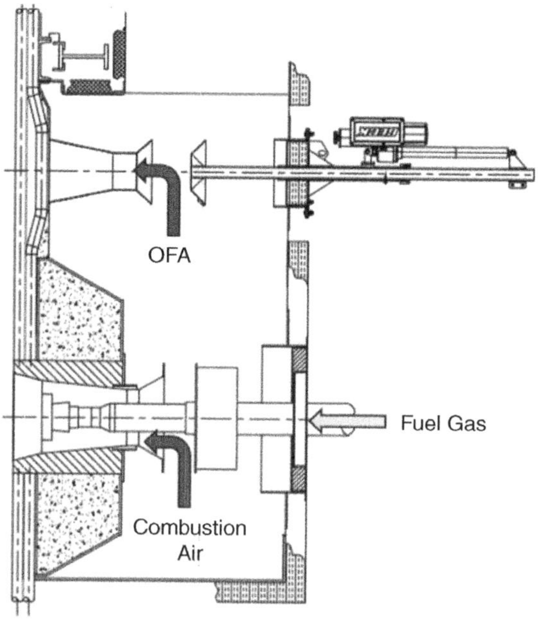

Figure 18.3 OFA and combustion air diagram.

available pressure, the better the penetration and mixing, and the better the OFA performance. In cases where a combustion air source or pressure is limited, boosted over-fired air systems can be used to enhance NO_x performance, but the cost benefit analysis should be considered for these applications.

Proper modeling of the furnace provides calculation and visualization of OFA mixing with the furnace gases, CO produced by incomplete combustion, and NO_x reduction as a result of port location, flow, and sizing.

OFA ports can include design features that includes flow straightening and a split port design option to provide better mixing and performance throughout the load range.

4. Low NO_x Burners (LNB)

Advancements in burner design have led to the advancement of burner nozzles and hardware to reduce NO_x emissions through fuel/air staging, low excess air operation, and aerodynamic analysis. An example of a typical Low-NO_x burner is shown in Figure 18.4. Fuel/Air staging is a proven technique used to reduce thermal NO_x emissions. Gas nozzles can be designed to establish fuel "rich" and "lean" zones at the burner outlet (staged combustion). During combustion, the fuel burns in the "rich" reducing environment and secondary air mixes further downstream to complete burnout. Operating the burner in fuel rich/lean zones avoids peak flame temperatures and thereby reduces thermal NO_x emissions.

Low NO_x burners are able to operate in the 2–3% excess O_2 range. Low excess air operation not only reduces peak flame temperatures but has been shown to reduce the fuel-bound nitrogen conversion rate as well as prompt NO_x generation.

The ultra-low NO_x burner expands upon fuel staging by injecting the staged gas outside of the combustion air stream. This allows the staged fuel to mix with inert flue gases before combustion, mimicking the effect of external FGR. This mixing reduces the peak flame temperature and further reduces thermal NO_x emissions.

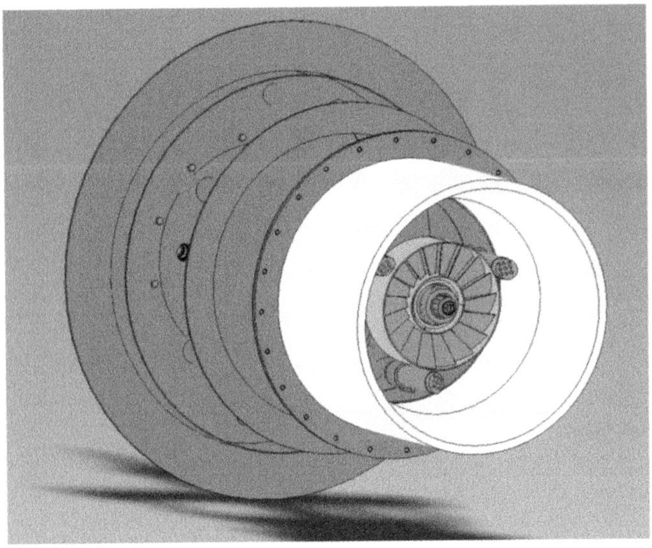

Figure 18.4 Falcon combustion axial flow, staged-fuel, low NO_x burner.

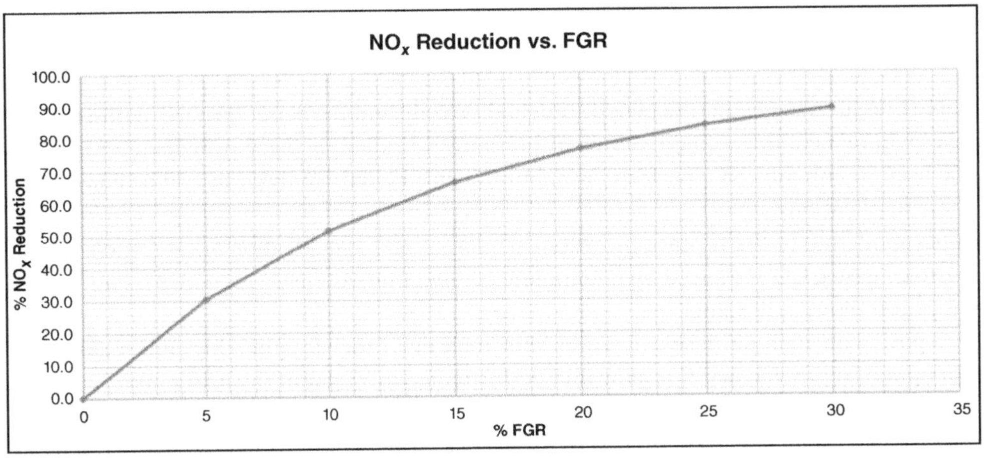

Figure 18.5 Percent FGR vs. NO_x reduction for gas fired applications.

5. Flue Gas Recirculation (FGR)
   Flue gas recirculation is a very effective method of reducing thermal NO_x emissions on industrial gas firing applications. FGR works by circulating relatively cool, low excess O_2 exhaust gas back into the combustion air. This dilutes the burner excess O_2 from the ambient concentration of 20.9% to 16–19% depending on the application. This reduces available O_2 for fuel or prompt NO_x formation in the primary combustion zone while also lengthening and cooling the flame for significant reductions in thermal NO_x (see Figure 18.5).

   There are two basic methods of implementing FGR: forced; and, induced. Forced FGR (see Figure 18.6) will utilize an additional fan (booster fan) to recirculate the flue gases back into the combustion air downstream of the forced draft fans. A damper controls the recirculation rate, based on the firing rate of the burners. Induced FGR (see Figure 18.7) utilizes the existing combustion air fan to recirculate the flue gases back into the combustion air stream. A portion

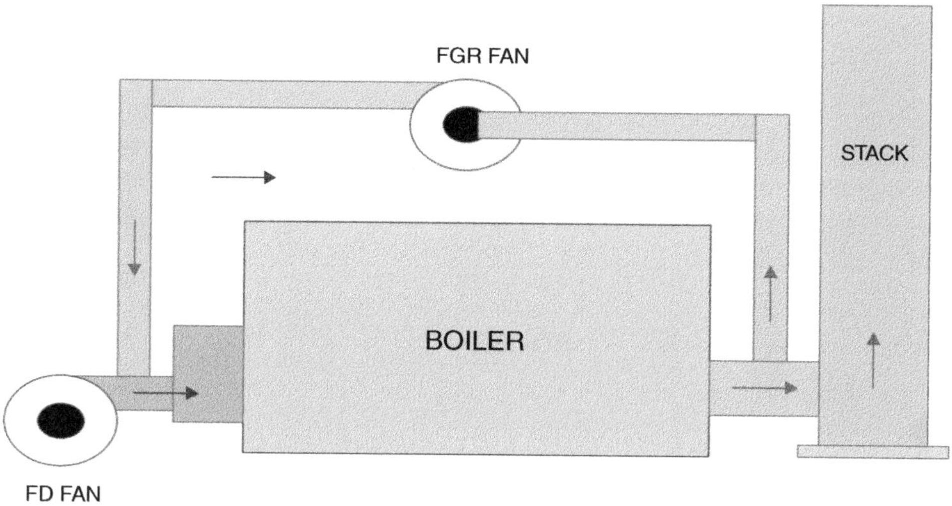

Figure 18.6 Diagram of forced FGR system.

of the flue gases are routed by duct work to the inlet of the existing FD fans, where they are premixed with the combustion air and introduced into the flame through the burner. The FD fans will need to be assessed to determine whether forced FGR is required, or the existing fans can accommodate the additional FGR flow. The assessment of the existing FD fans is completed to determine if the boiler nameplate rated capacity can be derated and still meet the normal operating performance of the boiler.

6. Water/Steam Injection
   Injection of water/steam dilutes the mixture and cools combustion. If flame temperature is sufficiently reduced, thermal NO_x will not be formed in as great a concentration. If fuel oil is used as a backup fuel in gas fired application, steam injection is readily available through the oil burner.
7. Reduced Air Preheat
   Some boilers implement an efficiency device called an air preheater. The air preheater is a gas-gas heat exchanger that transfers heat from the exhaust flue gas into the combustion air. Primarily an efficiency device, the air preheater is designed to retain heat destined to leave the stack unutilized, and recirculates it back into the unit. These devices were primarily utilized before NO_x emissions were of significant importance. It has since been discovered that the elevated combustion air temperature significantly increases NO_x emissions. Combustion air temperature effects NO_x emissions on different types of boilers differently. For tangentially fired boilers, the relationship is roughly described in Figure 18.8.

   By removing the air preheater, it will reduce NO_x emissions, but also reduce boiler efficiency. Therefore, removing the air preheater is typically accompanied by adding an economizer. An economizer performs a similar function to the air preheater relative to boiler efficiency. The economizer is a gas-liquid heater exchanger that transfers heat from the flue gas to the boiler feed water. The economizer can then be placed in the same location as the removed air heater so there are minimal duct work modifications. The economizer is more efficient than the air heater, so overall boiler efficiency will be improved.

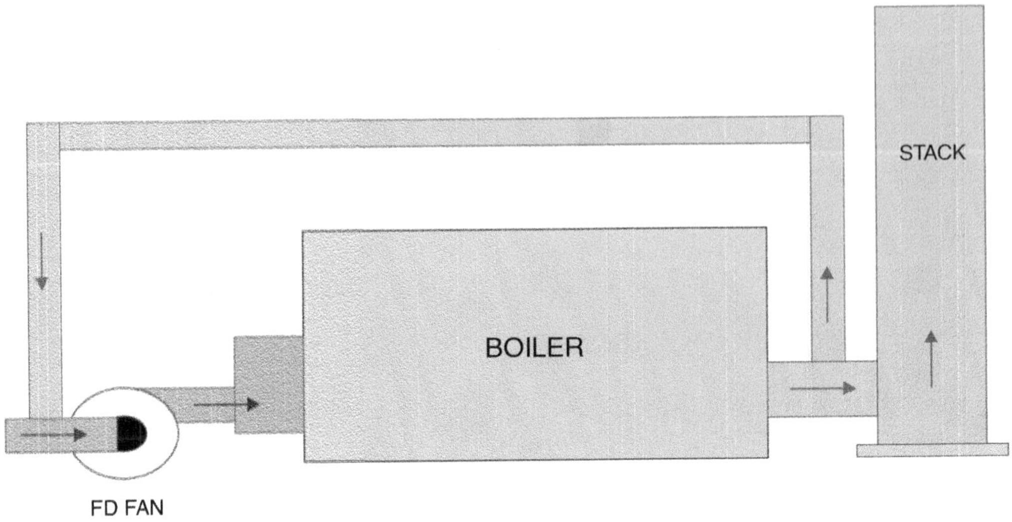

Figure 18.7 Diagram of induced FGR system.

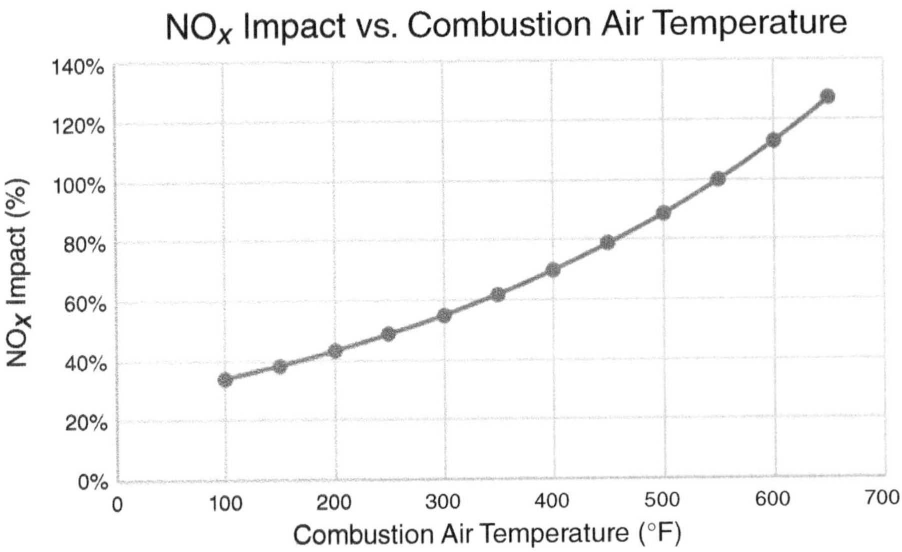

Figure 18.8 Combustion air temperature on thermax NO_x.

8. Fuel Reburning
   Recirculation of cooled flue gas with added fuel (this can be natural gas, pulverized coal, or even oil spray) causes a reduction in energy content similar to FGR, and primary combustion temperature can be lowered.
9. Combustion Optimization
   "You don't get NO_x out of the box." The first step in any NO_x reduction process is to take a complete set of baseline readings and perform combustion optimization tuning. As mentioned previously, thermal NO_x emissions account for essentially all of the NO_x emissions,

and thermal NO_x is primarily generated by elevated flame temperatures. The systems used to control industrial and utility combustion equipment are complex and sophisticated. They require in-depth tuning by qualified professionals to optimize performance and minimize emissions.

Gaseous fuel balancing can be achieved using burner mounted pressure gauges and manual balancing valves or orifices. If the gas spuds are in good condition, they can be effectively used as metering orifices and balancing can be achieved through providing uniform supply pressures at the burner fronts.

Airflow balancing is paramount to combustion optimization. Roughly 95% of the mass flow though the burner is combustion air. Poor airflow can decrease efficiency, increase emissions, and result in poor stability. Airflow balance testing can be conducted online with installed burner airflow measurement devices, but it is also recommended to perform periodic offline airflow balancing. The offline balancing is conducted while running combustion air fans only and can be done while the unit is hot. Balancing is typically completed in a single shift using an articulating probe. The airflow measurement probe is inserted into the burner and velocity measurements are taken at multiple circumferential points. Based on the calculated flows, balancing is performed using the existing air dampers.

10. Oxygen Instead of Air for Combustion
As mentioned previously, NO_x is formed by the breakdown of diatomic nitrogen in the air. If pure oxygen is used as an oxidant for the combustion process, all nitrogen (and therefore NO_x) is eliminated from the reaction. This is an extremely expensive approach compared to other methods and has numerous knock-on challenges.

11. Catalytic Combustion
Use of a catalyst or a ceramic mesh to stabilize combustion can occur below NO formation temperatures and can provide a suitable means of reducing combustion temperature. This technique is not used often because it is very sensitive. However, where it is used, catalytic combustion can achieve less than a 5 ppm concentration of NO_x in the flue gas.

18.5 Control of Flue Gas NO_x

In addition to the precombustion or in-combustion techniques mentioned above, NO_x can also be reduced in post combustion processes. This allows burners to operate in a more efficient, high intensity manner while still maintaining low NO_x emissions.

18.6 Selective Catalytic Reduction (SCR) Overview

Nitrogen oxides (NO_x) in combustion flues gases can be reduced though the process of selective catalytic reduction (SCR). In an SCR, NO_x reacts with a reducing agent, typically ammonia (NH_3), in the presence of a catalyst to form molecular nitrogen (N_2) and water (H_2O). The NH_3 is injected into the flue gas prior to the catalyst in either aqueous ammonia, anhydrous ammonia, or urea (which thermally decomposes into ammonia). Proper dosing and mixing is critical to SCR performance. Temperature and velocity must also be considered for proper NO_x reduction. The reaction occurs most effectively at temperatures between 500 °F (260 °C) and 1,050 °F (566 °C), with optimal performance based on catalyst design/makeup [5, 6].

The reaction occurs as NH_3 diffuses into the catalyst surface pore structure and is adsorbed. This active catalyst site then reacts with NO_x in the flue gas. The rate of and effectiveness of the reaction depends on the number of available active sites, flue gas temperature, and pollutant concentration. A well-balanced process will maintain the desired NO_x emissions with minimal NH_3 slip.

When designed and operated properly, the SCR system should have limited impact on upstream processes. The system will perform reliably and achieve high NO_x reduction under varying loads. Furthermore, a well operating SCR system has limited impact on operation of downstream equipment.

18.6.1 SCR Chemistry

The SCR reactions that occur to reduce NO and NO_2 with NH_3 are simplified as follows:

$$4NO + O_2 + 4NH_3 \rightarrow 4N_2 + 6H_2O \tag{18.4}$$

$$NO + NO_2 + 2NH_3 \rightarrow 2N_2 + 3H_2O \tag{18.5}$$

$$6NO_2 + 8NH_3 \rightarrow 7N_2 + 12H_2O \tag{18.6}$$

The first reaction is the primary reaction in this application. This reaction shows that, theoretically, only one mole of ammonia is required to remove each mole of NO from the flue gas. However, in practical applications, the process requires more ammonia to be injected than necessary for the desired NO removal. With variations in flue gas and imperfect mixing, more NH_3 must be available to react. Also, due to improper mixing, pollutant or reagent distributions, as well as temperature variations (to name a few), can lead to ammonia passing through the catalyst bed unreacted. This is referred to as ammonia slip.

Secondary reactions may also occur depending upon operating conditions and changes in flue gas composition. Examples of possible secondary reactions are as follows:

$$4NH_3 + 5O_2 \rightarrow 4NO + 6H_2O \tag{18.7}$$

$$SO_2 + O_2 \rightarrow 2SO_3 \tag{18.8}$$

$$2NH_3 + SO_3 + H_2O \rightarrow (NH_4)_2SO_4 \tag{18.9}$$

$$NH_3 + SO_3 + H_2O \rightarrow (NH_4)HSO_4 \tag{18.10}$$

Reaction 18.7 illustrates that ammonia oxidation can result when normal operating temperatures exceed 800 °F. As the flue gas temperature increases, the NH_3 oxidizes more readily, increasing the concentration of NO and decreasing the available NH_3 for the SCR process.

Reaction 18.8 is prevalent with higher sulfur fuels and not with low sulfur fuels or natural gas fuels. CO oxidation catalysts may also result in increased SO_2 to SO_3 conversion. Sulfur in the fuel generates SO_2 (predominantly) and SO_3 (SO_x) during combustion; the presence of both SO_3 and NH_3 in the flue gas can lead to the formation of ammonium sulfate and or ammonium bisulfate as indicated in reactions 18.9 and 18.10. Once formed, these substances may deposit on the catalyst and other system equipment, impacting and reducing the system performance.

18.6.2 Selective Noncatalytic Reduction (SNCR)

Similarly to SCR, selective noncatalytic reduction (SNCR) is a method of postcombustion NO_x reduction where an ammonia-based compound is injected into the furnace to react with NO_x to form N_2 and H_2O. SNCR systems can use anhydrous ammonia, aqueous ammonia, or aqueous urea. There are pros and cons to each reagent that should be evaluated for each specific site. The SNCR reaction occurs within a temperature range of approximately 1,600 to 2,100 °F. The optimum temperature within that range is dependent on residence time and carbon monoxide levels at the injection location, thus making the completion of a computational fluid dynamics (CFD) model of the boiler a requirement while evaluating the application of this technology. Lower temperatures require longer residence times. Elevated carbon monoxide levels shift the temperature curve such that better efficiencies are achieved at lower temperatures.

While NO_x conversion rates of 90% are possible in the laboratory, practical application of SNCR in conventional industrial and utility boilers currently limits this reduction to roughly 35–55%. This is due to temperature gradients, stratification, as well as droplet evaporation and distribution. Hotter furnaces will have lower NO_x reduction efficiencies, and boilers with furnace exit gas temperatures (FEGT) of 2,100 °F are not good candidates for SNCR (see Figure 18.9).

18.6.3 Absorption

While not commonplace practice in industry, the oxides of nitrogen can be absorbed by water, hydroxide and carbonate solutions, sulfuric acid, organic solutions, and molten alkali carbonates and hydroxides. When aqueous alkaline solutions such as NaOH and MgOH are used, complete removal requires that one-half of the NO must first be oxidized to NO_2. The best absorption occurs when the NO/NO_2 molar ratio is 1:1, which indicates that absorption of the combined oxide, N_2O_3, is the most favorable.

Absorption of nitric oxides by strong sulfuric acid has also been available. The compounds formed in this case are violet acid ($H_2SO_4 \cdot NO$) and nitrosyl sulfuric acid ($NOHSO_4$). The latter is very stable in concentrated acid. The reaction is:

$$NO + NO_2 + 2H_2SO_4 = 2NOHSO_4 + H_2O \quad (18.11)$$

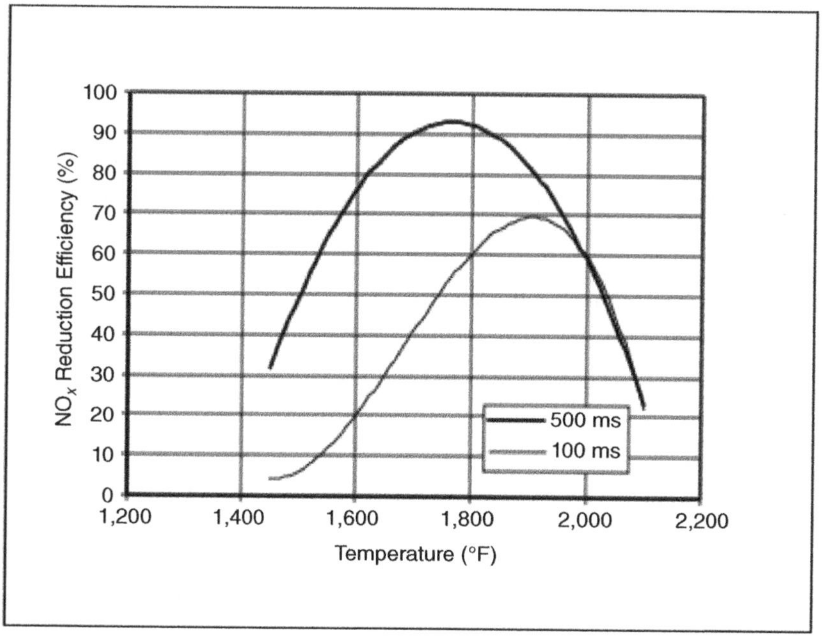

Figure 18.9 Selective noncatalytic reduction NO_x reduction as a function of temperature and residence time.

Aqueous scrubbing systems using aqueous alkaline solutions of concentrated sulfuric acid appear to offer the best potential for the control of both sulfur and nitrogen oxide emissions. Equimolar concentrations of NO and NO_2 (equivalent to N_2O_3) are required.

18.6.4 Adsorption

The use of char (activated carbon) to adsorb oxides of nitrogen has been studied [7]. Char has a high adsorption rate and capacity, compared to other materials. However, regeneration may be a problem. A potential fire and explosion hazard may be another difficulty with this material since O_2 is usually present in most stack gases. The technique does not show much promise at the present time. Adsorption may provide some limited means for the reduction of NO_2, emissions from flue gases associated with stationary combustion sources. Common adsorbents such as silica gel, alumina, molecular sieves, charcoal and ion exchange resins show some capacity for oxidizing NO to NO_2 and adsorbing the resultant NO_2. However, the capacity of these adsorbents is quite low at typical NO flue gas concentrations. Manganese and alkalized ferric oxides show some potential.

18.7 Operation, Maintenance, Inspection, and Optimization Considerations

The number of processes available for NO_x control is extremely large. In addition, the number of factors that can impact the processes is also very large. Because of this, it is not feasible to include any meaningful discussion on operation, maintenance, inspection, and optimization consideration for the various processes.

18.8 Conclusion

The initial push for large-scale NO_x reduction started in the 1990s and has continued to ramp up post 2010. NO_x has been a significant concern due to its greenhouse gas effect and formation of smog. The primary means of NO_x reduction is focused on reducing in-furnace flame temperature. There are extensive methods used in practice to accomplish this. Postcombustion NO_x reduction is also a viable alternative, though typically more costly. Design of NO_x reduction is typically left up to combustion experts (such as one of the authors of this text). Designing industrial scale combustion equipment can literally be playing with fire and must be maintained within certain bounds to prevent large-scale disaster.

References

1 J. Richardson and L. Theodore, *Optimizing Air Pollution Equipment Performance: Operation and Maintenance*, John Wiley & Sons, Hoboken, NJ, 2025.

2 L. Theodore. *Thermodynamics*. A Theodore Tutorial, East Williston, NY (originally published by USEPA/APTI, RTP, NC), 1983.

3 L. Theodore, F. Ricci and T. VanVleet. *Thermodynamics for the Practicing Engineer*, John Wiley & Sons, Hoboken, NJ, 2007.

4 L. Theodore. *Chemical Engineering: The Essential Reference*, McGraw Hill, New York City, NY, 2014.
5 L. Theodore and R. Nelsen. *Pollution Prevention*, A Theodore Tutorial, East Williston, NY, (originally published by USEPA/APTI, RTP, NC), 1994.
6 USEPA Technical Bulletin, Nitrogen Oxides (NO_x), Why and How they are Controlled, DARPS, EPA 456 1-99-006F, RTP, NC, 1999.
7 L. Theodore, R. Dupont and K. Ganesan. *Pollution Prevention: Sustainability, Industrial Ecology, and Green Engineering*. CRC Press Taylor & Francis Group, Boca Raton, FL, 2017.

19

Carbon Capture and Storage

Note: Parts of this chapter on the carbon capture and storage are an edited and updated version of an earlier chapter in the J. Richardson and L. Theodore publication titled, *Optimizing Air Pollution Control Equipment Performance: Operation and Maintenance.*

19.1 Introduction

Gases have performed varied functions in the economy. This is particularly true of carbon dioxide (CO_2). This chemical has been used in the liquid and solid phases for nearly two centuries. A half-century ago, over 1 million tons of CO_2 were produced in solid, liquid, and gaseous forms. The largest use of solid CO_2 was for refrigeration purposes; it also helped reduce bacteria spoilage and for carbonating some beverages. The main use of liquid CO_2 is for carbonating liquid beverages and as a fire extinguishing source. The gaseous state has been used in the production of urea, certain acids, and soda ash.

There are three main sources of CO_2 that have been deployed in the chemical process industries:

1. Flue gases (10–18% by volume) which result from burning fossil fuels.
2. By-products from fermentation industries (99%) produced from the reaction of dextrose breakdown into alcohol and carbon dioxide.
3. By-product of lime kiln operation (10–40%) where carbonates are calcined to oxides.

However, it appears that anthropogenic emissions of CO_2 may be causing global heating and climate change. It has been estimated that two-thirds of global CO_2 emissions are from the utility and transportation sectors. Since fossil fuels remain the least expensive source of energy, combustion of fossil fuels continues and will likely remain the energy source in the near future. Carbon capture and storage (CCS) is the process of capturing CO_2 from large point sources and producing concentrated and pressurized CO_2 for storage. However, it should also be noted that approximately 60% of CO_2 emissions can be attributed to dilute, distributed sources [1].

19.2 Properties of Carbon Dioxide

Carbon dioxide is a colorless, odorless, and lightly acid-tasting gas, sometimes called carbonic acid gas. The molecule consists of one atom of carbon joined to two atoms of oxygen (CO_2). Carbon dioxide is about 1.5 times as dense as air. It is soluble in water, with 0.9 volume of the gas dissolving in one volume of water at 20 °C (68 °F). The atmosphere contains carbon dioxide in variable amounts, usually 3–4 parts per 10,000. It is used by green plants in the process known as photosynthesis, by which carbohydrates are manufactured. It does not burn and does not support ordinary combustion, and because of these properties, it is used for extinguishing fires.

The CO_2 extinguisher is a steel cylinder filled with liquid carbon dioxide, which, when released, expands suddenly and causes so great a lowering of temperature that it solidifies into a powdery "snow." This snow volatilizes (vaporizes) on contact with the burning substance, producing a blanket of gas that cools and smothers the flame. Solid carbon dioxide, known as dry ice, is widely used as a refrigerant. Its cooling effect is almost twice that of ice water; its special advantages are that it does not melt as a liquid but turns into gas, and that it produces an inert atmosphere that reduces bacterial growth. The presence of carbon dioxide in the blood simulates breathing. For this reason, carbon dioxide is added to oxygen or ordinary air in artificial respiration and to the gases used in anesthesia.

19.3 Global Carbon Cycle

Carbon dioxide comprises only a very small portion of the Earth's atmosphere but has risen from about 280 to 380 ppmv (parts per million by volume) recently (approximately 0.03% by volume). However, it is undoubtedly one of the most essential molecules for life on Earth. Without this colorless, odorless gas, animals and plants would not be able to survive. Nevertheless, an excess of carbon dioxide in the atmosphere arising due to the burning of fossil fuels might prove to be harmful. The exchange of carbon throughout the world, termed the "global carbon cycle," is of particular importance for studying the greenhouse effect.

The transfer of carbon occurs naturally through both the terrestrial and oceanic cycles. The terrestrial cycle starts with photosynthetic plants, which use sunlight, water, carbon dioxide, and a pigment called chlorophyll to form glucose and oxygen. It has been estimated that this consumes approximately 500 billion tons of carbon dioxide each year, converting it into organic compounds and oxygen. Carbon dioxide is naturally returned into the atmosphere by respiration and decay. The oceans, which cover about 70% of the globe, consist of all the gases contained in the atmosphere. There is an exchange of carbon dioxide between the oceans and surrounding air until equilibrium is reached. Carbon dioxide is then transferred to deeper water by convective transport cycles, which lower the concentration at the surface, thereby allowing more of the gas from the atmosphere to diffuse in the ocean.

Surprisingly, about 25% of the increase in atmospheric carbon dioxide over the past century and a half has come from changes in land use, such as deforestation and soil cultivation. During the last 150 years, about 155 billion metric tons of carbon were released into the atmosphere from such actions. The amount released each year generally increased over the period, and by the 1990s, the rate of release averaged about two billion metric tons of carbon per year [2]. Deforestation increases carbon dioxide levels in two basic ways. First, when trees decay or are burned, they release carbon dioxide. Second, without the forest, carbon dioxide that would have been absorbed by photosynthesis remains in the atmosphere. For example, a rainforest can hold 1 or 2 kg of carbon/m^2/yr as compared to a field of crops, which can absorb less than 0.5 kg/m^2 every year [3].

In addition to deforestation and other such actions, the burning of hydrocarbons has caused a severe man-made influx of carbon dioxide into the global carbon cycle. The International Panel on Climate Change (IPCC) used knowledge of this natural process plus data on fuel consumption and land use in order to predict the concentration of carbon dioxide in the atmosphere. According to their model, approximately 3.3 gigatons more carbon dioxide is entering than leaving. This abundance of carbon dioxide has caused a disruption of the global carbon cycle. Without a "quick fix," some fear that this increase will result in catastrophic events, detailed in a later section.

Prior to 2024, the Environmental Protection Agency (EPA) appropriately replaced global warming with the term "climate change" to account for all of the environmental changes that are occurring in the world, not just the increase in temperature. Along with the federal government, the EPA has created a number of initiatives and incentives to encourage major polluters, particularly large corporations, to go "green" and reduce their environmental impacts. The U.S. government has also enacted many climate change programs to help curb greenhouse gas (GHG) emissions as well as prevent the release of other pollutants. A program introduced over a decade ago (2009) was the Climate Change Technology Program (CCTP), which will be discussed in more detail.

19.4 The Greenhouse Effect

The "greenhouse effect" is a phrase properly used to describe the increased warming of the Earth due to increased levels of carbon dioxide and other atmospheric gases, called GHGs. Just as the glass in a botanical greenhouse traps heat for growing plants, GHGs trap heat and warm the planet. The greenhouse effect, a natural phenomenon, has been an essential part of Earth's history for billions of years. The greenhouse effect is the result of a delicate and nonfixed balance between life and the environment. Yet, the greenhouse effect may be leading the planet to the brink of disaster. Since the Industrial Revolution, the presence of additional quantities of GHGs threatens to affect global climate and the predicted effects of this increase are still debated among scientists.

The greenhouse effect works as follows. The energy radiated from the sun to Earth is absorbed by the atmosphere and is balanced by a comparable amount of long-wave energy emitted back to space from the Earth's surface. Carbon dioxide molecules (and GHGs) absorb some of the long-wave energy radiating from the planet. Because of the greenhouse heat-trapping effect, the atmosphere itself radiates a large amount of long-wave energy downward to the surface of the Earth and makes the Earth warmer than if warmed by solar radiation alone. The GHGs trap heat because of their chemical makeup and, in particular, their triatomic nature. They are relatively transparent to visible sunlight, but they absorb long-wavelength infrared radiation emitted by the Earth [4].

As an environmental problem, climate change/global warming must be considered on an entirely different scale than that of most other environmental issues: the effects of climate change are long-term, global in magnitude, and largely irreversible. Because of the enormity of the problem and the uncertainties involved—it may take decades to determine with absolute certainty that global warming is underway—the difficult questions faced today are how and when one should react.

Fossil fuel burning and forestry and agricultural practices are likely responsible for most of the man-made contributions to the gases in the atmosphere that act like a greenhouse to raise the Earth's temperature; hence the term "greenhouse effect." Most of the processes that produce GHGs include common everyday activities such as driving cars, generating electricity from fossil fuels, using fertilizers, and using wood-burning stoves. Because so many of these activities are so ingrained in society, reducing emissions could be a difficult task.

The search for solutions has begun. However, there is a growing concern that the costs of reducing emissions may be too high. But to put cost concerns in proper perspective, one must ask what kind of future one wants on this planet and how much does one value the environment and the cultural heritage that depends on it [5].

A consensus has emerged in the scientific community that global warming has occurred. Most (but not all) scientists are certain that the concentrations of carbon dioxide and other GHGs in the atmosphere are increasing and these individuals generally agree that these gases will warm the Earth. Four questions remain to be answered:

1. How will the temperature rise?
2. When will the temperature rise?
3. Can the cause be attributed to man-made activities?
4. What are the potential consequences?

Recent estimates indicate that if the concentrations of these gases in the atmosphere continue to increase, the Earth's average temperature could rise by as much as 1.5–4.5 °C in the next century. While this may not sound like a tremendous increase, one must keep in mind that during the last ice age 18,000 years ago, when glaciers covered much of North America, the Earth's average temperature was only 5 °C cooler than today.

Certainly, global cooperation is an important consideration when addressing climate change issues. No single country contributes more than a fraction of GHGs, and only a concerted effort can reduce emissions. In the future, as developing nations grow and consume more energy, their share of GHG emissions will steadily increase. It is important for other nations to offer technological assistance so that these developing nations can grow in an energy-efficient manner [3]. Unlike other environmental problems that the EPA could address with the stroke of a regulation, potential climate change is a problem that needs innovative global solutions. Future trends of emissions will depend on a wide range of factors, from population and economic growth to technological development and policies to reduce emissions. Past trends show that all countries have been producing GHGs at an increasing rate, and many countries will continue to do so for years to come. Based on a careful study of the sources and trends of greenhouse emissions around the globe, countries can begin implementing measures for slowing down emissions while increasing economic development.

In the coming years, the technical community must reevaluate how emissions are likely to change. But given this preliminary picture of the future, it is important to take the next step of assessing the specific technologies and policy measures that can reduce emissions now at low costs. Each country will have to examine its unique situation and determine appropriate responses. However, only by acting together will the global community slow the trend toward high emissions in the next century.

19.5 Effects of Global Warming/Climate Change

A rise in average global temperature is expected by many to have profound effects. The need to study future situations has led the EPA and other environmental scientists to investigate what would happen to the planet after a 3–8 °F warming. These climate changes will have a significant effect on weather patterns. There will be changes in precipitation, storms, wind direction, etc. Rising temperatures are expected to increase tropical storm activity. The hurricane season in the

Atlantic and Caribbean oceans is expected to start earlier and last longer. Storms will be more severe. Some researchers believe that the planet will be a wetter place. Global circulation models (GCM) predicted that a doubling of carbon dioxide could increase humidity by 30–40%. However, such increases will not occur uniformly around the world. Perhaps, humid tropical areas will become wetter while some semiarid regions will become drier. It is possible that some farmers will experience a longer growing season, while others would suffer from more frequent droughts.

As the Earth gets warmer, the IPCC predicts that there will be a 0.5–3 ft rise in the average water level of oceans. As the ocean water is heated, it expands, or increases in volume. In addition, the polar ice caps will continue to diminish in size and contribute to the rising sea levels, even after taking into account the accumulation due to an increase in precipitation. Eventually, low-lying coastal areas are expected to flood. Beach erosions will be an increasing problem. The EPA has estimated that if the sea levels rise 3 ft (0.9 m), the nation will lose an area the size of Massachusetts, even if it spends more than $100 billion to protect critical shorelines. In addition to lost beaches, houses and other buildings that sit close to the water's edge will be engulfed and destroyed. This could bring havoc to the real estate and insurance industries, as well as the lives of many people.

If polar and temperate zones become warmer, there will be a poleward shift of ecological zones. Animal and plant species that now live in a particular area may no longer be able to survive there. The EPA predicts an increase in extinction rates as well as changes in migration patterns. As the ecological zones shift poleward, there may be a decrease in the amount of area suitable for forests, with a corresponding increase in grasslands and deserts. This means an overall loss in productive land, both for agriculture and for habitats of a broad range of plants and animals. For example, the United States' major crops grown—corn, wheat, and soybeans—are strongly affected by precipitation and temperatures. The warming trend may also cause changes in water quantity and quality in some areas. This will affect drinking supplies as well as the water needs of industry and agriculture. Rising sea levels may also contaminate water supplies as seawater migrates up rivers. For example, a foot (0.6 m) rise in sea levels would inundate Philadelphia's water intakes along the Delaware River, making the water too salty to drink.

The IPCC stated that "climate change is likely to have wide-ranging and mostly adverse impacts on human health, with significant loss of life." Heat waves take a physical toll on people. The heat puts a strain on the heart, as the body tries to cool itself. Studies have shown that the number of cases of heart and lung disease increases when temperature rises. Contagious diseases, such as malaria, influenza, and pneumonia, and allergic diseases, such as asthma, are also affected by the weather and become more prolific and are expected to spread among the public. The life cycles of mosquitoes and other disease-carrying insects are also extended in warmer weather. Warming is likely to allow tropical diseases, such as malaria, to spread northward in some areas of the world.

Climate changes, an increase in sea levels, and other direct effects of global warming can already be seen in many parts of the world. These changes are expected by many to have major social, economic, and political consequences for the future.

19.6 Carbon Dioxide Control Technologies

Currently, scientists are working hard to develop new technologies to safely reduce and dispose of carbon dioxide. There are a number of proposed advancements, many of which are still in the research and development stage. Absorption, adsorption, and reaction chemistry are the processes upon which these technologies are based. Some of these are discussed below.

One existing technology used to harvest carbon dioxide is a scrubber. Amine solutions extract carbon dioxide from flue gases through the processes of absorption. The spent amine solution is then sent to a regeneration process, where the CO_2 is collected, compressed, and liquefied [6].

Another process uses semipermeable membranes or catalyst-coated monoliths to collect carbon dioxide through adsorption. Membranes can either function according to the molecular size of the entrained molecules or ease of diffusion through the membrane. The latter poses an effective solution for CO_2 capturing. The membrane selectively allows certain gases that diffuse easily to pass through, while certain gases such as carbon dioxide are captured. In this case, the ease of diffusion of the flue gas through the membrane is dependent on the material of construction, which is polymer-based. Polyimide has some potential for this application [7].

A somewhat recent patent uses reaction chemistry for CO_2, removal. The process called accelerated weathering of limestone typically involves reacting CO_2 in a flue gas stream with water and calcium carbonate (limestone). Instead of releasing CO_2, a wastewater-rich stream of bicarbonate ions is produced, which can be directly released into the ocean. The coinventor of this project claims that carbon waste in the form of dissolved bicarbonate will have minimal adverse effects on the ocean and may even be beneficial for coral reefs. The Department of Energy has patented that technology, and several power plants along coastlines will serve as the testing grounds for this CO_2 sequestering method [8, 9].

In general, it is not very difficult to isolate carbon dioxide. However, safe disposal of the GHG has proven to be a daunting task. Many possibilities exist, but economic consequences are a major consideration when determining the feasibility of these technologies. Because industrial and commercial uses of CO_2 are minimal compared to amounts produced (approximately 1%), there needs to be safe, cost-effective disposal methods available. Such disposal methods that are likely to provide the first large-scale opportunity are direct injection to oil and gas reservoirs, deep, unmineable seams, or saline aquifers. These environments for injection have the ability to sequester large amounts of carbon dioxide as well as store and separate CO_2 from areas where injection would be harmful. Also, it could contribute to oil and methane recovery by the displacement method. Furthermore, ocean sequestering can be effective by the method of direct injection. This must be performed at 1,000–2,000 m deep, where the temperature gradient is great enough that it cannot mix with surface waters. It could also be injected over 3,000 m deep, where CO_2 becomes negatively buoyant which in turn would form a CO_2 lake. In this case, piping could be a major cost. For all of these methods, many technical, safety, liability, economic, and environmental issues remain unresolved.

Researchers in Australia conducted the biggest test of carbon disposal. They injected this material underground into a 1.3 mile hole in the ground, located at the Otway Basin. The base of this hole is a natural gas reservoir covered by several layers of impermeable rock. In 2008, 100,000 tons of carbon dioxide was introduced into the reservoir, and the hole was sealed. Scientists continue to monitor and test the surrounding areas to test for any leaks. A small leak could deem this project useless, while a large leak could potentially lead to disaster [10].

The direct air capture (DAC) is developing technology for removing CO_2 from the atmosphere that can, in principle, be built anywhere [11]. Unlike other CO_2 removal processes, it does not require a relatively concentrated source of CO_2 such as with flue gas. One method for DAC uses hydroxide solutions. The process consists of two main cycles: an air contactor, the CO_2 capture, and a sorbent regeneration cycle. In the first cycle, CO_2 from the atmosphere is absorbed by means of a chemical to form a carbonate (e.g., K_2CO_3). The carbonate solution then reacts with calcium hydroxide (Ca[OH]$_2$) to form calcium carbonate ($CaCO_3$). In the next cycle, the $CaCO_3$ is separated from the solution and calcined to lime (CaO) in an oxy-fired fluidized bed that operates at approximately 900 °C. During this stage, CO_2 is released while the CaO is recycled and rehydrated to Ca(OH). Finally, this hydroxide is used to react with the carbonate solution.

19.7 Carbon Dioxide Sequestration

Sequestration is to separate, seclude, and in this case, store something. Why the interest in sequestration? There is a simple answer: concern with carbon dioxide emissions into the atmosphere.

Preliminary estimates indicate that the amount of CO_2 captured from industrial processes, including combustion and chemical manufacturing, and produced from naturally occurring subsurface CO_2 reservoirs is approximately 44 million metric tons carbon dioxide equivalent [12]. Currently, more than 95% of this CO_2 supplied to the economy is injected underground for enhanced oil and gas recovery. However, CO_2 may be injected underground for geologic sequestration (GS). GS is the long-term containment of a CO_2 stream in subsurface geologic formations and is a key component of a set of climate change mitigation technologies known as carbon dioxide capture and geologic sequestration (CCS). CCS has the potential to enable large emitters of CO_2 such as coal-fired power plants to significantly reduce GHG emissions.

19.7.1 Carbon Capture Processes

19.7.1.1 Geologic Sequestration

Underground geologic formations that can be used for GS include deep saline formations, oil and natural gas reservoirs, and unmineable coal seams. In addition, CO_2 may be injected into other types of subsurface geologic formations, such as basalt formations.

GS occurs through a combination of structural and stratigraphic trapping, residual CO_2 trapping, solubility trapping, mineral trapping, and preferential adsorption trapping. These mechanisms are functions of the physical and chemical properties of CO_2 and the geologic formations into which the CO_2 is injected.

19.7.1.2 Enhanced Oil and Gas Recovery

CO_2 is currently being injected into subsurface geologic formations in the United States for enhanced recovery (ER). The CO_2 currently being used in ER is primarily produced from naturally occurring underground CO_2 reservoirs but also captured from industrial processes, including combustion and chemical manufacturing.

ER involves injecting CO_2 into oil or natural gas reservoirs via injection wells for the purposes of increasing crude oil production or to enhance recovery of natural gas. The crude oil and CO_2 mixture is produced from production wells and sent to a two-phase separator where the crude oil is separated from the gaseous hydrocarbons and CO_2. The gaseous CO_2-rich stream then is typically dehydrated, recompressed, and reinjected into the oil or natural gas reservoir to further enhance recovery. If the concentration of hydrocarbons in the CO_2 stream from the dehydrator is significant, then an acid gas recovery unit is used to separate the hydrocarbons from the CO_2. Injection of CO_2 into unmineable coal seams may result in displacement of methane (CH_4) from the coal seam and subsequent production of the methane as a product (or enhanced coal bed methane).

19.7.1.3 Other Industries That Use CO_2

EPA identified and considered a total of 22 commercial end-use sectors that use gaseous liquid, or solid CO_2, excluding ER and GS. These end-use sectors are available in the literature [13, 14].

19.7.1.4 Current Research

At the time of writing, there are several avenues of research into removing CO_2 from the air and converting it to carbonate so that it can be used as fertilizer, dissolved in the oceans, or locked in concrete. This process is largely in the infancy stage but seems promising.

19.8 Final Editorial Thoughts (of One of the Authors)

Regarding future trends, there appears to be three approaches to the response to this problem [15]. The first approach is the "wait-and-see" approach, where countermeasures may be ineffective if too much time is wasted. The second course of action is the "adaptation to incurable changes" attitude, which is based on the assumption that there will be plenty of time in which to decide and act on climate change, and the change is unavoidable. The third line of attack is the "act now" approach, which is the only one that demands an immediate legislative (and industrial) response. But the problem lies in the legal and economic systems which normally respond only to immediate and certain threats [16, 17]. Which course of action the world will adopt regarding the greenhouse problem is difficult to predict at this time. At this point in tune, the answer appears to be the third approach. Whether it will be wise or successful is another story.

The EPA earlier introduced the aforementioned CCTP in 2005. The purpose of this initiative was to attain the sustainable technology necessary to provide clean, safe, and reasonably priced energy sources on a global scale. The CCTP uses a Strategic Plan that outlines six major goals to be accomplished using the $3 billion in federal spending that was set aside for this program. These objectives are summarized as follows [17]:

1. Reduce emissions from energy end use and infrastructure.
2. Reduce emissions from energy supply, particularly by development and commercialization of no- or low-emission technologies.
3. Capture, store, and sequester carbon dioxide.
4. Reduce emissions of GHGs other than carbon dioxide.
5. Enhance the measurement and monitoring of GHG emissions.
6. Strengthen the contributions of basic science to climate change technology.

"It has always been a challenge for me to convince college graduates to join an oil and gas company. There is a strong desire for young professionals and academics to work with "green" technology. This is completely understandable, but what I always told them was that we can't forget to improve what exists while we keep an eye on the future. It would not be possible to turn off fossil fuels overnight, so the questions that need immediate action are: how do we improve coal combustion, what cleaner fuels such as natural gas can be substituted, how can we improve efficiency, how can we reduce emissions? Overhauling our energy system to renewable fuels is potentially feasible, but not in the short term. While it is important to look to the future, it is imperative to also give some attention to improving existing equipment and systems in a cost-effective manner [15]."

19.9 Final Editorial Thoughts (of the Other Author)

The literature that has appeared over the last decade on the influence of carbon on global warming has been extensive as noted earlier [18]. Most technical individuals strongly believe that human behavior is unalterably changing the world's climate by continuing to release carbon dioxide into the atmosphere, carbon that was previously bound for millennia on millennia in fossil fuels. These concerns have fostered new initiatives and concepts such as the Kyoto Protocol, carbon caps and trading, carbon sequestration, carbon footprint analyses, and restrictions on coal-fired utility boilers. Engineers and scientists believe there are lesser or insignificant impacts of human influence citing other phenomena such as sunspot activity, methane releases, and cosmic radiation, as examples of climate-forcing events. The difficulty with any opinion on climate change is that the subject is so broad and complex, that it is difficult for any one individual to be truly knowledgeable of all competing factors and analyses used to be able to reach conclusions on the impact of each process and their ultimate effect on climate. As an example, the Earth may be entering a period of significantly reduced solar radiation from sunspot activity in the immediate future. Such events have coincided with mini-ice ages in the past. Depending on the magnitude and duration of the reduction of solar activity in the near term, warming caused by increased carbon dioxide concentrations in the atmosphere may be counterbalanced by the cooling effects of minimal solar activity. The hypotheses on human's role in global warming are compelling but competing theories continue to emerge. Until independent, nonbiased research and analyses are performed to resolve the current differences in the theories on climate change, it may be best for the subject to remain outside the scope of this book. The reader is directed to the current literature to form their own opinion on human's role in global warming but is cautioned because there are many biased opinions that are being promoted by individuals with specific agenda, making it difficult even for a trained engineer or scientist to separate reality from hyperbole [19].

Here is an excerpt of part of a recent article by this same author [20].

You are no doubt aware that lawyers are adept at creating problems while engineers are noted for their ability to solve problems. In the days of old, some lectures to my chemical engineering students emphasized how to solve problems. Here is basically what I had to say. First, clearly define the problem. Then apply my three Cs concept: Cause, Consequence, and Cost. In effect, what is the cause, what are the consequences, and what will it cost to remedy/eliminate the problem? Here is what we have when this simple concept is applied to global warming and climate change: We really don't know the cause and we really don't know the consequences. Some of my critics claim that I often complain about problems but never offer any solutions. Well, here are seven possible solutions to this potential environmental problem by your favorite author that will have little to no adverse effect on society.

1. Plant 500,000,000 trees. The prestigious journal Science claimed a trillion trees would do the job in the most efficient manner.
2. Convert totally from a coal energy economy to one totally based on natural gas. I have previously shown that it would halve (actually 46%) CO_2 emissions. This partial energy conversion in the United States accounts for our achieving the goals of the 2015 Paris Climate Account.
3. Convert to a natural gas/oil energy economy. Oil emits approximately 25% less CO_2 than coal per unit of energy generated.

4. Embark on more aggressive domestic and industrial energy conservation programs. This can include converting to electric cars, capping unused wells, and controlling CO_2 and CH_4 (methane) emissions from fracking operations and landfills.
5. Convert to nuclear power. It's simple, safe, and economically attractive and there would be no CO_2 emissions. This is likely the cheapest and most effective solution.
6. Halt World Bank funding to China for potential coal-fired power plants since China is no longer (2015) classified as a developing nation.
7. Any combination of the above.

Bottom line: For the past 4.5 billion years, the Earth's temperature has varied from year to year, decade to decade, century to century, millennium to millennium, etc., and the changes, at times, were really significant. The same can be said about Earth's climate. None of these changes were man-made, and only some of these changes have been satisfactorily explained. Yet, here we are, not only having survived but also prospered. At the present time, the cause(s) for today's changes are unknown, the consequences are unknown, and therefore, any attempt to address these phenomena with massive economic expenditures and altering present lifestyles would be unreasonable [21].

References

1 D. Raynaud, J. Jouzel, J. M. Barnola, J. Chappellaz, R. J. Delmas and C. Lorius, The ice record of greenhouse gases, *Science*, 259(5097), 926–934, 1993.
2 *Global Carbon Cycle*, The Woods Hole Research Center. http://www.whrc.ore/carbon/index.htm
3 F. Lyman, *World Resources Institute: The Greenhouse Trap*, Beacon Press, West Palm Beach, FL, 1991.
4 S. Thompson, *The Greenhouse Effect*, Lucent Books, New York, 1993.
5 L. Theodore, *Personal Notes*, East Williston, NY, 1991.
6 J. Shao, *Amine Purification System (AmiPur) - Continuous Heat Stable Salts Removal from Amine Solutions*. Oil & Gas, Eco-Tec. Inc., Canada. http://www.eco-tec.com/techpapers/TP%20166%20AmiPur.pdf
7 D. Dortmundt and D. Kishore, *Recent Developments in CO_2 Removal Membrane Technology*, UOPLLC, Des Plaines, IL, 1999.
8 PhysOrg.com, *Patented Technology Captures Dioxide from Power Plants*, June 1, 2005.
9 R. Narula, H. Wen, K. Himes and B. Power, *Technical and Economic Comparison of CO_2 Reducing Technologies for Power Plants*, CEPSI, November 5, 2002.
10 P. Tesler, Air repair, *Curr. Sci.*, 93, 6–7, 2008.
11 M. Samari et al., *Direct Capture of Carbon Dioxide from Air via Lime-Based Sorbents*, Springer, New York City, NY, 2019.
12 US EPA, Inventory of U.S. Greenhouse Gas Emissions and Sinks, 1990–2008, EPA 430-R-10-006, RTP, NC, 2010.
13 US EPA, Vulnerability Evaluation Framework for Geologic Sequestration of Carbon Dioxide, EPA 430-R-08-009, RTP, NC, 2008.
14 R. Boyle, *Dead Heat*, Basic Books, New York, 1990.
15 J. Richardson, *Personal Notes*, 2023.

16 S. Schneider, *Global Warming*, Sierra Club Books, New York, 1993.

17 U.S. EPA, Climate Change Technology Program, http://www.epa.gov/climatechange/policy/cctp.html. Accessed April 28, 2009.

18 L. Theodore, *Personal Notes*, East Williston, New York, 2001

19 K. Skipka and L. Theodore, *Energy Resources: Availability, Management, and Environmental Impacts*, CRC Press/Taylor & Francis Group, Boca Raton, FL, 2014.

20 L. Theodore, *On Solving the Greenhouse Gasses Problem*, Queens Gazette, Astoria, NY, August 1, 2020.

21 J. Richardson and L. Theodore, *Optimizing Air Pollution Equipment Performance: Operation and Maintenance*, John Wiley & Sons, Hoboken, NJ, 2025.

20

Stacks

20.1 Introduction

Note: Parts of this chapter are an edited and updated version of an earlier chapter written by Eldon Dille for a 1981 Prentice-Hall publication titled "Air Pollution Control Equipment: Selection, Design, Operation, and Maintenance," which was edited by L. Theodore and A.J. Buonicore. The rights of the book were later transferred to Theodore Tutorials during the 1985–1986 time period.

After a waste gas has been treated, its release to the atmosphere is usually by means of a stack. Stacks discharging to the atmosphere have long been the most common industrial method of disposing waste gases. The concentrations to which humans, plants, animals, and structures are exposed at ground level can be reduced significantly by emitting waste gases from a process at great heights. This permits the gaseous pollutants to be dispersed over a much larger area and will be referred to as control by dilution. Although tall stacks may be effective in lowering the ground-level concentration of pollutants, they still do not in themselves reduce the amount of pollutants released into the atmosphere. However, in certain situations, it can be the most practical and economical way of dealing with an air pollution problem [1].

This chapter focuses on the practical considerations of stack design for the control of gaseous pollutants by dilution in the atmosphere. The emphasis will be on the determination of the "effective" stack height required to meet the maximum ground-level concentrations specified by regulation at various distances from a gaseous pollutant emitting source. To determine the acceptability of a stack as a means of disposing of waste gases, the acceptable ground-level concentration of the gaseous pollutant(s) must first be determined. The topography of the area must also be considered so that the stack can be properly located with respect to buildings and hills which might introduce a factor of air turbulence into the operation of the stack. Awareness of the meteorological conditions prevalent in the area, such as the prevailing winds, humidity, and rainfall, is also essential. Finally, an accurate knowledge of the constituents of the waste gas and its physical and chemical properties is equally important.

20.2 Description of Control Device

A typical stack consists of a concrete shell designed to withstand the stresses and loadings associated with a 500- to 600-ft vertical structure, plus a lining of independent brick, steel or concrete, corbel-supported brick, or gunned concrete. The type of lining used depends on the temperature tolerances and corrosive resistance required and the seismic zone in which the stack is located. There is a cap made of metal, ceramic, or fiberglass on the stack top.

Because of the stack environments caused by clean air regulations, most modem stacks were built in a period of design flux. Stacks built for power plants with flue gas desulfurization systems are all state-of-the-art designs. As such, some will perform better than others.

Stack designers use the American Concrete Institute's *Specification for Design and Construction of Reinforced Concrete Chimneys* (ACI 307-69) to plan for material, construction, and design requirements. (Note that the words "chimney" and "stack" are used interchangeably in this chapter. Individuals involved in the construction of this unit usually refer to it as a chimney; individuals involved in its use, particularly those in air pollution, refer to the unit as a stack.) This specification spells out the basic design standards needed to fit the right type of stack to a particular power plant. Some designers criticize the specification as being too conservative in some areas, but deviations from it are rare. ACI 307-69 sets forth recommended loadings for the design of reinforced concrete chimneys and recommended methods for determining the stresses in the concrete and reinforcement resulting from these loadings.

Although the specification is on firm ground about its structural standards, ACI 307-69 admits that the sections on lining materials "do not guarantee the adequacy of these materials to resist temperatures, abrasion or corrosion resulting from chimney gases." As you might expect, the area where engineering experience is least defined is where most maintenance problems occur, the liner is usually a maintainer's major concern.

Information on the stack foundation, shell, and liners follows:

20.2.1 The Stack Foundation

Stacks are often supported on pile foundations that support a concrete mat footing, usually a circular slab. A lead baseplate is located between the footing and the concrete slab. In most stack foundations, the slab provides a wearing surface and a drain for water that might collect at the stack base.

ACI 370-69 (3.10) states:

> *In chimneys conducting heated gases, consideration shall be given to the protection against the radiant heat of the gases for any part of the foundation which is exposed within the chimney and concrete floors either at the base of the chimneys or supported from the chimney column.*
>
> *It has been demonstrated that radiated heat from gases causes high temperatures at top surface of foundation or floor if not insulated. Even in modern power plants with moderate flue gas temperatures, this radiant heat may be sufficient to crack the chimney foundation or floor although flue openings enter the chimney an appreciable distance above the foundation.*
>
> *A satisfactory protection for the foundation floor is a ventilated insulation in the form of hollow tile covered with insulating brick or fire brick depending upon the temperature involved, with the voids of the hollow tile communicating directly to the atmosphere by means of air ducts when the chimney is lined or by openings in the shell when the chimney is unlined.*

Lead pans at the bottom of the flue(s) can be eaten away by acid attack. Since these pans are located below 1–1½ ft of concrete, an attack of the lead pan by the acid will occur after the concrete has been attached. When a flue begins to leak, a subverted lead pan beneath the concrete may be the cause.

In brick-lined stacks, some mortars in the lining are designed to contend with acid attack but not water. During the annual maintenance shutdown or whenever there is an extended boiler outage, the top of the stack should be covered if there is a possibility that rain may enter the stack. Water in the stack without flue gas present can result in a corrosive attack of the mortar. Water can change the acid resistant mortar into a paste that has no structural or cohesive strength. This is especially true at the base of the stack if water is allowed to collect. This explanation of the way the mortar fails is continuously reevaluated by the industry.

20.2.2 The Shell

All concrete stack shells built with ACI 370-69 recommendations are reinforced with vertical and circumferential reinforcing steel. The size, spacing, and amount of reinforcing bars used in construction depend on loading factors and the wall thickness required. Shell walls that are 18 inches or thicker have inner-face reinforcement of No. 4 bars on 24-inch centers vertically, and No. 4 bars on 12-inch centers horizontally. Some older stacks may not have inner-face reinforcing, so they are more subject to thermal and wind-load cracking than reinforced inner faces.

ACI 370-69 states: "The minimum shell wall thickness for any chimney with an internal diameter of 20 feet or less shall be seven inches. When the internal diameter exceeds 20 feet, the minimum thickness shall be increased ¼ inch for each two feet increase in internal diameter."

Since the external shell is concrete, it is resistant to most forms of mechanical attack (e.g., precipitation and wind). But eventually, every concrete shell will experience spalling. Thermal changes from normal day and night exposures can cause small cracks in the concrete. These will gradually increase in size until spalling occurs. In colder climates, water driven into these cracks can freeze and cause physical damage more rapidly. With the advent of FGD systems, there can be acid downwash along the top of the shell. This can result in chemical attack on the concrete. (Consider coating the outside of the concrete shell with a coating that is resistant to sulfuric acid attack. A coating based on polyester or epoxy resins could prove adequate in this environment. Fluoroelastomeric coatings also have the necessary acid resistance plus the advantage of bridging small hairline cracks in the concrete.)

Sulfuric acid in hairline cracks from downwash or internal infiltration can form salts and accelerate spalling. Structural failure may occur if the acid and water are allowed to reach the reinforcing bars. If the attack is allowed to go this far without detection, a major sectional repair or total replacement may be required.

Wind can also cause a cracking problem, or make an existing problem worse, especially in unreinforced areas of concrete. A timely system of crack inspections and a policy of instant repair will help you avoid the difficulty and expense of repairs or reconstruction.

Another form of mechanical attack is seismic disturbance. The shell design in earthquake-prone areas is usually based on the minimum lateral seismic force data given in ACI 370-69. A dynamic analysis based on the local seismic-response spectrum may also be a part of the design planning. A dynamic analysis gives insights into the local geology which may directly affect design decisions [2].

20.2.3 The Liner

Unlike the concrete shell, the brick liner is not built to precise engineering specifications. The type of liner used in the stack was chosen according to seismic conditions, the type of service (constant or intermittent peaking), gas temperature and gas composition in the flue, and the degree of reliability required. Structural requirements were, of course, a part of the liner selection.

The function of a liner is to accept the thermal stress in the stack column and to help protect the shell from the corrosive acids of coal-fired combustion. There are six basic types of liners: independent brick, steel, gunned concrete (gunite), fiberglass-reinforced plastic (more commonly known as FRP), alloys, and cast-in place concrete. The use of high-alloy metal for liners is nascent; cast-in-place concrete liners are all but extinct. A liner that was popular in the past—sectional brick liners supported by corbels in the concrete shell—is rarely found today because acid penetration can subvert the support corbels.

Each modern-era liner will be discussed separately, but the environments a stack may encounter will be addressed first.

The major concern in liner design is acid formation which leads to corrosion. Acids generated in the coal combustion process include sulfuric acid, sulfurous acid (an early stage of sulfuric acid), hydrofluoric acid, hydrochloric acid, and nitric acid. The dew points of sulfurous, hydrochloric, and hydrofluoric acids are well below the dew point of sulfuric acid. Prior to the advent of FGD systems, these did not pose a threat to the stack. Now, temperatures in the stack can be within or below the dew points of sulfurous, hydrochloric, and hydrofluoric acids.

Hydrofluoric acid is the Achilles' heel of inorganic monolithic gunned linings, potassium silicate mortars, bricks, and fiberglass fillers. Concentrations of this acid are beginning to be detected in linings and are being monitored to determine if they are reaching problem concentration levels.

Nitric acid, an oxidizing acid, will readily attack most components of the stack. All the corrosion products of nitric acid are water-soluble, so corrosion will proceed until the nitric acid is used up. Only scattered reports have been made of nitric acid presence in power plant stacks. It is expected that new boiler designs that minimize the formation of nitrogen oxide compounds will result in minimizing the production of stack-bound nitric acid.

There are several categories of liners, including: brick liners, independent brick liners, steel liners, fiberglass liners, and alloy liners.

20.2.4 Brick Liners

Independent brick liners, the most widely used type, evolved from corbel brick liners, the standard design of the 1930s and 1940s. The corbel liner consisted of 30- to 40-ft lengths of brickwork, usually 4 inch thick, that were supported by corbels (ledges) built into the liner face of the concrete shell. A space called an annulus, 2 inches or more, separated the lining from the column. The annulus was often filled with insulation to offset thermal stresses within the shell.

20.2.5 Independent Brick Liners

Brick liners in plants with FGD units should be built with ASTM Specifications C-279 acid-resistant brick. If a wet scrubber is used, the mortar should be sulfuric acid and sulfonation resistant. Expect stacks made of standard brick and nonacid resistant mortars to fail from acid attack of the mortar once an FGD system becomes operational. Since a brick liner is subject to positive pressure from the FGD system, there has to be some type of protection for the metal portions of the stack shell. Annular space pressurization and metal coatings that are acid and sulfonation resistant help prevent this acid attack. The stack cap and the exterior of the shell can also be damaged by corrosive acids. Condensed gases can be carried down the outside of the shell by a draft and attack the concrete shell, the protective system, the navigational lights, access ladders, and service platforms.

20.2.6 Steel Liners

Steel liners were introduced around 1960 to help power plant owners cope with corrosion problems then experienced with corbel brick linings. The stated benefits of a steel liner were a lower first cost than brick liners, resistance to earthquakes, and high resistance to corrosive forces [3]. Prior to passage of the 1972 Clean Air Act, the performance of steel liners met most expectations. There were some cases of liner buckling, but steel liner failures were very rare.

20.2.7 Fiberglass Liners

The lining materials of construction for power plant liners alternated between carbon steel and brick during the pre-FGD era. With the lowering of the flue gas to around 120 °F without reheat and to 180 °F with reheat, a new lining material was needed. In an attempt to find a new material in the 1960s, a coating of fiberglass-reinforced plastic coating was applied to steel liners. This composite material experienced some failures. However, the coating did withstand the corrosive liquids found in the liner.

20.2.8 Alloy Liners

An increasing interest is being shown in the use of high-corrosion resistant alloys for the fabrication of stack liners. Since each coal and FGD system results in differing corrosive liner environments, the precise alloy being used varies. The most often used materials are nickel and chromium alloys with a 6% or higher molybdenum content. These materials are known for their resistance to sulfuric acid corrosion and a varying resistance to chloride pitting. The precise chloride resistance of each alloy is dependent on the pH together with the temperature and chloride concentrations the alloy will experience.

20.3 Process Design Considerations

The atmosphere is the medium in which air pollution is carried away from a stack and diffuses. Meteorological factors have a considerable influence over the frequency, length of time, and concentrations of effluents to which the general public may be exposed. The variables that affect the severity of an air pollution problem at a given time and location are wind speed and direction, insolation (amount of sunlight), lapse rate (temperature variation with height), mixing depth, and precipitation. Unceasing change is the predominant characteristic of the atmosphere; for example, temperatures and winds vary widely with latitude, season, and surrounding topography [4, 5].

Atmospheric dispersion depends primarily on horizontal and vertical transport. Horizontal transport depends on the turbulent structure of the wind field. As the wind velocity increases, the degree of dispersion increases with a corresponding decrease in the ground-level concentration of the contaminant at the receptor site. This is a result of the emissions being mixed into a larger volume of air. The dilute effluents may, depending on the wind direction, be carried out into essentially unoccupied terrain away from any receptors. Under different atmospheric conditions, the wind may funnel the diluted effluent down a river valley or between mountain ranges. If an inversion (temperature increases with height) is present aloft that would prevent vertical transport, the pollutant concentration may build up continually.

For a given stack height, the calculational sequence begins by first estimating the "effective" height of the emission employing an applicable plume rise equation. The maximum GLC may then be determined using an appropriate atmospheric diffusion equation. A simple comparison of the calculated maximum GLC for the particular gaseous pollutant with the maximum GLC permitted by the local air pollution codes dictates whether the stack is operating satisfactorily. GLC standards, a stack can be properly designed which will satisfy these standards.

Dispersion of air contaminants from stacks is strongly dependent on the local meteorology of the atmosphere into which the pollutants are emitted. The mathematical formulation for the design of pollutant dispersal is associated with the open-ground terrain free of obstructions. Either natural or man-made obstructions alter the atmospheric circulation and with it the dispersion of pollutants. In addition are the effects of mountain valley terrain, hills, lakes, shorelines, and buildings.

20.3.1 Plume Rise

A plume of hot gases emitted vertically has both a momentum and a buoyancy. As the plume moves away from the stack, it quickly loses its vertical momentum (owing to drag by and entrainment of the surrounding air). As the vertical momentum declines, the plume bends over in the direction of the mean wind. However, quite often, the effect of buoyancy is still significant, and the plume continues to rise for a long time after bending over. The buoyancy term is due to the less-than-atmospheric density of the stack gases and may be temperature or composition induced. In either case, as the plume spreads out in the air (all the time mixing with the surrounding air), it becomes diluted by the air.

Modeling the rise of the plume of gases emitted from a stack into a horizontal wind is a complex mathematical problem. Plume rise depends not only on such stack gas parameters as temperature, molecular weight, and exit velocity but also on such atmospheric parameters as wind speed, ambient temperature, and stability conditions [6].

The behavior of plumes emitted from any stack depends on localized air stability. Effluents from tall stacks are often injected at an effective height of several hundred feet to several thousand feet above the ground because of the added effects of buoyancy and velocity on the plume rise. Other factors affecting the plume behavior are the diurnal variations in the atmospheric stability and the long-term variations that occur with changing seasons [7].

Details regarding a list of plume rise models and calculations of procedures are available in the literature.

20.3.2 Effective Stack Height

Reliance on atmospheric dispersion as a means of reducing ground-level concentrations is not foolproof. Inversions can occur with a rapid increase in ground-level pollutant concentrations. One solution to such situations is the tall stack concept. The goal is quite simple: Inject the effluent above any normally expected inversion layer. This approach is used for exceptionally difficult or expensive treatment situations because tall stacks are quite expensive. To be effective, they must reach above the inversion layer to avoid local plume fallout. The stack itself does not have to penetrate the inversion layer if the emissions have adequate buoyancy and velocity. In such cases, the effective stack height will be considerably greater than the actual stack height [5]. The effective stack height (equivalent to the effective height of emission) is usually considered the sum of the actual stack height, the plume rise due to velocity (momentum) of the issuing gases,

and the buoyancy rise which is a function of the temperature of the gases being emitted and the atmospheric conditions.

The effective stack height depends on a number of factors. The emission factors include the gas flow rate, the temperature of the effluent at the top of the stack, and the diameter of the stack opening. The meteorological factors influencing plume rise are wind speed, air temperature, shear wind speed with height, and atmospheric stability. No theory on plume rise presently considers all these variables, and it appears that the number of equations for calculating plume rise varies inversely with one's understanding of the process involved. Even if such a theory was available, measurements of all the parameters would seldom be available. Most of the equations that have been formulated for computing the effective height of an emission stack are semiempirical in nature. When considering any of these plume rise equations, it is important to evaluate each in terms of assumptions made and the circumstances existing at the time the particular correlation was formulated. Depending on the circumstances, some equations may be more applicable than others.

The effective height of an emission rarely corresponds to the physical height of the source or the stack. If the plume is caught in the turbulent wake of the stack or of buildings in the vicinity of the source or stack, the effluent will be mixed rapidly downward toward the ground. If the plume is emitted free of these turbulent zones, several emission factors and meteorological factors will influence the rise of the plume. The influence of mechanical turbulence around a building or stack can significantly alter the effective stack height. This is especially true with high winds when the beneficial effect of the high stack gas velocity is at a minimum and the plume is emitted nearly horizontally.

Details regarding a host of plume rise models and calculation procedures are available in literature [8, 9].

20.3.3 Atmospheric Dispersion Models

The Pasquill–Gifford model has been the basis of many models developed and accepted today [10–12]. This model has served as an atmospheric dispersion formula from which the path downwind of emissions can be estimated after obtaining the effective stack height. There are many other dispersion equations (models) presently available, most of them semiempirical in nature. Calculation details regarding the use of this equation are available in the literature [8–9].

The problem of having several models is that various predictions can be obtained. In order to establish some reference, a standard was sought by the government. The Guideline on Air Quality Models [13, 14] is used by the Environment Protection Agency (EPA), by the states, and by private industries in reviewing and preparing prevention of significant deterioration (PSD) permits and in state implementation plans (SIP) revisions. The guideline serves as a means by which consistency is maintained in air quality analyses. On September 9, 1986 (51 FR 32180), the EPA included four different changes to this guideline: (a) addition of specific version of the rough terrain diffusion model (RTDM) as a screening model, (b) modification of the downwash algorithm in the industrial source complex (ISC) model, (c) addition of the offshore and coastal dispersion (OCD) model to EPA's list of preferred models, and (d) addition of the AVACTA II model as an alternative model in the guideline. In industries today, the ISC models are the preferred models for permitting and therefore are used in many applications involving normal or "after the fact" releases, depending on which regulatory agency must be answered to.

The EPA originally published the Guideline in April 1978 (EPii-450/2-78-027), and it was incorporated by reference in the regulations in June 1978. The EPA revised the Guideline in 1986

(51 FR 32176) and updated it with supplement A in 1987 (53 FR 3208 1), supplement B in July 1993 (58 FR 38816), and supplement C in August 1995 (60 FR 40465).

Other effects to be considered include evaporative cooling, aerodynamics downwash, multiple stacks, and economics are briefly discussed below.

20.3.4 Evaporative Cooling

When effluent gases are washed to absorb certain constituents prior to emission, the gases are cooled and become saturated with water vapor. Upon release of the gases from the absorption tower, further cooling due to contact with cold surfaces of duct work or stack is likely. This cooling causes condensation of water droplets in the gas stream. Upon release of the gases from the stack, the water droplets evaporate, withdrawing the latent heat of vaporization from the air and cooling the plume. The resulting negative buoyancy reduces the effective stack height. The result may be a plume (with a greater density than that of the ambient atmosphere) which will fall to the ground. If any pollutant remains after scrubbing, its full effect will be felt on the ground in the vicinity of the stack.

20.3.5 Aerodynamic Downwash

Should the stack exit velocity, v_s, be too low as compared to the speed of the crosswind, u, some of the effluent can be pulled downward by the low pressure on the lee side of the stack. This phenomenon, known as downwash, can be minimized in one way by keeping v_s greater than u. [15] presents some general rules of thumb that correlate the wind velocity with the stack gas exit velocity for avoiding downwash. Another way suggested to minimize downwash is by fitting the top of the stack with a flat disc that extends for at least one stack diameter outward from the stack. If it becomes necessary to increase the stack gas exit velocity to avoid downwash, it may be necessary to redesign the stack exit. A venturi nozzle design has been found to be most effective. This design also keeps pressure losses to a minimum.

20.3.6 Multiple Stacks

The effect of multiple-stack sources has been handled in the past by simply treating each as a distinct source and the resulting concentrations added to obtain the total concentration. At the present time probably, the best method available for estimating the effect of multiple sources is that suggested by the TVA. Their procedure is as follows: For the case of a series of stacks all having the same height, estimate the concentration for a single stack, and then multiply this result by the factor indicated in the literature which corresponds to the total number of stacks of this size. For example, if at a large coal-fired power plant with four units (each connected to a single stack of the same size) the atmospheric dispersion calculations on a single stack (unit) resulted in a downstream pollutant concentration of $100\,\mu g/m^3$, then to account for the multiple source effect (the other three stacks), multiply the $100\,\mu g/m^3$ by a factor of to get the total concentration $(300\,\mu g/m^3)$. If the multiple-stack source consists of stacks of approximately one height and also stacks of considerably different heights, these groups should be treated as distinct groups of sources and the resulting concentrations from each group added to obtain the total concentration.

20.3.7 Rules of Thumb in Stack Design

As experience in designing stacks has accumulated over the years, several rules of thumb have evolved:

1. Stack heights should be at least 2½ times the height of any surrounding buildings or obstacles so that significant turbulence is not introduced by these factors.
2. The stack gas exit velocity should be greater than 60 ft/sec so that stack gases will escape the turbulent wake of the stack. In many cases, it is good practice to have the gas exit velocity on the order of 90 or 100 ft/sec.
3. A stack located on a building should be located in a position which will assure that the exhaust escapes the wakes of nearby structures.
4. Gases from stacks with diameters less than 5 ft and heights less than 200 ft will hit the ground part of the time, and ground concentrations may be excessive. In this case, the plume becomes unpredictable.
5. The maximum ground concentration of stack gases subjected to atmospheric diffusion occurs about five to ten effective stack heights downwind from the point of emission.
6. When stack gases are subjected to atmospheric diffusion and building turbulence is not a factor, ground-level concentrations on the order of 0.001–1% of the stack concentration are possible for a properly designed stack.
7. Ground concentrations can be reduced by the use of higher stacks. The ground concentration varies inversely as the square of the effective stack height.
8. Average concentrations of a contaminant downwind from a stack are directly proportional to the discharge rate. An increase in discharge ram by a factor increases ground-level concentrations at all points by the same factor.
9. In general, increasing the dilution of stack gases by the addition of excess air in the stack does not affect ground-level concentrations appreciably. Practical stack dilutions are usually insignificant in comparison to later atmospheric dilution by plume diffusion. Addition of diluting air will increase the effective stack height, however, by increasing the stack exit velocity. This effect may be important at low wind speeds. On the other hand, if the stack temperature is decreased appreciably by the dilution, the effective stack height may be reduced. Stack dilution will have an appreciable effect on the concentration in the plum close to the stack.

The nine rules of thumb listed above represent the basic design elements of a pollution control system. The engineering approach suggests that each element be evaluated independently and as part of the whole control system. However, the engineering design and evaluation must be an integrated part of the complete pollution control program.

20.3.8 Economics

Stack costs vary widely with geographical location, gas conditions, materials, allowable construction time, and height (cost per foot of height increases with increasing height by a factor of 1.25–1.5 per unit increase). Although tall stacks have greater dispersion/dilution effects, these self-supporting refractory lined structures can become considerably expensive. Originally, stacks consist of a carbon steel shell lined with either castable material or firebrick. The diameter of the stack is often dictated by the load-bearing requirements.

20.4 Equipment Design Considerations

Welding nickel-based alloys can be a major problem because of "hot cracking." Hot cracking is caused by stress and the presence of a "strain-intolerant microstructure." Since stress of the metal is a natural occurrence during welding, care must be taken to minimize the stress by proper welding design and proper welding technique. The minimization of the strain intolerant microstructure can be achieved by minimizing the heat generated during the welding process. This is done by keeping the current required on the welding machine as low as possible and using a proper sized electrode to maintain the minimum current load. Excessive heat input will increase the amount of metal exposed to the crack-sensitive temperature range and increase the duration of the elevated temperature exposure. A common source of excessive heat input is from "weaving" the electrode instead of using the preferred stringer bead welding technique. The weaving should not exceed three times the diameter of the electrode being used; however, the best precaution is to rely on sound welding practice and good common sense. A welder has described the proper welding current for alloys as the current required to weld so that it feels like you are taking the metal from the electrode and putting it on the metal.

20.4.1 Fusion-zone Hot Cracking

Bend contour and shape are important factors in minimizing fusion-zone (weld metal) hot cracking, especially when using butt welding plate of pipe which requires multiple passes. The individual stringer passes or beads should be deposited with a distinct crown. Weld beads that are deposited with a convex or flat surface are more susceptible to centerline cracking. If coated electrodes are used to make the welds, all the slag should be removed between passes so that there will not be any slag inclusion in the weldment. All flaws in the weld should be removed before another pass is placed in the welding area.

Cracks in the alloy can also occur at the edge of the heat-affected zone. These cracks are caused by excessive welding heat input coupled with stresses associated with welding. Some manufacturers recommend that no welding of any kind occur if the metal temperature is below 32 °F. Ideal welding temperature is 72 °F. Interpass temperatures should not exceed 200 °F. No cooling techniques should be used between passes that can introduce oil, grease, dirt, or contamination into the weld. This includes shop air, oil-soaked rags, hard-water deposits, and so on. For best results, allow the metal to cool naturally.

The surfaces of all areas adjacent to the welding joint should be cleaned to bright metal prior to welding. This means that all grease, oil, crayon marks, dirt, magic marker marks, and so on, should be removed by wiping the area with the recommended solvent and lint-free rags. Failure to do this can result in porosity and increased susceptibility to corrosion at the weldment.

Coated electrodes should be stored in sealed moisture-resistant containers until they are used. Electrodes that are removed from the original containers should be heated at least 15 °F above the ambient temperature to assure that the electrodes stay dry. If the electrodes get wet, they should be dried as recommended by the manufacturer. One manufacturer recommends that the electrodes be dried at 250–300 °F for at least 2 hours before they are used.

Porosity can occur in shielded metal arc (SMA) welding when welded vertically. This probably occurs because gravitational forces prevent the molten slag from properly protecting the weld pool. Another cause may be the breaking of the molten slag layer by improper rod movements. As much as possible, the SMA welding should be done in the horizontal position. Check with the supplier of the alloy and the welding rod to determine the proper rod for vertical welding if it is required.

Two other techniques are also used in welding alloys. These are gas tungsten arc (GTA) and gas metal arc (GMA). In both processes, the gas shielding the molten pool should precede the arc initiation by several seconds. Also, the gas flow should not be excessive. Excessive gas flow can result in porosity of the resulting weld. This excessive gas flow causes air to be brought into the weld puddle region. Should the gas flow be too low, there is insufficient gas flow to properly protect the molten pool, and air is brought into the molten pool. When welding with either GTA or GMA, care should be taken to keep the diffuser screen clean and free of splatter.

20.4.2 Other Shielding-related Problems

Another problem related to proper shielding is the inability to deposit a successful root pass, especially when butt welding pipe. The underside of a root pass should be adequately shielded with inert gas or else the weld bead will not "wet" and "flow" to produce the desired penetration effect. Proper shielding should also include the use of purge dams. When initiating the root pass, a temporary covering of tape over the remainder of the joint will minimize the loss of purge gas through the root opening.

The surface of weld beads deposited by the gas tungsten arc process and gas metal arc process will occasionally appear to have very thick oxide rafts, sometimes referred to as "slag spots." Although these artifacts may be questionable from a cosmetic viewpoint, they very rarely interfere with the subsequent weld passes or with overall integrity of the welded joint.

20.4.3 Repair of Welds

All rejected welds should be removed by a suitable grinding method. Remelting of the weld beads or depositing additional weld beads over the defect is definitely not recommended.

In joining a high alloy to carbon steel, an alloy welding filler material should be used that will give a final metal composition that is equal to or better than the parent metal.

One of the alloy manufacturers has provided a suggested list of fabricating tips when working with corrosion-resistant alloys. These tips are guides for maintaining the corrosion-resistant properties of the alloys [16].

DO:

- Follow acceptable welding practices as set forth in the welding section of the ASME Boiler and Pressure Vessel Code.
- Use filler metal of the same composition as the alloy base metal.
- Achieve complete penetration of the weld to avoid crevices which might lead to accelerated corrosion and stress raisers in the fabrication.
- Remove all traces of foreign material from the alloy before heating to above the black heat range when welding or for any other reason.
- 100% dye-check all welds to assure crack-free joints for corrosion applications as required by ASME Boiler and Pressure Vessel Code.
- 10% X-ray where required, to conform to code.
- Solution-heat-treat after fabrication if the alloy and the corrosion application warrant it.
- When heat treating, use fuels of low sulfur content to avoid sulfur penetration and subsequent detrimental effect on mechanical properties and corrosion resistance.

DO NOT:

- Allow direct flame impingement on the alloys.
- Overheat when solution-heat-treating or hot working. Overheating of some of the alloys by as little as 25–50 °F above the recommended solution heat treatment temperature can cause incipient melting.
- Age the alloys for prolonged periods of time in the 1,100–1,800 °F range. Aging causes carbide precipitation in the grain boundaries, which leads to loss of ductility and greater susceptibility to corrosive attack. If aging does occur, solution heat treatment at the recommended temperature will restore much of the original ductility and corrosion resistance to the alloy.
- Use oxyacetylene for welding the alloys unless recommended by the manufacturer.
- Use brazing alloys that contain copper.
- Make intermittent cuts when machining these alloys. They have a tendency to work-harden, which makes subsequent cuts much more difficult.
- Use shot blasting to clean the alloys. Iron particles that are embedded in the surface of the alloys can cause serious product contamination and/or severe corrosion problems in some media.

20.5 Sulfuric Acid Attack

Sulfuric acid is primarily responsible for the corrosion attack on any material used in the stack liners. It is not the only acid eating away at the liner, but it is the most plentiful and potentially the most destructive. This acid comes from the sulfur trioxide and sulfur dioxide produced in the boiler during fuel combustion. Other acids that can increase the corrosion rate of sulfuric acid are hydrochloric and hydrofluoric acids.

The Stationary Source Performance Standard of 1979 allows 10% of the sulfur dioxide emissions from solid and solid-derived fuels combustion to pass through the FGD system into the stack. Since this sulfur dioxide is saturated with water when it leaves the wet FGD system, there will be sufficient moisture to make an acidic solution in the stack.

There is also sulfur trioxide present in the flue gas when it comes from the boiler. Depending on the sulfur content of the coal, oil, or other fuel, the boiler design, vanadium content of the ignitor oil, vanadium pentoxide deposits on the boiler tubes, pattern of varying loads on the boiler, and so on, various amounts of sulfur trioxide can be produced. The Shawnee Pilot Plant tests have shown that about 58% of the sulfur trioxide present in the unsaturated flue gas is removed by the FGD system. Some sulfur trioxide is also removed in the cold-side electrostatic precipitator. At some plants, if the fly ash being collected in the electrostatic precipitator has a very high electrical resistivity, it is sometimes the practice to add sulfur trioxide before the precipitator to improve the removal efficiency of the unit.

Sulfur trioxide that passes through the precipitator and FGD system goes through the demisters and into the stack, where it is available to make sulfuric acid. The quantity of sulfuric acid will depend on the precise amount of sulfur trioxide present in the treated flue gas.

Sulfur trioxide has a high affinity for water. When the sulfur trioxide concentration goes below its dew point, sulfur trioxide reacts with water vapor in less than 1 second to form minute droplets of sulfuric acid. These droplets will be very concentrated, approximately 98% sulfuric acid, but they will continue to react with water until the concentration on the liner is lowered to its boiling point equilibrium concentration. This medium-concentrated acid will either react with components of the flue lining or continue to absorb water and form more dilute sulfuric acid. The pH value of dilute sulfuric acid that finally forms after a wet FGD system is in the 0.5–2.2 range.

Sulfuric acid can be found whenever the flue gas temperature drops below 400 °F. In the past, flue gases from the preheater were around 350 °F. Even though this temperature is below the maximum acid condensation temperature, there were only minimal corrosion problems. In the contemporary stack environment of about 130 °F, copious amounts of sulfuric acid are formed when sulfur trioxide gas goes below its dew point.

There has been considerable discussion about what the dew point of sulfuric acid is. It can be defined as the temperature at which liquid sulfuric acid forms in equilibrium with the sulfuric acid in the vapor state. Research to develop some instrument to accurately determine the sulfuric acid dew point has produced several mechanical techniques. These instruments generally determine the sulfuric acid dew point by the use of electrical conductivity probes. These probes are put into the flue gas stream and then slowly cooled. At the point that sulfuric acid and water condense on the probe, a sudden change will be seen on the meter, indicating the conductivity change across the probe. The temperature at which this sudden change in conductivity occurs is the dew point of the flue gas. Since sulfuric acid is an unusual compound, this dew-point temperature does not mean that liquid sulfuric acid will be seen on the walls of the stack.

Care must be taken when using these instruments to assure that a value is obtained indicative of what is happening in the liner. Gibson and Kufta suggest that the corrosion rate with sulfuric acid increases from the theoretical dew point of the acid to a temperature about 85 °F below the sulfuric acid dew point [16].

From this temperature to the dew point of water, the corrosion rate decreases. Below the water dew point, the corrosion rate rapidly increases. This lower condition is the primary area of interest for scrubbed flue gases.

To illustrate the effect of trace amounts of sulfur trioxide on the dew point of sulfuric acid, here is an example taken from a paper by Robert R. Pierce:

> *Thus, if a combustion gas containing as much as 10 percent water vapor by volume also contains as little as 40 parts per million sulfur trioxide by volume (0.004 percent), it has a weight percentage of sulfuric acid (based on sulfur trioxide and water alone) of only 0.22 percent. The vapor phase in equilibrium with the liquid phase has a concentration of 82.5 percent by weight sulfuric acid at its boiling point at 0.1 atmosphere pressure which is 148 degrees C (298 degrees F). Thus, as soon as the temperature of such a mixture of 10 percent water vapor and 40 parts per million sulfur trioxide by volume drops below 148 degrees C, liquid must form because the gas becomes supersaturated with 82.5 percent sulfuric acid. This powerfully acidic liquid will condense on any solid surface which is cool enough, which is why it is called the dew point or sometimes called the acid dew point. If the gas is cooled below the dew point temperature by radiation, etc., a mist of acid droplets will be formed. It may seem strange that a gas containing a mixture of sulfur trioxide and water corresponding to only 0.22 percent by weight of sulfuric acid will form a droplet or a film on a surface containing 82.5 percent by weight of sulfuric acid, but such is the nature of this acid. Of course, only a very small amount of the water in the gas is condensed out with the acid* [17].

As was mentioned earlier, considerable work has been done to develop some type of instrument to accurately determine the sulfuric acid dew point. But in the literature reviewed, there seems to be no agreement about the accuracy of the available mechanical techniques. There is, however, a method for determining the sulfur trioxide concentration in the flue gas. Method 8 from the Federal Register, Appendix A-32, can be used to determine the sulfur trioxide concentration. The concentration determined by the EPA Method 8 is in grams per dry standard cubic meter or pounds per dry standard cubic foot.

One of the operating conditions that has become more apparent with the advent of flue gas desulfurization systems is the formation of acid smut in the stacks. Gills and Lees [18] reported:

Based on many years' of study of this problem, including examinations of the inside of chimneys, microscopic examination and analysis of the deposits, and examination of the smuts, it has been concluded that acidic smuts consist of large agglomerates of black spheres bonded together with sulfuric acid or acidic salts.

Smuts form where chimney surfaces are below 250 °F and occur chiefly with metal stacks, but they have also been observed with concrete and brick stacks. It is possible that they can also form in ducts at the base of stacks.

It is believed that the four stages in the formation of smuts are:

1. Sulfuric acid solution is first deposited from the flue gas stream onto the cool surface where it reacts with the metal and forms an acidic salt.
2. This is a sticky layer and consequently soot particles adhere to it. Careful control of combustion conditions and use of good burners will reduce the concentration of these carbonaceous particles, but they will not be entirely eliminated. Hence, smut emission will be reduced but not eliminated by these measures.
3. Part of the acidic layer containing carbonaceous matter then flakes off due to a change in load or temperature.
4. The detached flakes break up into agglomerates or smuts. Some fall to the bottom of the chimney (where they form a thick acidic sooty layer) and some are carried out by the gases into the atmosphere. Owing to their large size and weight they fall near to the stack. These appear as large black smuts, often shaped like a cone up to one-half-inch diameter. Because of their acidic nature they will:
   a. Damage the cellulose finish on cars.
   b. Damage fabrics, particularly synthetic fibers, such as nylon.
   c. Cause a brown stain on fabrics and asbestos roofs due to the acidic iron salts present in the smuts.
   d. Damage leaves of plants.
5. Appear as sticky black smudges on surfaces on which they fall. This is often mistakenly described as an oil-smudge. It is not oily, but acidic [18].

Up until the advent of FGD systems, the basic technique to minimize the formation of acid smut was to operate the stack above the sulfuric acid dew point. Now that the flue gas is cooled by the FGD system, there is a good possibility that acid smut will be formed in the stacks and ducts. Fortunately, as previously mentioned, a FGD system does remove part of the sulfur trioxide from the flue gas.

Another technique for the control of the acid smut mentioned by Gills and Lees [18] is the neutralization of the sulfuric acid by an additive. The injection of finely divided magnesium hydroxide gives effective results in some cases. In this country, a number of companies offer additives for reduction of sulfur trioxide. Most help decrease sulfuric acid formation in the stack.

20.6 Inspection and Repair of Liners

With the advent of FGD systems, the inspection of the stack liner has taken on increased importance. It is suggested to contact a stack inspection firm to make a formal inspection of your stack liner after its first year of operation. The frequency of additional inspections should be determined after consultations with the stack inspectors and the engineer.

Inspection and repair information in this section will concentrate on five liner types:

1. Independent brick
2. Steel
3. Gunned concrete
4. Fiberglass
5. Alloy

20.6.1 Independent Brick Liners

Some of the inspection criteria for brick liners have already been discussed in the section on design of brick liners. One of the more frequent defects in the brick liner is cracking. There are basically three types of cracks. The first is the crack caused by a settling foundation. These cracks generally run upward or zigzag through mortar joints. Next is the crack caused by excessive wind loads. These cracks usually go through the mortar joints at 45 ° angles. The last is the thermal crack that extends upward through the brick and mortar. If any of these cracks are observed, special attention should be given to the metal tension bands holding the bricks together. It is possible that acid has migrated through the cracks in the bricks and attacked the metal bands. Repair cracks based on the recommendations of the stack designer; replace all damaged bands.

20.6.2 Steel Liners

Steel liners are found in two basic forms in the power industry. The first is the bare carbon steel liner that is used on a conventional power plant that does not have a FGD system. The second is the carbon steel liner coated with some organic or inorganic material used after a FGD system. Inspection of the bare carbon steel liner has been a standard procedure for a number of years. The main areas of concern are the places where the temperature is allowed to drop below the acid dew point for sulfuric acid. At these locations, there will be corrosion of the steel. Areas susceptible to this form of corrosion are the entry doors into the liner, sample ports, inlet breechings, and the liner that extends beyond the cap. The basic repair procedure is to remove the corroded steel and replace it with new metal.

20.6.3 Gunned Concrete

Problems encountered with gunned concrete liners are the same as those experienced with gunned coatings on steel liners. The repair and inspection procedures are also the same.

20.6.4 FRP Liners

Specifications used as part of the specifications for fiberglass liners include:

- ASTM Standard D3299
- NBS Voluntary Product Standard PS15-69
- ASTM Standard D2393-1972
- ASTM Standard D2471-1971
- ASTM Standard D2583-1975
- ASTM Standard D2584-1972
- ASTM Standard B84-1975

All resins used in fiberglass liners should contain a fire suppressant (antimony trioxide—less than 5% concentration) and an ultraviolet absorber to increase resistance to the destructive effect of the sun on the resin. Where a coating that is resistant to the chemical environment in the absorber is applied over the fiberglass liner, the life of the fiberglass is increased. (These coatings should contain a high ultraviolet-reflecting pigment such as titanium dioxide.)

20.6.5 Alloy Liners

Alloy liners are constructed of various metals. The precise alloy selected for a liner depends on the chemical environment it will face. To get maximum corrosion resistance from alloys, it is important that they be kept clean. That is true for all types of alloys. There is a saying in the alloy field that "stainless steel to be stainless must be clean." To a certain extent, this applies to nickel alloys as well as to stainless steels. To examine the alloys, it is necessary to remove the dust and scale deposits from the metal surface. The engineer should recommend the type of tools and cleaning devices needed for the alloy type. Because there are a number of alloys being used in the fabrication of liners, it is hard to generalize about the precise modes of failure that might occur in the alloy. There are several generalizations that can be made, however. For example, the use of steel shovels to clean an austenitic stainless steel can leave iron on the surface of the stainless steel. These iron scrapings will rust when the liner is put back into service and possibly leave pits in the stainless steel. These pits could then become points where chloride pitting might occur. Therefore, it is standard practice to use stainless steel tools when working with stainless steel.

20.7 Recent Advances

The basic design, construction, and operation of stacks have remained relatively unchanged over their first use. However, there are new designs and construction materials used in state-of-the-art stacks. In theory, each design will give satisfactory service for its life. However, until they actually have been in service for 30 years, the state-of-the-art designs will have to be considered "new" designs. Three relatively "new" design techniques used in power industry stacks will be described below:

1. One method that has proven its ability to handle all types of acid condensation while permitting complete flexibility of operation with any type of fuel is a double-column construction using a concrete outer column and a completely separate inner brickwork column. Proper brick and the use of sodium-free potassium silicate mortars can make the inner column resistant to acid condensation. The space between the two columns can be narrow or, preferably, sufficiently wide to permit inspection and repair. In this design, the acid-resisting requirements of the inner column and the physical requirements of the outer column can be separated. In addition, several methods can be used to prevent acid condensation from reaching and damaging the foundation.
2. Another variation of the double-column design is a concrete outer column and a steel inner column. The carbon steel column is insulated on its outer surface (facing the concrete column) to keep the inside surface of the steel above the acid condensation temperature. When this method is used, the temperatures at the inside steel surface must not be permitted to drop below the acid condensation point. Should the steel surface temperature drop below its acid dew point, a corrosive attack of the carbon steel will occur at a rapid rate. This corrosive attack could occur at a rate of 1/4 inch of steel per month of operation.
3. Still another variation is the fiberglass liner inside a concrete outer column. These fiberglass liners have to be joined to the concrete outer column with metals that are resistant to the corrosive liquids inside the liner. This assures the structural integrity of the liner if the acid should permeate the liner. Special precautions must also be taken to allow for the high thermal expansion of the fiberglass liner. Should these expansion characteristics not be taken into account, there will probably be structural damage to the liner. Also, since the fiberglass is flammable, special fire protection systems must be installed and maintained to assure fire protection. In addition, the sulfuric acid concentrations and flue gas temperatures that will be on the surface of the FRP liner must be considered.

20.8 Conclusion

From the day a stack is built, both external and internal forces are at work to destroy it. External villains are much easier to design for than internal ones, so exterior inspection and maintenance procedures can usually be based on known design parameters and realistic life expectations of construction materials. Despite this relative stability, however, internal corrosion may affect external performance. It cannot be overstressed that programmed inspections, timely crack repairs, and scheduled maintenance procedures are the keys to continuing stack shell integrity.

Inside the stack, knowing the strengths and weaknesses of a particular type of liner is just as important as knowing the kind of environment a liner must face. Independent brick liners and the emerging family of alloy liners have the best record and highest potential for success in the battle against low-temperature, wet stack environments, but even with these design Goliaths, there may be a David hiding in the post-FGD or bypass effluent. One must determine effluent temperature at various stack heights, dew-point condensation, and a myriad of other stack environment data. Only then can one tell what a liner can tolerate and for how long.

Continuing research and development portend optimism for increased stack longevity and performance reliability. Someday, there may be a stack that "can take it," but that day is not here yet.

References

1 L. Theodore and A. J. Buonicore, *Air Pollution Control Equipment: Selection, Design, Operation, and Maintenance*, Theodore Tutorials, East Williston, NY (originally published by Prentice-Hall, NJ), 1981.

2 B. Cooley, *Consider Dynamic Wind and Earthquake Loads When Designing Reinforced-Concrete Stacks*, Power, 24, October 1977.

3 R. Snook, *Interim Report of the ASCE Task Committee on Steel Chimney Liners*, American Power Conference, Chicago, IL, 1, April 20, 1972.

4 H. Hesketh, *Understanding and Controlling Air Pollution*, Ann Arbor Science Publishers, Ann Arbor, MI, 33–70, 1972.

5 A. Gilpin, *Control of Air Pollution*, Butterworth, NY, 326–333, 1963.

6 L. Theodore, *Air Pollution Control Equipment Calculations*, John Wiley & Sons, Hoboken, NJ, 2008.

7 R. Bethea, *Air Pollution Control Technology*, Ivan Nostrand Reinhold, NY, 39–59, 1987.

8 J. Mycock, J. McKenna and L. Theodore, *Handbook for Air Pollution Control Engineering and Technology*, CRC Press/Taylor & Francis Group, Boca Raton, FL, 1995.

9 J. Reynolds, J. Jeris and L. Theodore, *Handbook for Chemical and Environmental Engineering Calculations*, John Wiley & Sons, Hoboken, NJ, 2004.

10 F. Pasquill, *Meteorol. Mag.*, 90(33), 1063, 1961.

11 F. Gifford, *Nucl. Saf.*, 2(4), 47, 1961.

12 H. Cota, *J. Air Pollut. Control. Assoc.*, 31(8), 253, 1984.

13 U.S. EPA, *Guideline on Air Quality Models*, Publication No. EPA-450/2-78-027, RTP, NC, 1978 (OAQPS No. 1.2-08).

14 U.S. EPA, Industrial Source Complex (ISC) Dispersion Model User's Guide, 2nd edition, Vols. 1 alto 2, Publication Nos. EPA-450/4-86-005a and EPA-450/4-86-005b. RTP, NC 1986 (NRTIS PB 86 234259 and 23467).

15 G. Read Jr., *Corrosion-Resistant Linings for Chimneys, Scrubbers and Ducts of Flue Gas Systems*, American Power Conference, 40th Annual Meeting, Chicago, 2, April 24–26, 1978.

16 G. Gibson and R. Kufta, *Cold-End Corrosion Monitoring Methods*, Power Eng., 71, April 1977.

17 R. Pierce, *Acid Dew Point Estimating Method*, 1976 Chimney Design Symposium and Exhibition, TH 2281.C55, Quarto Vol. 3, 143, 1976.

18 B. Gills and B. Lees, *Insulated Steel Chimneys Avoid Corrosion and Smut Emission, American Society of Mechanical Engineers Winter Annual Meeting*, p. 2, Chicago, November 7–11, 1965.

Part V

Epilogue

The epilogue contains one chapter that is titled "Comparing, Selecting, and Purchasing Air Pollution Control Equipment." This chapter's sections are shorter than the earlier chapters, but still worth including. References are provided that allow the reader to delve deeper into the subject matters. Their accompanying sections are listed below:

21

Comparing, Selecting, and Purchasing Air Pollution Control Equipment

21.1 Introduction

This concluding chapter introduces the reader to several reference tables that are related directly or indirectly with the design of air pollution control equipment. The earlier work of Theodore and Theodore [1] and Theodore and Buonicore [2] provides detailed review of these topics.

21.2 Factors in Control Equipment Selection

There are a number of factors to be considered prior to selecting and operating particular piece of air pollution control hardware. In general, they can be grouped into three categories: environmental, engineering, and economic.

Environmental

1. Equipment location
2. Available space
3. Ambient conditions
4. Availability of adequate utilities (i.e., power, water, etc.) and ancillary system facilities (i.e., waste treatment and disposal, etc.)
5. Maximum allowable emission (air pollution codes)
6. Aesthetic considerations (i.e., visible steam or water vapor plume, etc.)
7. Contribution of air pollution control system to wastewater and land pollution
8. Contribution of air pollution control system to plant noise levels

Engineering

1. Contaminant characteristics (i.e., physical and chemical properties, concentration, particulate shape and size distribution—in the case of particulates, chemical reactivity, corrosivity, abrasiveness, toxicity, etc.)
2. Gas stream characteristics (i.e., volume flow rate, temperature, pressure, humidity, composition, viscosity, density, reactivity, combustibility, corrosivity, toxicity, etc.)

3. Design and performance characteristics of the particular control system (i.e., size and weight, fractional efficiency curves—in the case of particulates, mass transfer and/or contaminant destruction capability—in the case of gases or vapors, pressure drop, reliability and dependability, turndown capability, power requirements, utility requirements, temperature limitations, maintenance requirements, flexibility toward complying with more stringent air pollution codes, etc.)

Economic

1. Capital cost (equipment, installation, engineering, etc.)
2. Operating cost (utilities, maintenance, etc.)
3. Expected equipment lifetime and salvage value

The usual problems can be further compounded by any one or combination of the following:

1. Unfamiliarity of process engineers with air pollution engineering
2. New and changing air pollution codes
3. New suppliers with frequently unproven equipment
4. Lack of industry standards in some key areas
5. Interpretations of control agency field personnel
6. Compliance schedules that are too tight
7. Vague specifications
8. Weak guarantees for the new control equipment
9. Unreliable delivery schedules
10. Process unreliability problems

Proper selection of a device for a specific application can be both difficult and complicated. It is in the best interest of the prospective user to work closely with regulatory officials as early in the process as possible. Finally, previous successful experience on a similar application cannot be overemphasized.

21.3 Comparing Control Equipment Alternatives

The final choice in equipment selection is usually dictated by that equipment capable of achieving compliance with regulatory codes at the lowest uniform annual cost. In order to compare specific control equipment alternatives, knowledge of the particular application and site is essential. A preliminary screening, however, may be performed by reviewing the advantages and disadvantages of each type of air pollution control equipment. For example, if water or a waste treatment system is not available at the site, this may preclude use of a wet scrubber system and instead focus particulate removal on dry systems, such as cyclones, baghouses, and/or electrostatic precipitators. If auxiliary fuel is unavailable on a continuous basis, it may not be possible to combust organic pollutant vapors in an incineration system. If the particulate size distribution in the gas stream is relatively fine, cyclone collectors most probably would not be considered. If the pollutant vapors can be reused in the process, control efforts may be directed to adsorption systems. There are many more situations where knowledge of the capabilities of the various control options, combined with common sense, will simplify the selection procedure.

General advantages and disadvantages of the most popular types of air pollution control equipment for gases and particulates are presented in Tables 21.1–21.9 [3–6]:

Table 21.1 Advantages and disadvantages of cyclone collectors.

Advantages

1. Low cost of construction.
2. Relatively simple equipment with few maintenance problems.
3. Relatively low operating pressure drops (for degree of particulate removal obtained) in the range of approximately 2 to 6 in. water column.
4. Temperature and pressure limitations imposed only by the materials of construction used.
5. Dry collection and disposal.
6. Relatively small space requirements.

Disadvantages

1. Relatively low overall particulate collection efficiencies, especially on particulates below 10 μm in size.
2. Inability to handle tacky materials.

Table 21.2 Advantages and disadvantages of wet scrubbers.

Advantages

1. No secondary dust sources.
2. Relatively small space requirements.
3. Ability to collect gases as well as particulates (especially "sticky" ones).
4. Ability to handle high-temperature, high-humidity gas streams.
5. Capital cost is low (if wastewater treatment system is not required).
6. For some processes, the gas stream is already at high pressures (so pressure drop considerations may not be significant).
7. Ability to achieve high collection efficiencies on fine particulates (however, at the expense of pressure drop).

Disadvantages

1. May create water disposal problem.
2. Product is collected wet.
3. Corrosion problems are more severe than with dry systems.
4. Steam plume opacity and/or droplet entrainment may be objectionable.
5. Pressure drop and horsepower requirements may be high.
6. Solids buildup at the wet–dry interface may be a problem.
7. Relatively high maintenance costs.

Table 21.3 Advantages and disadvantages of electrostatic precipitators.

Advantages

1. Extremely high particulate (coarse and fine) collection efficiencies can be attained (at a relatively low expenditure of energy).
2. Dry collection and disposal.
3. Low-pressure drop (typically less than 0.5 inch water column).
4. Designed for continuous operation with minimum maintenance requirements.
5. Relatively low operating costs.
6. Capable of operation under high pressure (to 150 psi) or vacuum conditions.
7. Capable of operation at high temperatures (to 1,300 °F).
8. Relatively large gas flow rates can be effectively handled.

Disadvantages

1. High capital cost.
2. Very sensitive to fluctuations in gas stream conditions (in particular, flows, temperatures, particulate and gas composition, and particulate loadings).
3. Certain particulates are difficult to collect due to extremely high or low resistivity characteristics.
4. Relatively large space requirements required for installation.
5. Explosion hazard when treating combustible gases and/or collecting combustible particulates.
6. Special precautions are required to safeguard personnel from the high voltage.
7. Ozone is produced by the negatively charged discharge electrode during gas ionization.
8. Relatively sophisticated maintenance personnel required.

Table 21.4 Advantages and disadvantages of fabric filter systems.

Advantages

1. Extremely high collection efficiency on both coarse and fine (submicron) particulates.
2. Relatively insensitive to gas stream fluctuation. Efficiency and pressure drop are relatively unaffected by large changes in inlet dust loadings for continuously cleaned filters.
3. Filter outlet air may be recirculated within the plant in many cases (for energy conservation).
4. Collected material is recovered dry for subsequent processing or disposal.
5. No problems with liquid waste disposal, water pollution, or liquid freezing.
6. Corrosion and rusting of components are usually not problems.
7. There is no hazard of high voltage, simplifying maintenance and repair, and permitting collection of flammable dusts.
8. Use of selected fibrous or granular filter aids (precoating) permits the high-efficiency collection of submicron smokes and gaseous contaminants.
9. Filter collectors are available in a large number of configurations, resulting in a range of dimensions and inlet and outlet flange locations to suit installation requirements.
10. Relatively simple operation.

Disadvantages

1. Temperatures in excess of 550 °F require special refractory mineral or metallic fabrics that are still in the developmental stage and can be very expensive.
2. Certain dusts may require fabric treatments to reduce dust seeping or, in other cases, assist in the removal of the collected dust.
3. Concentrations of some dusts in the collector (~50 g/m^3) may represent a fire or explosion hazard if a spark or flame is introduced by accident. Fabrics can burn if readily oxidizable dust is being collected.
4. Relatively high maintenance requirements (bag replacement, etc.).
5. Fabric life may be shortened at elevated temperatures and in the presence of acid or alkaline particulate or gas constituents.
6. Hygroscopic materials, condensation of moisture, or tarry adhesive components may cause crusty caking or plugging of the fabric or require special additives.
7. Replacement of fabric may require respiratory protection for maintenance personnel.
8. Medium-pressure drop requirements, typically in the range 4 to 10 inch water column.

Table 21.5 Advantages and disadvantages of absorption systems (packed and plate columns).

Advantages

1. Relatively low pressure drop.
2. Standardization in fiberglass-reinforced plastic (FRP) construction permits operation in highly corrosive atmospheres.
3. Capable of achieving relatively high mass-transfer efficiencies.
4. Increasing the height and/or type of packing or number of plates can improve mass transfer without purchasing a new piece of equipment.
5. Relatively low capital cost.
6. Relatively small space requirements.
7. Ability to collect particulates as well as gases.

Disadvantages

1. May create water (or liquid) disposal problem.
2. Product collected wet.
3. Particulates deposition may cause plugging of the bed or plates.
4. When FRP construction is used, it is sensitive to temperature.
5. Relatively high maintenance costs.

Table 21.6 Comparison of plate and packed columns.

Packed column

1. Lower pressure drop.
2. Simpler and cheaper to construct.
3. Preferable for liquids, with high foaming tendencies.

Plate column

1. Less susceptible to plugging.
2. Less weight.
3. Less of a problem with channeling.
4. Temperature surge will result in less damage.

Table 21.7 Advantages and disadvantages of adsorption systems.

Advantages

1. Product recovery may be possible.
2. Excellent control and response to process changes.
3. No chemical disposal problem when pollutant (product) recovered and returned to process.
4. Capability of systems for fully automatic, unattended operation.
5. Capability to remove gaseous or vapor contaminants from process streams to extremely low levels.

Disadvantages

1. Product recovery may require an exotic, expensive distillation (or extraction) scheme.
2. Adsorbent progressively deteriorates in capacity as the number of cycles increases.
3. Adsorbent regeneration requires a steam or vacuum source.
4. Relatively high capital cost.
5. Prefiltering of gas stream may be required to remove any particulate capable of plugging the adsorbent bed.
6. Cooling of the gas stream may be required to get to the usual range of operation (less than 120 °F).
7. Relatively high steam requirements to desorb high-molecular-weight hydrocarbons.

Table 21.8 Advantages and disadvantages of combustion systems.

Advantages

1. Simplicity of operation.
2. Capability of steam generation or heat recovery in other forms.
3. Capability for virtually complete destruction of organic contaminants.

Disadvantages

1. Relatively high operating costs (particularly associated with fuel requirements).
2. Potential for flashback and subsequent explosion hazard.
3. Catalyst poisoning (in the case of catalytic incineration).
4. Incomplete combustion can create potentially worse pollution problems.

Table 21.9 Advantages and disadvantages of condensers.

Advantages

1. Pure product recovery (in the case of indirect-contact condensers).
2. Water used as the coolant in an indirect contact condenser (i.e., shell-and-tube heat exchanger) does contact the contaminated gas stream and can be reused after cooling.

Disadvantages

1. Relatively low removal efficiency for gaseous contaminants (at concentrations typical of pollution control applications).
2. Coolant requirements may be extremely expensive.

21.4 Selecting Control Equipment for Specific Industries

The basic types of emission control devices are mechanical collectors, wet scrubbers, baghouses, electrostatic precipitators, combustion systems, condensers, absorbers, and adsorbers. All of these have been used to some extent to control emissions from a nearly infinite variety of processes, with the selection procedure almost always dictated by experience. This section provides information on suggested control measures for a number of industries. Details are provided in Table 21.10 [3–7].

Table 21.10 Selecting control equipment for specific industries.

Industry	Process or operation	Air pollutions emitted	Control techniques
Aluminum reduction plants	Materials handling: buckets and belt conveyor or pneumatic conveyor	Particulates (dust)	Exhaust system and baghouse
	Anode and cathode electrode preparation Cathode: baking Anode: grinding and blending	Hydrocarbon emissions from binder Particulates (dust)	Exhaust systems and mechanical collectors
	Baking	Particulates (dust), CO, SO_2, hydrocarbons, and fluorides	High-efficiency cyclone, electrostatic precipitators, scrubbers, catalytic combustion or incinerators, flares, and baghouse
	Pot charging	Particulates (dust), CO, HF, SO_2, CF_4, and hydrocarbons	High-efficiency cyclones, baghouses, spray towers, floating-bed scrubbers, electrostatic precipitators, chemisorption, and wet electrostatic precipitators
	Metal casting	Cl_2, HCl, CO, and particulates (dust)	Exhaust systems and scrubbers
Asphalt batch plants	Materials handling, storage and classifiers: elevators, chutes, vibrating screens	Particulates (dust)	Local exhaust systems with a cyclone precleaner and a scrubber or baghouse
	Drying: rotary oil- or gas-fired	Particulates and smoke	Proper combustion controls, fuel-oil preheating where required local exhaust system, cyclone, and a scrubber or baghouse
Cement plants	Truck traffic	Dust	Wetting down truck routes
	Quarrying: primary crusher, secondary crusher, conveying, storage	Particulates (dust)	Wetting and exhaust systems with mechanical collectors
	Dry processes: materials handling, air separator (hot-air furnace)	Particulates (dust)	Local exhaust system and mechanical collectors and baghouse
	Grinding	Particulates (dust)	Local exhaust system, with cyclones and baghouse
	Pneumatic conveying and storage	Particulates (dust)	
	Wet process: materials, handling, grinding, and storage	Wet materials, no dust	

(*Continued*)

Table 21.10 (Continued)

Industry	Process or operation	Air pollutions emitted	Control techniques
	Kiln operations rotary kiln	Particulates (dust), CO, SO_2, NO_2, hydrocarbons, aldehydes, ketones	Electrostatic precipitators and baghouses, scrubber, flare
	Clinker cooling: materials handling	Particulates (dust)	Local exhaust system and mechanical collectors
	Grinding and packaging: air separator, grinding, pneumatic conveying, materials handling, packaging	Particulates (dust)	Local exhaust systems and mechanical collectors
Coral preparation plants	Materials handling: conveyors, elevators, chutes	Particulates (dust)	Local exhaust systems and cyclones
	Sizing: crushing, screening, and classifying	Particulates (dust)	Local exhaust systems and cyclones
	De-dusting	Particulates (dust)	Local exhaust system, cyclone precleaners, and baghouse
	Storing coal in piles	Blowing particulates (dust)	Wetting, plastic spray covering
	Refuse piles	H_2S, particulates, and smoke from burning storage piles	Digging out fire, pumping water onto fire area, blanket with incombustible material
	Coal drying: rotary, screen, suspension, fluid bed, and cascade	Dust, smoke, particulates, sulfur oxides, and H_2S	Exhaust systems with cyclones and venturi scrubbers
Coke plants	By-product ovens charging	Smoke, particulates (dust)	Pipeline charging, careful charging techniques; portable hooding and scrubber or baghouses
	Pushing	Smoke, particulates (dust), SO_2	Minimize green coke pushing, scrubbers, and baghouses
	Quenching	Smoke, particulates (dust and mists), phenols, and ammonia	Baffles and spray tower
	By-product processing	CO, H_2S, methane, ammonia, H_2, phenols, hydrogen cyanide, N_2, benzene, xylene, etc.	Electrostatic precipitator, scrubber, flaring
	Material storage (coal and coke)	Particulates (dust)	Wetting, plastic spray, fire-prevention techniques
Fertilizer industry (chemical)	Phosphate fertilizers: crushing, grinding, and calcining	Particulates (dust)	Exhaust system, scrubber, cyclone, and baghouse

Industry	Process or operation	Air pollutions emitted	Control techniques
	Hydrolysis of P_2O_5	PH_3, P_2O_5, H_3PO_4, mist	Scrubbers, flare
	Acidulation and curing	HF, SiF_4	Scrubbers
	Granulation	Particulates (dust) (product recovery)	Exhaust system, scrubber, or baghouse
	Ammoniation	NH_3, NH_4Cl, SiF_4, HF	Cyclone, electrostatic precipitator, baghouse, and high-energy scrubber
	Nitric acid acidulation	NO_x, gaseous fluoride compounds	Scrubber, addition of urea
	Superphosphate storage and shipping	Particulates (dust)	Exhaust system, cyclone, or baghouse
	Ammonium nitrate reactor	NH_3, NO_x	Scrubber
	Prilling tower	NH_4, NO_3	Proper operation control, scrubbers
Foundries	Melting (cupola):		
Iron	Charging Melting Pouring Bottom drop	Smoke and particulates Smoke particulates, and fume Oil, mist, and CO Smoke and particulates	Closed top with exhaust system, CO afterburner, gas-cooling device and scrubbers, baghouse or electrostatic precipitator, and wetting to extinguish fire
Brass bronze	Melting:		
	charging melting pouring	Smoke particulates, & oil Zinc oxide fume, particulates, and smoke zinc oxide Fume and lead oxide fume	Low-zinc-content red Brass: use good combustion controls and slag cover; high-zinc-content brass: use good combustion controls, local exhaust system, and baghouse or scrubber
Aluminum	Melting: charging, melting, and pouring	Smoke and particulates	Charge clean material (no paint or grease) and proper operation should be required, no air pollution control equipment if no fluxes are used and degassing is not required, and dirty charge requires exhaust system with scrubbers and baghouses
Zinc	Melting Charging	Smoke and particulates	Exhaust system with cyclone and baghouse; charge clean material (no paint or grease)
	Melting	Zinc oxide fume	Careful skimming of dross

(Continued)

Table 21.10 (Continued)

Industry	Process or operation	Air pollutions emitted	Control techniques
	Pouring	Oil mist and hydrocarbons from die-casting machines	Use low-smoking die-casting lubricants
	Sand handling shake-out	Particulates (dust), smoke, organic vapors	
	Magnetic pulley, conveyors and elevators, rotary cooler, screening, and crusher mixer	Particulates (dust)	Exhaust system, cyclone, and baghouse
	Coke-making ovens	Organic acids, aldehydes, smoke, hydrocarbons	Use of binders that will allow ovens to operate at less than 400 °F, or exhaust systems and afterburners
Galvanizing operations	Hot-dip galvanizing tank kettle, dipping material into the molten zinc, and dusting flux onto the surface of the molten zinc	Fumes, particulates (liquid), vapors: NH_4Cl, ZnO, $ZnCl_2$, Zn, NH_3, oil, and carbon	Close-fitting hoods with high indraft velocities (in some cases the hood may not be able to be close to the kettle, so that the in-draft velocity must be very high), baghouses, electrostatic precipitators
Kraft pulp mills	Digesters: batch and continuous	Mercaptans, methanol (odors)	Condensers and use of lime kiln, hog fuel boiler, or furnaces as afterburners
	Multiple-effect evaporators	H_2S, other odors	Caustic-scrubbing and thermal oxidation of noncondensables
	Recovery furnace	H_2S, mercaptans, organic sulfides, and disulfides	Paper combustion controls for fluctuating load and unrestricted primary and secondary air flow to furnace and scrubber or electrostatic precipitator
	Weak and strong black liquor oxidation	H_2S	Packed tower and cyclone
	Smelt tanks	Particulates (mist or dust)	Demisters, venturi, packed tower, or impingement-type scrubbers
	Lime kiln	Particulates (dust), H_2S	Venturi scrubbers
Municipal and industrial incinerators	Single-chamber incinerators	Particulates, smoke, volatiles, CO, SO_x, ammonia, organic acids, aldehydes, NO_x, hydrocarbons, odors, and HCl	

Industry	Process or operation	Air pollutions emitted	Control techniques
	Flue-fed		Settling chambers, scrubbers, afterburner, bypass flue, and ash cleanout
	Multiple-chamber incinerators: retort, inline	Particulates, smoke, and combustion contaminants	Operating at rated capacity, using auxiliary fuel as specified and good maintenance, including timely cleanout of ash
	Flue-fed	Particulates, smoke, and combustion contaminants	Use of charging gates and automatic controls for draft
	Wood waste	Particulates, smoke, and combustion contaminants	Continuous-feed systems, operate at design load and excess air, and limit charging of oily material
	Municipal incinerators: 50–100 tons/days	Particulates, smoke, volatiles, CO, ammonia, organic acids, aldehydes, NO_x, hydrocarbons, SO_x, hydrogen chloride, and odors	Preparation of materials, including weighing, grinding, shredding, control of tipping area, furnace design with proper automatic controls, proper startup techniques, maintenance of design operating temperatures, use of electrostatic precipitators, scrubbers and baghouses, and proper ash cleanout
	Pathological incinerators	Odors and hydrocarbons	Proper charging
	Wood waste and industrial waste	Particulates, smoke, and combustion contaminants	Modified fuel feed, auxiliary fuel, and dryer systems
	Box type	Particulates, smoke, and combustion contaminants	Allow proper startup, and charge material slowly, without overload
Nonferrous smelters, primary copper	Roasting	SO_2, particulates, and fume	Exhaust system, settling chambers, cyclones or scrubbers and electrostatic precipitators for dust and fumes, and sulfuric acid plant for SO_2
	Reverberatory furnace	Smoke, particulates, fume, SO_2	Exhaust system, settling chambers, cyclones or scrubbers and electrostatic precipitators for dust and fumes and sulfuric acid plant for SO_2

(*Continued*)

Table 21.10 (Continued)

Industry	Process or operation	Air pollutions emitted	Control techniques
	Converters: charging, slag skim, pouring, air or oxygen blow	Smoke, fume, and SO_2	Exhaust system, settling chambers, cyclones or scrubbers, electrostatic precipitators for dust and fumes and sulfuric acid plant for SO_2
Lead	Sintering	So_2, particulates, and smoke	Exhaust system, cyclones and baghouse or precipitators for dust and fumes, sulfuric acid plant for SO_2
	Blast furnace	SO_2, CO, particulates, lead oxide, and zinc oxide	Exhaust system, settling chambers, afterburner and cooling device, cyclone, and baghouse
	Dross reverberator furnace	SO_2, particulates, and fume	Exhaust system, settling chambers, cyclone and cooling device, and baghouse
	Refining kettles	SO_2, particulates	Local exhaust system, cooling device, baghouse, or precipitator
Cadmium	Roasters, slag, fuming furnaces, and de-leading kilns	Particulates	Local exhaust system, baghouse, or precipitator
Zinc	Roasting	Particulates (dust) and SO_2	Exhaust system, humidifier, cyclone scrubber, electrostatic precipitator, and acid plant
	Sintering	Particulates (dust) and SO_2	Exhaust system, humidifier, electrostatic precipitator, and acid plant
	Calcining	Zinc oxide fume, particulates, SO_2, and CO	Exhaust system and baghouse
	Retorts: electric arc		
Nonferrous smelters, secondary	Blast furnaces and cupolas recover metal from scrap and slag	Dust, fumes, particulates, oil vapor, smoke, and CO	Exhaust systems, cooling devices, CO burners and baghouses or precipitators
	Reverberatory furnaces	Dust, fumes, particulates, smoke, and gaseous fluxing materials	Exhaust systems and baghouses, or precipitators or venturi scrubbers
	Crucible furnaces	See nonferrous smelters	
	Sweat furnaces	Smoke, particulates, and fumes	Precleaning metal and exhaust systems with afterburner and baghouse
	Wire reclamation and autobody burning	Smoke and particulates	Scrubbers and afterburners

Industry	Process or operation	Air pollutions emitted	Control techniques
Paint and varnish manufacturing	Resin manufacturing: closed reaction vessel	Acrolein, other aldehydes and fatty acids (odors), and phthalic anhydride (subl.)	Exhaust systems with scrubbers and fume burners
	Varnish cooking-open or closed vessels	Ketones, fatty acids, formic acids, acetic acid, glycerin, acrolein, other aldehydes, phenols and terpenes, from tall oils, hydrogen sulfide, alkyl sulfide, butyl mercaptan, and thiofene (odors)	Exhaust system with scrubbers and fume burners—close-fitting hoods are required for open kettles
	Solvent thinning	Olefins, branched chain aromatics and ketones (odors), solvents	Exhaust system with fume burners
Rendering plants	Feedstock storage and housekeeping	Odors	Quick processing, washdown of all concrete surfaces, pave dirt roads, and proper sewer maintenance, packed towers
	Cookers and percolators	SO_2, mercaptans, ammonia, odors	Exhaust system, condenser, scrubber, or incinerator
	Grinding	Particulates (dust)	Exhaust system and scrubber
Roofing plants (asphalt saturators)	Felt or paper saturators: spray section, asphalt tank, and wet looper	Asphalt vapors and particulates (liquid)	Exhaust system with high inlet velocity at hoods (>200 ft/min) with either spray scrubbers, baghouses, or two-stage low-voltage electrostatic precipitators
	Crushed rock or other minerals handling	Particulates (dust)	Local exhaust system, cyclone or multiple cyclones
Steel mills	Blast furnaces: charging, and pouring	CO, fumes, smoke, and particulates (dust)	Good maintenance, seal leaks, use of higher ratio of pelletized or sintered ore, CO burned in waste heat boilers, stoves or coke ovens, cyclone, scrubber, electrostatic precipitator, or venturi scrubber
	Electric steel furnaces: charging, pouring, and oxygen blow	Fumes, smoke, particulates (dust), and CO	Segregate dirty scrap, and proper hooding, baghouses, venturi scrubbers, or electrostatic precipitator

(Continued)

Table 21.10 (Continued)

Industry	Process or operation	Air pollutions emitted	Control techniques
	Open-hearth furnaces: oxygen blow, and pouring	Fumes, smoke, SO_x, particulates, (dust), CO, and NO_x	Proper hooding, settling chambers, waste-heat boiler, baghouse, electrostatic precipitator, or venturi scrubber
	Basic oxygen furnaces: oxygen blowing	Fumes, smoke, CO, particulates, (dust)	Proper hooding (capture emissions and dilute CO), scrubbers, or electrostatic precipitator
	Raw material storage	Particulates (dust)	Wetting or application of plastic spray
	Pelletizing	Particulates (dust)	Proper hooding, cyclone, and baghouse
	Sintering	Smoke, particulates (dust), SO_2, and NO_x	Proper hooding, cyclones, venturi scrubbers, baghouse, or precipitator

Source: [7]/U.S. Government Printing Office/Public Domain.

21.5 Purchasing Equipment

Prior to purchase of control equipment, experience has shown that the following points should be emphasized:

1. Refrain from purchasing any control equipment without reviewing certified independent test data on its performance under a similar application. Request the manufacturer to provide performance information and design specifications.
2. In the event that sufficient performance data are unavailable, request that the equipment supplier provide a small pilot model for evaluation under existing conditions.
3. Request participation of the local control authorities in the decision-making process.
4. Prepare a good set of specifications. Include a strong performance guarantee from the manufacturer to ensure that the control equipment will meet all applicable local, state, and federal codes at specific process conditions.
5. Closely review your process and economic fundamentals. Assess the possibility for emission trade-offs (offsets) and/or applying the "bubble concept." The bubble concept permits a plant to find the most efficient way to control its emissions as a whole, rather than the Environmental Protection Agency regulating the emissions from individual sources. Reductions at a source where emissions can be lessened for the least cost can offset emissions of the same pollutant from another source in the plant.
6. Make a careful material balance study before authorizing an emission test or purchasing control equipment.
7. Refrain from purchasing any equipment until firm installation cost estimates have been added to the equipment cost. Escalating installation costs are the rule rather than the exception.
8. Give operation and maintenance costs high priority on the list of equipment selection factors.

9. Refrain from purchasing any equipment until a solid commitment from the utility supplier(s) is obtained. Make every effort to ensure that the new system will utilize fuel, controllers, filters, motors, and so on, that are compatible with those already available at the plant.
10. The specification should include written assurance of prompt technical assistance from the equipment supplier. This, together with a complete operating manual (with parts list and full schematics), is essential and is too often forgotten in the rush to get the equipment operating.
11. Schedules, particularly on projects being completed under a court order or consent judgment, can be critical. In such cases, delivery guarantees should be obtained from the manufacturers and penalties identified.
12. The air pollution equipment should be of fail-safe design with built-in indicators to show when performance is deteriorating.
13. Withhold 10 to 15% of the purchase price until compliance is clearly demonstrated.

References

1 M. K. Theodore and L. Theodore, *Introduction to Environmental Management*, 2nd edition, CRC Press/Taylor & Francis Group, Boca Raton, FL, 2021.

2 L. Theodore and A. J. Buonicore, Chapter 6, *Industrial Air Pollution Control Equipment for Particulates*, CRC Press, Inc., Boca Raton, FL, 1976.

3 L. Theodore and R.R. Dupont, *Environmental Health Risk and Hazard Risk Assessment: Principles and Calculations*, CRC Press/Taylor & Francis Group, Boca Raton, FL, 2013.

4 L. Theodore, *Air Pollution Control Equipment Calculations*, John Wiley & Sons, Hoboken, NJ, 2008.

5 L. Theodore, R. R. Dupont and K. Ganesan, *Unit Operations in Environmental Engineering*, John Wiley & Sons, Hoboken, NJ, 2017.

6 L. Theodore, *Chemical Engineering: The Essential Reference*, McGraw Hill, New York City, NY, 2012.

7 M. Weisburd, *Field Operations and Enforcement Manual for Air Pollution Control*, Vol. 2: Control Technology and General Source Inspection, EPA APTD-1101, 1972.

Appendix

SI Units

The Metric System

The need for a single worldwide coordinated measurement system was recognized over 500 years ago. Gabriel Mouton, Vicar of St. Paul in Lyons, proposed in 1670 a comprehensive decimal measurement system based on the length of 1 minute of arc of a great circle of the earth. In 1671, Jean Picard, a French astronomer, proposed the length of a pendulum beating seconds as the unit of length. (Such a pendulum would have been fairly easily reproducible, thus facilitating the widespread distribution of uniform standards.) Other proposals were made, but over a century elapsed before any action was taken.

In 1790, in the midst of the French Revolution, the National Assembly of France requested the French Academy of Sciences to "deduce an invariable standard for all the measures and all the weights." The Commission appointed by the Academy created a system that was, at once, simple and scientific. The unit of length was to be a portion of the Earth's circumference. Measures for capacity (volume) and mass (weight) were to be derived from the unit of length, thus relating the basic units of the system to each other and to nature. Furthermore, the larger and smaller versions of each unit were to be created by multiplying or dividing the basic units by 10 and its multiples. This feature provided a great convenience to users of the system by eliminating the need for such calculations as dividing by 16 (to convert ounces to pounds) or by 12 (to convert inches to feet). Similar calculations in the metric system could be performed simply by shifting the decimal point. Thus, the metric system is a "base-10" or "decimal" system.

The Commission assigned the name "metre" (which is now spell meter) to the unit of length. This name was derived from the Greek word "metron," meaning "a measure." The physical standard representing the meter was to be constructed so that it would equal one ten-millionth of the distance from the north pole to the equator along the meridian of the Earth running near Dunkirk in France and Barcelona in Spain.

The metric unit of mass, called the "gram," was defined as the mass of one cubic centimeter (a cube that is 1/100 of a meter on each side) of water at its temperature of maximum density. The cubic diameter (a cube 1/10 of a meter on each side) was chosen as the unit of fluid capacity. This measure was given the name "liter."

Although the metric system was not accepted with enthusiasm at first, adoption by other nations occurred steadily after France made its use compulsory in 1840. The standardized character and decimal features of the metric system made it well suited to scientific and engineering work. Consequently, it is not surprising that the rapid spread of the system coincided with an age of rapid technological development. In the United States, by Act of Congress in 1866, it was made "lawful throughout the United States of America to employ the weights and measures of the metric system in all contracts, dealings, or court proceedings."

By the late 1860s, even better metric standards were needed to keep pace with scientific advances. In 1875, an international treaty, the "Treaty of the Meter," set up well-defined metric standards for length and mass and established permanent machinery to recommend and adopt further refinements in the metric system.

This treaty, known as the Metric Convention, was signed by 17 countries, including the United States.

As a result of the Treaty, metric standards were constructed and distributed to each nation that ratified the Convention. Since 1893, the internationally agreed-to metric standards have served as the fundamental weights and measures standards of the United States.

By 1900, a total of 35 nations—including the major nations of continental Europe and most of South America—had officially accepted the metric system. With the exception of the United States and a few small countries, the entire world today predominantly uses the metric system or is committed to such use. In 1971, the Secretary of Commerce, in transmitting to Congress the results of a 3-year study authorized by the Metric Study Act of 1968, recommended that the United States change to predominant use of the metric system through a coordinated national program.

The International Bureau of Weights and Measures located at Sevres, France, serves as a permanent secretariat for the Metric Convention, coordinating the exchange of information about the use and refinement of the metric system. As measurement science develops more precise and easily reproducible ways of defining the measurement units, the General Conference of Weights and Measures—the diplomatic organization made up of adherents to the Convention—meets periodically to ratify improvements in the system and the standards.

The SI System

In 1960, the General Conference adopted an extensive revision and simplification of the system. The name "Le Systeme International d'Unites" (International System of Units), with the international abbreviation SI, was adopted for this modernized metric system. Further improvements in and additions to SI were made by the General Conference in 1964, 1968, and 1971.

The basic units in the SI system are the kilogram (mass), meter (length), second (time), Kelvin (temperature), ampere (electric current), candela (the unit of luminous intensity), and radian (angular measure). All are commonly used by the engineer. The Celsius scale of temperature ($0\,°C = 273.15$ K) is commonly used with the absolute Kelvin scale. The important derived units are the newton (SI unit of force), the joule (SI unit of energy), the watt (SI unit of power), the pascal (SI unit of pressure), and the hertz (unit of frequency). There are a number of electrical units: coulomb (charge), farad (capacitance), henry (inductance), volt (potential), and weber (magnetic flux). One of the major advantages of the metric system is that larger and smaller units are given in powers of ten. In the SI system, a further simplification is introduced by recommending only those units with multipliers of 10^3. Thus, for lengths in engineering, the micrometer (previously micron), millimeter, and kilometer are recommended, and the centimeter is generally avoided.

A further simplification is that the decimal point may be substituted for by a comma (as in France, Germany, and South Africa), while the other numbers, before and after the comma, will be separated by spaces between groups of three, i.e., 1 million dollars will be "$1,000,000.00." More details are provided below.

Seven Base Units

Length—meter (m). The meter (common international spelling, metre) is defined as 1,650,763.73 wavelengths in vacuum of the orange-red line of the spectrum of krypton-86. The SI unit of area is the square meter (m^2). The SI unit of volume is the cubic meter (m^3). The liter (0.001 cubic meter), although not an SI unit, is commonly used to measure fluid volume.

Mass—kilogram (kg). The standard for the unit of mass, the kilogram, is a cylinder of platinum-iridium alloy kept by the International Bureau of Weights and Measures at Paris. A duplicate in the custody of the National Bureau of Standards serves as the mass standard for the United States. This is the only base unit still defined by an artifact. The SI unit of force is the newton (N). One newton is the force which, when applied to a 1-kilogram mass, will give the kilogram mass an acceleration of 1 (meter per second) per second. Thus $1\ N = 1\ kg{\cdot}m/s^2$. The SI unit for pressure is the pascal (Pa). Thus $1\ Pa = 1\ N/m^2$. The SI unit for work and energy of any kind is the joule (J). Thus $1\ J = 1\ N{\cdot}m$. The SI unit for power of any kind is the watt (W). Thus $1\ W = 1\ J/s$.

Time—second (s). The second is defined as the duration of 9,192,631,770 cycles of the radiation associated with a specified transition of the cesium-133 atom. It is realized by tuning an oscillator to the resonance frequency of cesium-133 atoms as they pass through a system of magnets; a number of periods or cycles per second is called frequency. The SI unit for frequency is the hertz (Hz). One hertz equals one cycle per second. The SI unit for speed is the meter per second (m/s). The SI unit for acceleration is the (meter per second) per second (m/s^2).

Electric Current—ampere (A). The ampere is defined as that current which, if maintained in each of two long parallel wires separated by 1 meter in free space, would produce a force between the two wires (due to their magnetic fields) of 2×10^{-7} Newtons for each meter of length. The SI unit of voltage is the volt (V). Thus $1\ V = 1\ W/A$. The SI unit of electrical resistance is the ohm (Ω). Thus $1\ \Omega = 1\ V/A$.

Temperature—kelvin (K). The kelvin is defined as the fraction 1/273.16 of the thermodynamic temperature of the triple point of water. The temperature 0 K is called "absolute zero." On the commonly used Celsius temperature scale, water freezes at about 0 °C and boils at about 100 °C. The °C is defined as an interval of 1 K, and the Celsius temperature 0 °C is defined as 273.15 K. 1.8 Fahrenheit degrees are equal to 1.0 °C or 1.0 K; the Fahrenheit scale uses 32°F as a temperature corresponding to 0 °C.

Amount of Substance—mole (mol). The mole is the amount of substance of a system that contains as many elementary entities as there are atoms in 0.012 kilogram of carbon 12. When the mole is used, the elementary entities must be specified and may be atoms, molecules, ions, electrons, other particles, or specified groups of such particles. The SI unit of concentration (of amount of substance) is the mole per cubic meter (mol/m^3).

Luminous Intensity—candela (cd). The candela is defined as the luminous intensity of 1/600,000 of a square meter of a blackbody at the temperature of freezing platinum (2045 °K). The SI unit of light flux is the lumen (lm). A source having an intensity of 1 candela in all directions radiates a light flux of 4π lumens.

Two Supplementary Units

Plane Angle—radian (rad). The radian is the plane angle with its vertex at the center of a circle that is subtended by an arc equal in length to the radius.

Solid Angle—steradian (sr). The steradian is the solid angle with its vertex at the center of a sphere that is subtended by an area of the spherical surface equal to that of a square with sides equal in length to the radius.

Table A.1 Multiples and prefixes that may be applied to all SI units.

Multiples		Prefixes	Symbols
1,000,000,000,000	10^{12}	tera	T
1,000,000,000	10^{9}	giga	G
1,000,000	10^{6}	mega	M
1,000	10^{3}	kilo	k
100	10^{2}	hecto	h
10	10^{1}	deka	da
Base Unit 1	10^{0}		
0.1	10^{-1}	deci	d
0.01	10^{-2}	centi	c
0.001	10^{-3}	milli	m
0.000 001	10^{-6}	micro	μ
0.000 000 001	10^{-9}	nano	n
0.000 000 000 001	10^{-12}	pico	p
0.000 000 000 000 001	10^{-15}	femto	f
0.000 000 000 000 000 001	10^{-18}	atto	a

Table A.2 Common conversions (accurate to six significant figures).

Symbol	When you know	Multiply by	To find	Symbol
in	inches	25.4	millimeters	mm
ft	feet	0.3048	meters	m
yd	yards	0.9144	meters	m
mi	miles	1.60934	kilometers	km
yd^2	square yards	0.836127	square meters	m^2
	acres	0.404686	hectares	ha
yd^3	cubic yards	0.764555	cubic meters	m^3
qt	quarts (lq)	0.946353	liters	l
oz	ounces (avdp)	28.3495	grams	g
lb	pounds (acdp)	0.453592	kilograms	kg

Symbol	When you know	Multiply by	To find	Symbol
°F	Fahrenheit	(°F − 32) × 5/9	Celsius	°C
mm	millimeters	0.039370	inches	in
m	meters	3.28084	feet	ft
m	meters	1.09361	yards	yd
km	kilometers	0.621371	miles	mi
m^2	square meters	1.19599	square yards	yd^2
ha	hectares	2.47105	acres	
m^3	cubic meters	1.30795	cubic yards	yd^3
l	liters	1.05669	quarts (lq)	qt
g	grams	0.035274	ounces (avdp)	oz
kg	kilograms	2.20464	pounds (acdp)	lb
°C	Celsius	°C × 9/5 + 32	Fahrenheit	°F

Table A.3 Conversion constants for quantities most often employed in air pollution control.

Quantity	Multiply by	To obtain
ft	0.3048	meters (m)
in	0.0254	meters (m)
miles	1.609×10^{-3}	meters (m)
ft/s	0.3048	meters/sec (m/s)
ft/min	5.08×10^{-3}	meters/sec (m/s)
ft^3/s	28.32×10^{-3}	m^3/s
ft^3/min	0.472×10^{-3}	m^3/s
gal/min	75.77×10^{-6}	m^3/s
pounds	0.4536	kilograms (kg)
ounces	28.35	kilograms (kg)
grains	64.8×10^{-6}	kilograms (kg)
lbs/ft^3	16.02	kg/m^3
$grains/ft^3$	2.29	g/m^3
lb/s	0.4536	kg/s
lb/min	7.56×10^{-3}	kg/s
lb/hr	0.126×10^{-3}	kg/s
ft^2	92.9×10^{-3}	m^2
ft^3	0.0283	m^3
gal	4.546	m^3
lb/in^2	6,895	Pa (N/m^2)
atm	101,300	Pa (N/m^2)

(Continued)

Table A.3 (Continued)

Quantity	Multiply by	To obtain
in W.C.	249.1	Pa (N/m^2)
in Hg	3,386	Pa (N/m^2)
mm Hg	133.3	Pa (N/m^2)
Poise	0.1	Pa (N/m^2)
lbs · s/ft^2	47.88	Pa (N/m^2)
kg · s/m^2	9.807	Pa (N/m^2)
Stoke (cm/s)	10^{-4}	m^2/s
ft^2/hr	25.81×10^{-6}	m^2/s
ft^2/s	92.9×10^{-3}	m^2/s
Btu	1,055	J
therm	105.5×10^6	J
kWh	3.60×10^6	J
calorie	4.1888	J
ft · lb/s	1.358	W
horsepower	745.7	W
Btu/lb	2,326	J/kg
Btu/ft^3	37,210	J/m^3
Btu//lb	4,187	J/kg · K
Btu/ft^2 · hr	3.155	W/m^2

Index

Printed and bound by CPI Group (UK) Ltd, Croydon, CR0 4YY
17/09/2026
14964237-0002